# *Monopulse Radar*

## The Artech House Radar Library

*Radar System Analysis* by David K. Barton

*Introduction to Electronic Warfare* by D. Curtis Schleher

*Electronic Intelligence: The Analysis of Radar Signals* by Richard G. Wiley

*Electronic Intelligence: The Interception of Radar Signals* by Richard G. Wiley

*Principles of Secure Communication Systems* by Don J. Torrieri

*Multiple-Target Tracking with Radar Applications* by Samuel S. Blackman

*Solid-State Radar Transmitters* by Edward D. Ostroff et al.

*Logarithmic Amplification* by Richard Smith Hughes

*Radar Propagation at Low Altitudes* by M.L. Meeks

*Radar Cross Section* by Eugene F. Knott, John F. Shaeffer, and Michael T. Tuley

*Radar Anti-Jamming Techniques* by M.V. Maksimov et al.

*Introduction to Synthetic Array and Imaging Radars* by S.A. Hovanessian

*Radar Detection and Tracking Systems* by S.A. Hovanessian

*Radar System Design and Analysis* by S.A. Hovanessian

*Radar Signal Processing* by Bernard Lewis, Frank Kretschmer, and Wesley Shelton

*Radar Calculations Using the TI-59 Programmable Calculator* by William A. Skillman

*Radar Calculations Using Personal Computers* by William A. Skillman

*Techniques of Radar Reflectivity Measurement,* Nicholas C. Currie, ed.

*Monopulse Principles and Techniques* by Samuel M. Sherman

*Receiving Systems Design* by Stephen J. Erst

*Designing Control Systems* by Olis Rubin

*Advanced Mathematics for Practicing Engineers* by Kurt Arbenz and Alfred Wohlhauser

*Radar Reflectivity of Land and Sea* by Maurice W. Long

*High Resolution Radar Imaging* by Dean L. Mensa

*Introduction to Monopulse* by Donald R. Rhodes

*Probability and Information Theory, with Applications to Radar* by P.M. Woodward

*Radar Detection* by J.V. DiFranco and W.L. Rubin

*Synthetic Aperture Radar,* John J. Kovaly, ed.

*Infrared-to-Millimeter Wavelength Detectors,* Frank R. Arams, ed.

*Significant Phased Array Papers,* R.C. Hansen, ed.

*Handbook of Radar Measurement* by David K. Barton and Harold R. Ward

*Statistical Theory of Extended Radar Targets* by R.V. Ostrovityanov and F.A. Basalov

*Radar Technology,* Eli Brookner, ed.

*MTI Radar,* D. Curtis Schleher, ed.

# *Monopulse Radar*

## A.I. Leonov
## K.I. Fomichev

## Translated by
## William F. Barton

**International Standard Book Number: 0-89006-217-X**
**Library of Congress Catalog Card Number: 86-71715**

Translation of *Monoimpul' snaya radiolokatsiya,* originally published in Russian by Radio i Svyaz, Moscow. Copyright © 1984.

10   9   8   7   6   5   4   3   2   1

## Contents

# *Preface*

Since the initial publication of *Monopulse Radar,* interest in the monopulse method has continued to grow, due to the advantages it offers in angle sensing and consequent widespread application of monopulse systems in various fields. This interest has been further encouraged by improvements in monopulse techniques realized through recent scientific and technological achievements.

The most important developments in monopulse include its improved interference immunity, resolution, and angular accuracy. New methods of radar signal processing have been introduced in the West in recent years, the statistical theory of space-time filtering has progressed, and computer simulations are being used more widely in the development and testing of monopulse systems.

As a result, a significant revision of this book was required. The general areas of revision are as follows:

1. The first chapter was completely rewritten. The basic principles of modern monopulse radars are introduced in a separate chapter. Added to that chapter are descriptions of monopulse Doppler radar, digital signal processing for monopulse systems, and monopulse radars with conical scanning.
2. Greater emphasis is placed on phased antenna arrays, and additional treatment is given to spiral antennas.
3. Material is presented on the operating principles of modern multi-processor computers and their use in automating radar operation. Digital signal processing techniques employing the fast Fourier transform (FFT) are examined.
4. A statistical approach to the analysis of angular resolution is introduced, and some possible methods for improving angular resolution are considered.
5. In analyzing the accuracy of monopulse systems, greater emphasis has been placed on the effects of microwave reflection from the

earth's surface, and on methods of minimizing this degrading factor in angle measurements on low-flying targets.

6. Crosspolarization of the antenna radiation is analyzed by studying sources of instrumental errors. The polarization structure of reflector antennas is examined, and we study the nature of angle errors arising from the reception of signals varying in phase, along with some means for reducing the crosspolarization of antennas.

7. The chapter on interference immunity of monopulse radars has been completely rewritten. Since a good estimate of such tolerance is impossible without knowledge of the interference conditions under which the radar will operate, modern methods of radar countermeasures are reviewed, as are methods for protecting against the more dangerous forms of jamming. Given is a detailed analysis of distortion of the angle-sensing characteristic as a function of the polarization error in the received signals. The effects of interference on target detection and selection are also examined.

8. The chapter on the application of simulations to the assessment of operating characteristics has been revised and expanded. Algorithms are presented for simulating antenna-feed structures and for signal processing in the receiver. Some simulation results are included.

9. In considering practical application of monopulse techniques, information on modern US systems is presented.

Chapters 1, 2, 3, 4, 9, and 10 were written by A. I. Leonov. In writing Chapter 9, the author made use of material offered by N. I. Antropov, V. N. Vasenev, and F. V. Nagulinko. K. I. Fomichev wrote Chapters 5, 6, 7, and 8.

The authors express deep thanks to Professor S. I. Krasnogorov for valuable observations offered during the review of the manuscript.

# Chapter 1

# The Principles and Methods of Monopulse Radar

## 1.1 THE PRINCIPLE OF MONOPULSE ANGLE SENSING

Target angle sensing, the process of determining the direction to a target, is one of the principal functions of radar. In early systems, conical (or linear) scanning or sequential lobing was used to achieve accurate automatic angle sensing of a signal source, a method implemented in single-channel trackers. In such systems angle sensing was typically achieved through comparison of signals received sequentially by an antenna with a varying pattern, and the angle measurement took the form of amplitude modulation of the received signals. The modulation level determined the value of the error signal, while comparison of the phases of the signal modulation and a reference signal gave the direction of the deviation of the axis from the target.

The formation of the angular error signal in single-channel trackers requires the processing of a sequence of reflected pulses received during one complete scan. This error signal, therefore, is sensitive to the amplitude fluctuations in the received signals resulting from random variations in the target cross section. This is the greatest shortcoming of single-channel angle sensing using conical and linear scanning or sequential lobing.

Recently, widespread use has been made of the monopulse method of target angle measurement, in which complete information on the angular position of the target may be obtained through the processing of a single reflected pulse (hence the term "monopulse"). In order to implement the monopulse approach, multichannel reception is required. Signals simultaneously reflected from the target are received by pairs of independent channels in the receiver, one pair for each tracking coordinate (e.g., two channels in azimuth and two in elevation). Because monopulse systems can perform angle sensing with one pulse, using two independent channels in each coordinate, amplitude fluctuations in the reflected signal do not have a significant effect on the accuracy of the target angle measurement. The variation in a given pulse from the mean return is canceled out by processing the pair of signals for each coordinate.

The monopulse method was originally developed to achieve accurate automatic target tracking. In the West, monopulse is now also used for 3D radar surveillance systems, which are monopulse systems performing angle measurement for all targets within the limits of the antenna pattern, the beam direction being programmable.

We will distinguish between three basic monopulse techniques, each characterized by the manner in which the angle measurement is extracted from the received signals: amplitude-comparison, phase-comparison, and combination.

In monopulse systems employing amplitude-comparison angle sensing, two identical overlapping patterns are formed, offset $\pm \theta_0$ from the equi-signal axis (the axis along which the amplitudes of the two patterns are equal). When the target is offset by an angle $\theta$ from this axis (see Fig. 1.1(a)), the signal received through the lower pattern has a greater amplitude than that received through the upper. The amplitude of this difference signal determines the magnitude of the angular offset from the equisignal axis to the target. The sign of the difference indicates the target direction in the measured coordinate (e.g., up or down from the axis for the case of elevation). More precisely, the direction of the target is given by the cosine of the difference signal's phase relative to a reference signal, this phase difference being restricted to 0 or 180 degrees in the ideal case. When the target is located on the axis, the amplitudes of the target echoes received through both patterns are equal, and the difference signal will ideally have zero magnitude.

In phase-comparison monopulse systems, the target angle in a given coordinate is determined by comparing the phases of the signals received by a pair of antennas (see Fig. 1.1(b)). In the far field, the two patterns virtually coincide and, as a result, echoes received at the two antennas from a target in the far field have identical amplitudes, but different phases. This situation is illustrated in Fig. 1.1(b), with the phase centers of the antennas separated by the distance $l$.

The line of sight to the target forms an angle $\theta$ with the joint axis of the antennas, which passes through the midpoint of the line joining their phase centers. The distance to the target from antenna 1 is

$$R_1 = R + (l/2)\sin \theta$$

and from antenna 2

$$R_2 = R - (l/2)\sin \theta$$

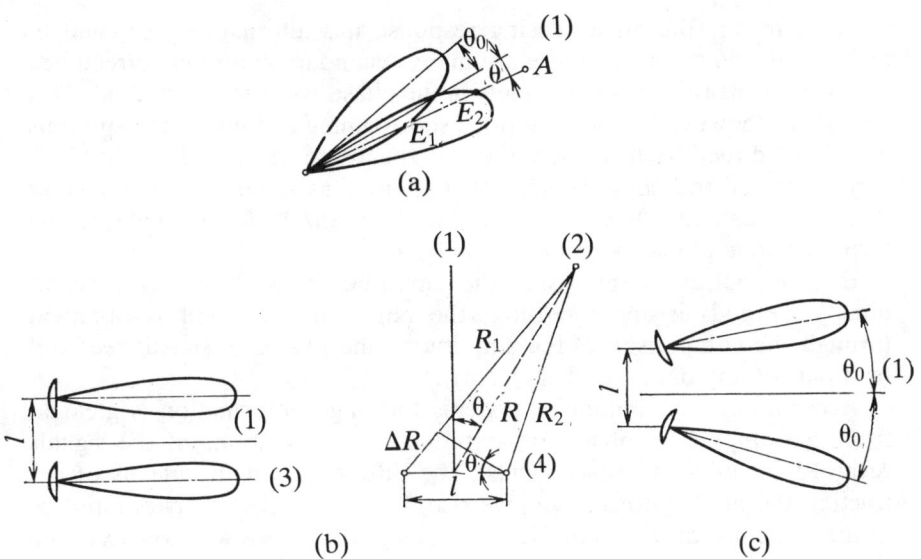

**FIGURE 1.1.** Partial antenna patterns and comparison of signals in amplitude-comparison (a), phase-comparison (b), and combination (c) monopulse systems.

| | |
|---|---|
| (1) equisignal direction | (3) antenna 1 |
| (2) target | (4) antenna 2 |

The difference between these displacements:

$$\Delta R = R_1 - R_2 = l \sin \theta$$

gives the phase difference:

$$\Delta\phi = 2\pi\Delta R/\lambda = 2\pi l \sin(\theta/\lambda) \qquad (1.1)$$

where $\lambda$ is the wavelength.

This is the relation which makes it possible to determine the target offset angle $\theta$ from the phase difference in signals arriving at two displaced antennas.

Equation (1.1) shows that this phase shift goes to zero, not only when $\theta = 0$, i.e., when the target is along the axis, but also when the target is at certain angular offsets satisfying the condition:

$$\theta = \arcsin(2n\pi/k_\lambda l), \qquad k_\lambda = \frac{2\pi}{\lambda}$$

As a result, the angle-sensing response has alternating sign, and in addition to the main axis there are many secondary equisignal directions. This causes measurement ambiguity in the phase-comparison method. This ambiguity, however, is not a serious shortcoming as long as the spurious equisignal directions lie outside the main beam, to which end the displacement between the phase centers of the antennas should be kept smaller than their diameter. Such a system may be realized, for example, in the form of two antennas positioned side by side.

In monopulse systems using the amplitude-phase-comparison (combined) method of angle sensing, the angle measurement is obtained through the comparison of the amplitudes and phases of signals received by a pair of antennas (see Fig. 1.1(c)).

Accordingly, in monopulse systems the angle information is gleaned from the amplitude, phase, or amplitude-phase relations of the signals received in independent channels. The differences in the means of extracting the angle information give rise, in turn, to particular differences in the structure of a monopulse system as a whole. We will now examine the basic structure of a monopulse system.

## 1.2 THE STRUCTURE OF A MONOPULSE SYSTEM

As has been explained, target angle measurement in monopulse systems is achieved in a processor which compares the received signals in pairs. A feature of this comparison is that the voltage at the output of the processor is not dependent on the received signal amplitudes, but rather is determined by the angle of arrival of the signals.

The angle-sensing response, i.e., the output of the initial monopulse signal processor, should indicate the value and sign of the angle of arrival of the received signals. In other words, the response for each measured coordinate should be an odd, real function of the angle of arrival projected onto that coordinate. Initial information about the angle of arrival contained in the pairs of signals is formed upon reception at the antenna, which is the angle sensor. At this point, such operations should be performed as necessary to obtain a function whose real part satisfies the above requirements for the angle-sensing response.

We will now examine some possible approaches to the problem [32, 35]. The field intensity at the point of reception may be written as a complex function of time:

$$\underline{E}(t) = E_m \exp i(\omega t - \phi_0) \tag{1.2}$$

where $E_m$ is the amplitude, $\omega$ is the angular frequency, and $\phi_0$ is the initial phase.

The beam patterns of the antenna are complex functions of the angle of arrival and may be expressed as

$$f_1(\theta) = F_1(\theta)\exp i\phi_1(\theta)$$
$$f_2(\theta) = F_2(\theta)\exp i\phi_2(\theta)$$

(1.3)

where $F_1(\theta)$ and $F_2(\theta)$ are the amplitude patterns and $\phi_1(\theta)$ and $\phi_2(\theta)$ are the phase patterns.

We will consider the phase patterns to be mirror reflections of one another across the equisignal axis, and the amplitude patterns to be identical with maxima displaced from the axis by $\pm\theta_0$, i.e.,

$$\phi_1(\theta) = \tfrac{1}{2}\phi(\theta); \qquad \phi_2(\theta) = -\tfrac{1}{2}\phi(\theta)$$
$$F_1(\theta) = F(\theta_0 - \theta); \qquad F_2(\theta) = F(\theta_0 + \theta)$$

(1.4)

where $\phi(\theta)$ is the antenna phase difference pattern.

Reflected signals received by the spatially separated antennas may be written as

$$\underline{E}_1(t, \theta) = \underline{E}(t)f_1(\theta), \qquad \underline{E}_2(t, \theta) = \underline{E}(t)f_2(\theta)$$

(1.5)

Then, for the ratio of the two received signals, we have

$$r_m(\theta) = \frac{\underline{E}_1(t, \theta)}{\underline{E}_2(t, \theta)} = \frac{f_1(\theta)}{f_2(\theta)} = \frac{F_1(\theta)\exp i\phi_1(\theta)}{F_2(\theta)\exp i\phi_2(\theta)}$$

(1.6)

We shall call this the multiplicative ratio, and under certain conditions it can serve as the initial ratio in forming the angle-sensing response.

A different response can be formed by using the ratio of the difference to the sum signal:

$$r_a(\theta) = \frac{\underline{E}_1(t, \theta) - \underline{E}_2(t, \theta)}{\underline{E}_1(t, \theta) + \underline{E}_2(t, \theta)} = \frac{f_1(\theta) - f_2(\theta)}{f_1(\theta) + f_2(\theta)}$$

(1.7)

which is usually called the additive ratio of the received signals. The relation between the two ratios may be expressed by the following expressions:

$$r_a(\theta) = \frac{r_m(\theta) - 1}{r_m(\theta) + 1}, \qquad r_m(\theta) = \frac{1 + r_a(\theta)}{1 - r_a(\theta)}$$

(1.8)

Aside from the multiplicative and the additive ratios there are others which satisfy the requirements for angle sensing in the monopulse processor. With multiplication of the multiplicative and additive ratios by the complex constant:

$$\underline{a} = a_0 \exp i\psi_0$$

we obtain the linearly transformed quantities

$$\underline{r}_m(\theta) = \underline{a}r_m(\theta), \qquad \underline{r}_a(\theta) = \underline{a}r_a(\theta) \tag{1.9}$$

Devices in which the sum and difference signals are formed or in which multiplication of the received signal by the complex coefficient $\underline{a}$ occurs are called comparators; they are also called converters, because in general it is possible to transform amplitude relations into phase relations and *vice-versa*. These transformations are usually accomplished at RF, using passive elements which feature relative simplicity and stability.

It is not possible to compute the ratio of two signals, converted or not, without first amplifying them. A device making use of active elements, including amplifiers and a comparison circuit, which produces the multiplicative or additive ratio and the angle-sensing response, is called an angle discriminator.

In amplitude-comparison monopulse the angle information is contained in the amplitude patterns $F_1(\theta)$ and $F_2(\theta)$. In this case, with identical phase patterns, the multiplicative and additive ratios may only be expressed through the amplitude patterns:

$$\begin{aligned} r_m(\theta) &= \frac{F_1(\theta)}{F_2(\theta)} = \rho(\theta) \\[2mm] r_a(\theta) &= \frac{F_1(\theta) - F_2(\theta)}{F_1(\theta) + F_2(\theta)} = \frac{\rho(\theta) - 1}{\rho(\theta) + 1} \end{aligned} \tag{1.10}$$

In phase-comparison monopulse, when the amplitude patterns are identical, the angle information is contained in the phase difference:

$$\Delta\phi = \phi_1(\theta) - \phi_2(\theta) \tag{1.11}$$

The multiplicative and additive ratios may be expressed only through the phase patterns:

$$\begin{aligned} r_m(\theta) &= \frac{\exp i\phi_1(\theta)}{\exp i\phi_2(\theta)} = \exp i\Delta\phi \\[2mm] r_a(\theta) &= \frac{\exp i\phi_1(\theta) - \exp i\phi_2(\theta)}{\exp i\phi_1(\theta) + \exp i\phi_2(\theta)} = i \tan \frac{\Delta\phi}{2} \end{aligned} \tag{1.12}$$

The ratio of the amplitudes $\rho(\theta)$ and the phase shift $\Delta\phi$ will be referred to as multiplicative angle functions, and the ratios $[\rho(\theta) - 1]/[\rho(\theta) + 1]$ and $\tan(\Delta\phi/2)$ as additive angle functions.

Angle discriminators using multiplicative angle functions and responding only to the amplitude relations in forming the angle-sensing response are called amplitude angle discriminators [they are noncoherent processors because they remove the phase information from the input signals]. Angle discriminators responding only to the phase relations are called phase angle discriminators. Discriminators which react to both the amplitude and phase relations and which use an additive angle function to form the response are called sum and difference angle discriminators.

Thus there are only three possible methods of angle measurement (three types of discriminators): amplitude, phase, and sum-and-difference. Each of these may be applied in amplitude-comparison, phase-comparison, or combination monopulse systems. Resulting are nine basic classes of monopulse systems as classified in Table 1.1. Throughout the remainder of the text, monopulse systems will be characterized on the basis of this classification. The first term will describe the type of angle sensing (i.e., the nature of the information in the signal prior to any processing); the second will identify the type of angle discriminator used (i.e., the nature of the processing used to extract this information). For example, an amplitude-comparison monopulse system using amplitude angle discrimination will be referred to as an amplitude-amplitude (AA) system, and, with the same discriminator, a phase-comparison system would be a phase-amplitude (PA) system. The other seven classes are named similarly (see Table 1.1).

**Table 1.1**

| method of measurement (type of angle discriminator) | basic classes of monopulse systems for three types of angle sensing | | |
|---|---|---|---|
| | amplitude (A) | phase (P) | combination (C) |
| amplitude (A) | AA | PA | CA |
| phase (P) | AP | PP | CP |
| sum and difference (SD) | ASD | PSD | CSD |

Of the nine classes, four are used the most in practice: amplitude-amplitude (AA), phase-phase (PP), amplitude-sum-and-difference (ASD), and phase-sum-and-difference (PSD). These monopulse systems will be considered in more detail in later chapters.

Stemming from the complete sequence of operations, the block diagram outlining a monopulse system should have the following basic elements (see Fig. 1.2):

- an angle sensor which forms signals containing the target angle information;
- a comparator or converter, which transforms the target angle information into some combination of amplitude and phase relations between the signals in two independent channels; and
- an angle discriminator, which forms a real function of the ratio of signal parameters, the function having a single-valued relation to the angle of arrival of the signal.

The antenna of the monopulse system is the angle sensor. It is the most important element of the system and has several characteristic features, which will be considered in detail in Chapter 2.

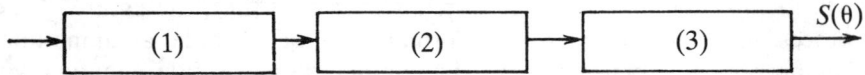

**FIGURE 1.2.** Block diagram of a monopulse system.

| (1) angle<br>    detector | (2) comparator<br>    (converter) | (3) angle<br>    discriminator |

The converter may be a 90-degree phase shifter which multiplies by $\pm i$, or a sum-and-difference comparator employing a hybrid ring junction or magic tee. A sum-and-difference device is examined in Chapter 3.

## 1.3 ANGLE DISCRIMINATORS IN MONOPULSE SYSTEMS

The structure of angle discriminators is presented in Figs. 1.3. to 1.5. In order to preserve the symmetry of the signals and maintain the phase coherence between them, a single local oscillator (LO) is used to generate the intermediate frequency for both channels. The ratio of the received signals $r_m(\theta)$ or $r_a(\theta)$, characterizing the arrival angle, is generated as a result of normalized amplification within the circuits.

In the amplitude angular discriminator shown in Fig. 1.3(a), the ratio $r_m(\theta)$ results from the subtraction of the logarithms of the two signal amplitudes. Since this is equivalent to the logarithm of their ratio, the output voltage is independent of the signal input level. The expression $\ln r_m(\theta)$ is equal to zero for the equisignal direction (the amplitudes are equal and $r_m(\theta) = 1$) and has an odd symmetry relative to this direction. Therefore, it may be used to obtain the angle-sensing response:

$$S(\theta) = \text{Re}[\ln r_m(\theta)] = \ln|u_1(t, \theta)| - \ln|u_2(t, \theta)| \qquad (1.13)$$

where $|u_1(t, \theta)|$ and $|u_2(t, \theta)|$ are the amplitudes at the output of the receiving-amplifying channels as functions of time and the angle error.

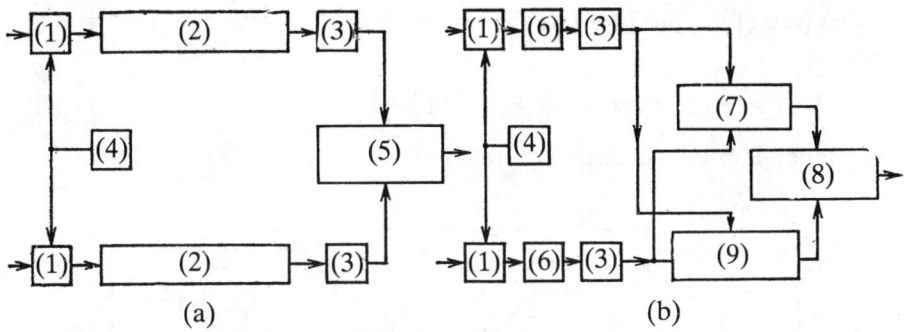

FIGURE 1.3. Amplitude angle discriminator:
   (a) with logarithmic IF amplifiers
   (b) with normalization by the sum signal at video frequency

   (1) mixer                    (6) IF amp
   (2) logarithmic IF amp       (7) adder
   (3) amplitude detector       (8) division circuit
   (4) local oscillator         (9) subtraction device
   (5) subtraction device

The phase difference between the signals present at the input to the amplifiers has no effect on the ratio $r_m(\theta)$, because the phase is lost in the amplitude detectors prior to subtraction. The amplitude responses of the two logarithmic amplifiers are unstable and nonuniform, leading to distortion in the angle-sensing response and hence the target angle measurement, which is the fundamental disadvantage of amplitude angle discriminators using logarithmic amplifiers.

Logarithmic amplifiers are not used in the amplitude angle discriminator shown in Fig. 1.3(b); the difference signal is normalized against the sum signal at video frequencies in the division circuit. In practice, with the sum signal in an amplitude-amplitude monopulse system having a dynamic range on the order of 80–100 dB, it may not be possible to use linear IF amplifiers. In this case, logarithmic amplifiers must be employed, minimizing the influence of their instability with the help of a periodic sequence of control signals.

In a phase angle discriminator, the amplitude modulation must be eliminated prior to normalization. One solution, shown in Fig. 1.4(a), is to use limiting amplifiers whose output amplitudes are independent of the input amplitudes; another possibility (see Fig. 1.4(b)) is to employ an independent AGC circuit in each channel. It is then possible to use the function $ir_m(\theta)$ (see (1.12)) to generate the angle-sensing response, because its real part:

$$S(\theta) = \text{Re}[-i \exp i\Delta\phi] = \sin \Delta\phi \qquad (1.14)$$

is independent of the amplitude relations.

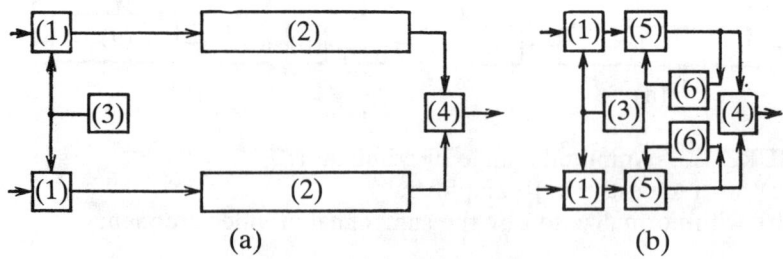

(a)                                                    (b)

**FIGURE 1.4.** Block diagrams for phase-angle discriminators achieving normalization with limiting amplifiers (a) and AGC (b).

(1) mixer                              (4) phase detector
(2) IF amp with limiter        (5) IF amp
(3) local oscillator               (6) AGC

In the sum-and-difference angle discriminators shown in Fig. 1.5(a), (b), the ratio $r_a(\theta)$ is formed through the use in each channel of AGC circuits, controlled by the sum signal voltage. Thus, both the sum-and-difference signals are normalized with respect to the sum signal. The additive ratio is used to form the angle-sensing response and, with identical receiving channels, for the case of amplitude-comparison monopulse we have

$$S(\theta) = \text{Re}\left[\frac{f_1(\theta) - f_2(\theta)}{f_1(\theta) + f_2(\theta)}\right] = \frac{F_1(\theta) - F_2(\theta)}{F_1(\theta) + F_2(\theta)} \qquad (1.15)$$

and for phase comparison monopulse:

$$S(\theta) = \text{Re}\left[\frac{\exp i\phi_1(\theta) - \exp i\phi_2(\theta)}{\exp i\phi_1(\theta) + \exp i\phi_2(\theta)}\right] = \tan\frac{\Delta\phi}{2} \qquad (1.16)$$

The sum-and-difference angle discriminators shown, in Fig. 1.5(a), (b), differ in the means of forming the magnitude and sign of the error signal, and are used in automatic tracking monopulse radars. In surveillance monopulse systems (see Fig. 1.5(c)), normalization in the sum-and-difference angle discriminator does not occur at IF with AGCs, but through division of the phase detector (PD) output by the sum signal at video frequencies. The time constant of AGC systems renders them unusable in surveillance radars, which must determine the coordinates of all resolved targets within the beam.

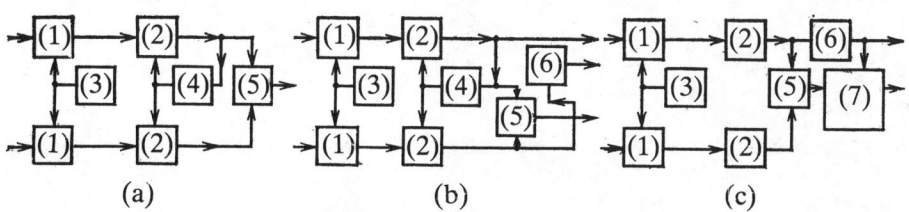

(a)                          (b)                          (c)

**FIGURE 1.5.** Sum-and-difference angle discriminators: with combined (a) and separate (b) formation of the magnitude and sign of the error signal, and with normalization by the sum signal at video frequency (c).

(1) mixer               (5) phase detector
(2) IF amp              (6) amplitude detector
(3) local oscillator    (7) division current
(4) AGC

# Chapter 2
## Monopulse Radar Antennas

### 2.1 PARABOLIC ANTENNAS

A diagram of a parabolic antenna, which consists of a feed and a metallic reflector in the form of a paraboloid of rotation, is shown in Fig. 2.1(a). The two basic properties of a parabolic antenna are as follows: (1) diverging beams radiated by a source at the focus become parallel upon reflection at the surface, and (2) the distance traveled by any beam from the focus to a plane perpendicular to the axis upon reflection from the surface is independent of the angle at which the beam left the focus. As a result, such antennas produce a plane wavefront with uniform phase.

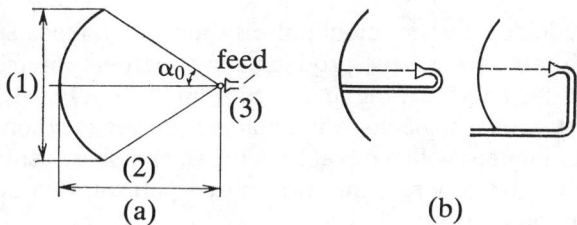

(a)  (b)

**FIGURE 2.1.** Parabolic antenna:
(a) illustration of basic parameters:
  reflector diameter $(d_p)$
  focal length $(f_p)$
  focal point $(F_p)$
  angular aperture dimension $(2_o)$
(b) feed supports
  (1) $d_p$  (2) $f_p$  (3) $F_p$

If the aperture angle $2\alpha_0 > \pi$, then the paraboloid is termed short-focus; if $2\alpha_0 < \pi$, it is long-focus. For a short-focus paraboloid the ratio of the focal length to the diameter $f_p/d_p < 0.25$, and for a long-focus $f_p/d_p > 0.25$. (Western engineering tradition sets the threshold at $2\alpha_0 = \pi/2$, which corresponds to $f_p/d_p = 0.6$.) Parabolic antennas used in monopulse systems are designed with $f_p/d_p$ lying in the range 0.5 to 1.0, so that the patterns overlap at the desired level.

The type of feed employed most often in parabolic antennas is the waveguide horn, which may be mounted by either method shown in Fig. 2.1(b). In both cases, the aperture is partially shadowed by the feed, waveguide, and supporting structure, which alters the effective antenna pattern. In addition, some energy is reflected back into the feed and waveguide, degrading the performance of the transmitter.

If the feed is displaced by an angle $\alpha_1$ from the focus, and moved in a circle about the axis, then the main beam will be offset from the axis by an angle $\alpha_2$, which is opposite to and somewhat smaller than the feed displacement $\alpha_1$, as shown in Fig. 2.2.

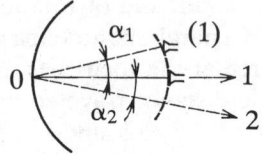

**FIGURE 2.2.** Moving the beam by displacing the feed:

(1) displaced feed

In amplitude-comparison monopulse systems, two feeds are positioned symmetrically about the axis, producing symmetrical intersecting amplitude patterns for angle sensing in one coordinate (see Fig. 1.1). In phase-comparison systems, single-coordinate angle sensing is accomplished with two separate antennas with feeds at each focus, and phase centers separated by a distance $l$. To generate the transmitted pattern, one or both of the antennas may be used.

In order to realize the benefits considered below, parabolic antennas with subreflectors (Cassegrain antennas) are widely used in monopulse systems. The feed in such antennas is placed at the main reflector, and between the feed and the focus of the paraboloid is a hyperbolic subreflector. One focus of the hyperboloid is coincident with the paraboloid focus at the point $F'_p$. The second hyperboloid focus is coincident with the center of the feed along the joint axis of the reflectors. Using the first-order approximations of geometric optics, it may be shown that all rays emitted by the feed leave the Cassegrain antenna in a plane wave with uniform phase.

The parameters of the Cassegrain antenna shown in Fig. 2.3(a) are related as follows:

$$\tan(\alpha_0/2) = d_p/4f_r \tag{2.1}$$

$$1/\tan \alpha_0 + 1/\tan \alpha_r = 2/f_r/d \tag{2.2}$$

$$\sin[(\alpha_0 - \alpha_r)/2]/\sin[\alpha_0 + \alpha_r)/2] = 2f_0/f_r \tag{2.3}$$

In practice, the required antenna dimensions and characteristics indicate the necessary values for the parameters $d_p$, $f_p$, $f_r$, and $\alpha_r$, from which are derived the values of $\alpha_0$, $d_s$, and $f_0$.

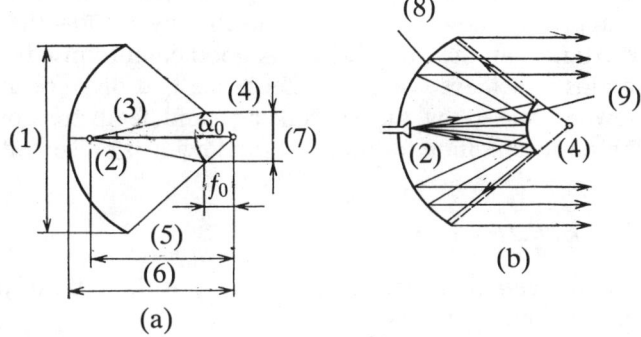

**FIGURE 2.3.** Parabolic antenna with subreflector:
(a) antenna geometry
(b) antenna operation
   real focus ($F_r$)
   imaginary (paraboloid) focus ($F_p$)

     (1) $d_p$  (4) $F'_p$  (7) $d_s$
     (2) $F_r$  (5) $f_r$  (8) parabolic reflector
     (3) $\alpha_r$  (6) $f_p$  (9) hyperbolic subreflector

If the subreflector is considered to be a hyperbolic mirror producing a mirror image of the feed at the focal point of the paraboloid $F'_p$, then the Cassegrain antenna (see Fig. 2.3) may be viewed as an ordinary parabolic antenna with the advantage of a focal length increased by the factor

$$K_h = (l_h + 1)/(l_h - 1) \tag{2.4}$$

where

$$l_h = \sin[(\alpha_0 + \alpha_r)/2]/\sin[(\alpha_0 - \alpha_r)/2]$$

is the eccentricity of the hyperboloid.

The increase in effective focal length is also given by the ratio of the distance between the subreflector and the real focus to the distance from the subreflector to the imaginary focus:

$$K_h = (f_r - f_0)/f_0 \tag{2.5}$$

On the other hand, the subreflector shadows the aperture, decreasing the antenna gain and raising the level of the sidelobes. These effects may be reduced by decreasing the dimensions of the subreflector while increasing the directivity of the feed or placing it closer to the main reflector. With this approach, however, the feed itself may shadow the aperture to a greater extent than the subreflector. A good compromise between these considerations is achieved when the feed size and distance are such that the shadowing caused by the subreflector and feed are approximately equal. The basic condition for minimal shadowing is presented in [42] (see Fig. 2.4):

$$f_r/f_p \approx k_f d_f^2/2F_r\lambda \approx d_f/d_s \tag{2.6}$$

where $d_f$ is the feed-aperture diameter, and $k_f$ is the ratio of the effective and geometric feed apertures.

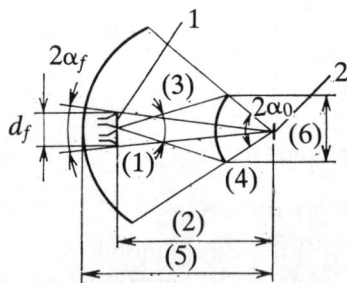

**FIGURE 2.4.** Geometry for minimizing shadowing:
   1   real feed
   2   imaginary feed

(1) $F_r$     (4) $F_r'$
(2) $f_r$     (5) $f_p$
(3) $2\alpha_r$   (6) $d_s$

The approximation (2.6) is obtained with the assumption that the angles $\alpha_r$ and $\alpha_f$ are small and that the subreflector is closer to the imaginary focus than to the feed.

The minimum diameter of shadowing is then given by

$$d_{s\ min} \approx \sqrt{(2/k_f)f_p\lambda} \tag{2.7}$$

The above relations concerning minimum shadowing apply for any polarization. With linear polarization it is possible to reduce the effects of shadowing further by the use of polarization twisting. This method employs a horizontal grating as a subreflector (Fig. 2.5), which reflects horizontally polarized radiation while passing vertically polarized waves. Conversion of the beam from horizontal to vertical polarization takes place at the main reflector, after which the beam passes virtually unaffected through the subreflector, which now causes almost no blockage.

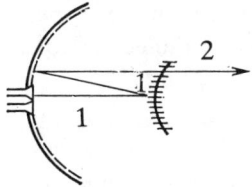

**FIGURE 2.5.** Polarization twisting:
1  horizontally polarized wave
2  vertically polarized wave

Due to the possible small dimensions of the feeds, the feed blockage is small, comparable to that for typical parabolic reflectors. Consequently, it is advantageous to use large subreflectors and small feeds in systems employing phase twisting. The separate patterns required for monopulse angle sensing are formed in Cassegrain antennas in the same manner as in parabolic antennas.

The greatest advantage to double-reflector antennas is the shortening of the waveguides supplying the feed, which is now at the surface of the main reflector. With shorter feed lines, smaller phase errors are introduced between the individual segments, thereby reducing the total angle error. Furthermore, it is possible to use low-noise amplifiers (transistor or parametric) in the receiver, because they can be placed directly next to the feeds. In an ordinary parabolic antenna with the feed at the focus, placing the amplifiers behind the reflector allows the significant losses in the feed line to degrade the sensitivity, while placing the amplifiers with the feed at the focus increases the aperture blockage.

Another important property of double-reflector antennas is that the effective focal length can exceed the axial dimension of the antenna; in other words, an antenna with a small $f_p/d_p$ ratio can produce the same effects as an antenna with a larger ratio. This ratio for double-reflector antennas can be as much as double that for ordinary parabolic antennas, allowing a Cassegrain antenna to be half the depth of an equivalent parabolic antenna.

A third advantage to double-reflector antennas is the ability to scan the beam by moving the subreflector, which removes the difficulties attending use of a rotating feed.

## 2.2 LENS ANTENNAS

Monopulse radars also employ lens antennas, which operate on the basis of the refraction of the beam at the interface of two media. From elementary optics it is known that if a beam falls on the plane surface between media of respective dielectric constants $\epsilon_1$ and $\epsilon_2$, the angle of refraction is given by the relation:

$$\sin \alpha_1 = (n_1/n_2)\sin \alpha_2 \tag{2.8}$$

where $\alpha_2$ is the angle of incidence, and $n_1$ and $n_2$ are the indices of refraction.

The index of refraction gives the ratio of the propagation speed of electromagnetic waves in a vacuum to the propagation speed in the given medium. In the absence of strong magnetic effects, it is given by the square root of the medium's dielectric constant:

$$n_1 = \sqrt{\epsilon_1}, \qquad n_2 = \sqrt{\epsilon_2}$$

Thus,

$$\sin \alpha_1 = \sqrt{\epsilon_1/\epsilon_2} \sin \alpha_2 \tag{2.9}$$

Using these relations, the physics of lens-antenna operation may be explained as follows (see Fig. 2.6). In the air ($n_1 = 1$) is a source $F_b$, emitting a bundle of diverging beams. The beams propagate through the dielectric lens more slowly than through the air, adding a phase shift proportional to the distance traveled by the beam through the lens. The lens decreases in thickness from the center to the edge, resulting in diminishing phase shifts towards the edge, which compensates for the longer path lengths from the source to points off the lens center. Furthermore, the lens profile is such that the incident spherical wavefront is "straightened"; as a result of the combination of these effects, a planar wavefront of uniform phase leaves the front face of the lens.

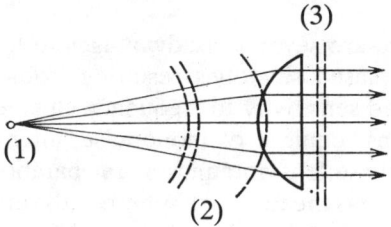

**FIGURE 2.6.** Dielectric lens:
(1) $F_1$   (2) spherical wavefront   (3) plane wave

As a result of their high cost and large mass, ordinary dielectric lenses are not often used in monopulse radar systems. Instead, metal-plate lenses and Luneberg lenses are most widely used.

The metal-plate lens (see Fig. 2.7(a)) consists of plates, parallel to the electric field vector and separated from one another by the distance $l_l$ ($\lambda/2 < l_l \leq \lambda$). The space between the plates acts as a waveguide, in which the phase velocity is higher than in free space, and has a corresponding index of refraction less than unity and given by the formula

$$n_2 = \sqrt{1 - \lambda^2/4l_l^2} \tag{2.10}$$

With the proper lens profile, all beams leaving the point $F_b$ arrive simultaneously and in phase at the aperture.

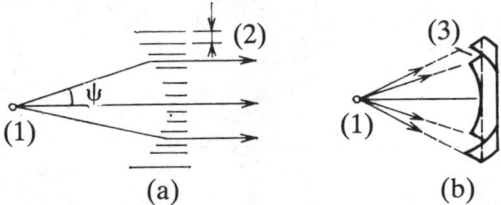

**FIGURE 2.7.** Metal plate lens:
(a) waveguide lens
(b) stepped lens
(1) $F_1$   (2) $l_1$   (3) $l_n$

With decreasing distance between plates, the index of refraction becomes smaller and the lens may be made thinner. But even for an index of refraction on the order of 0.5 ($l_l = 0.56\lambda$), the lens thickness is substantial. To decrease the lens thickness, stepped metal-plate lenses may be used. The path length in such lenses is decreased by the stepped profile. The depth of each step $l_n$ is chosen to produce a corresponding phase shift of 360 degrees, leaving the coherence at the front face of the lens undisturbed: $l_n = \lambda/(1 - n \cos \psi)$. The thickness of the lens is reduced to that at its center. There are several disadvantages to lens antennas, including energy losses, increased sidelobes resulting from the shadowing of the steps, and increased sensitivity to frequency changes.

The two patterns required of monopulse antennas are formed in the same manner for these lens antennas as for parabolic antennas.

A Luneberg lens has the form of a sphere with varying index of refraction (Fig. 2.8). Because most land-based and shipboard radars scan or track only in the upper hemisphere, hemispherical Luneberg antennas may be used, with a mirror surface at the base producing a reflection at $\theta_u'$ of a source at $\theta_u$. Due to its spherical symmetry, the focusing ability of such lenses is independent of the angle of arrival of the beams. The fundamental property of a Luneberg lens is that an incident plane wave is focused on a point on the opposite side of the surface and, conversely, beams emitted from a point source at $\theta_u$ on the surface of the sphere are transformed by the lens into plane waves.

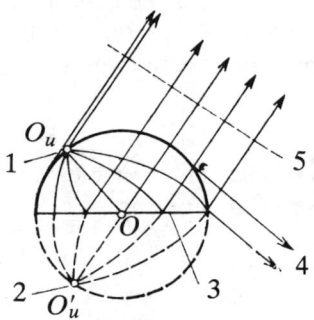

**FIGURE 2.8.**  Hemispherical Luneberg lens:
1  source
2  source image
3  reflecting surface
4  lost radiation
5  plane wave

The index of refraction in a Luneberg lens has the following form:

$$n_2 = \sqrt{2 - r_u^2/r_l^2} \tag{2.11}$$

where $r_l$ is the radius of the lens, and $r_u$ is the distance from the center to the point where the index of refraction is being calculated.

Equation (2.11) indicates that the index of refraction is greatest at the center and when equal to $\sqrt{2}$; it is equal to 1 on the surface of the lens.

Moving a source along the surface of a Luneberg lens causes the corresponding beam to move accordingly, in the opposite direction. The beam may be scanned in one of two ways: by moving the feed along the lens surface, or by using a large number of feeds distributed over the surface, switching the transmitter or receiver from one feed to another. To produce more than one beam, a corresponding number of feeds must be placed appropriately on the lens surface. Luneberg lens antennas may be used in systems in which the beam must be scanned rapidly over a wide angle, or when the antenna must be placed on an unstable platform, such as a ship. In this case, stabilization of the beam with ship motion may be achieved through compensating control of the feed location.

Lens antennas have several advantages over reflector antennas. One of the more important is the absence of aperture shadowing with the feed and waveguides lying outside the radiated field. Another advantage is the capability for rapid scanning over large sectors; a Luneberg lens, for example, can scan the entire upper hemisphere. In addition, lens antennas are subject to more lenient mechanical and electrical tolerances, and have low inertia.

The drawbacks of lens antennas include low efficiency resulting from losses in the lens material, large space requirements, and complexity of construction.

## 2.3 SPIRAL ANTENNAS

In monopulse systems operating over a wide frequency range, spiral antennas may be used. Several types of spiral antennas are encountered in practice (see Fig. 2.9): cylindrical, conical, flat, and others named according to the shape of the surface on which they are constructed. The parameters of a cylindrical spiral antenna are the radius $r_c$, step $S_c$, number of turns $N_c$, and length of the spiral $l_c$; a description of a conical spiral includes the angle $\alpha_c$ at the apex of the cone.

Geometrically, a flat spiral is the projection of a conical spiral onto a plane perpendicular to the cone's axis. The spirals shown in Fig. 2.9 are unifilar; multifilar spirals are also used. Figure 2.10 shows bifilar, trifilar, and quadrifilar spiral configurations. In the bifilar case, the two elements are driven 180 degrees out of phase. The phase shift between elements of a trifilar spiral is 120 degrees, and of a quadrifilar spiral 90 degrees.

Spiral antennas can generate several symmetrical patterns. According to Western literature [19], flat spiral antennas can be used in monopulse systems in the guidance heads of certain missiles, and also in aircraft radar systems. To obtain radiation in one direction, they are placed in front of a metal screen, which is usually mounted on a dielectric substrate.

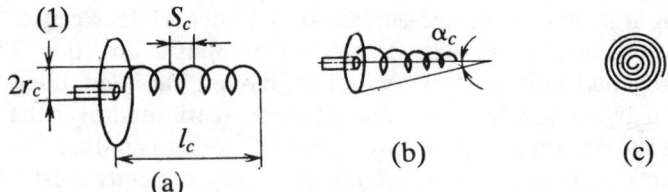

**FIGURE 2.9.** Spiral antennas:
  (a) cylindrical
  (b) conical
  (c) plane
    (1) screen

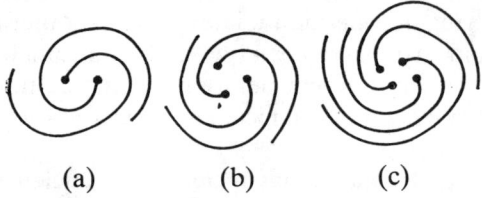

**FIGURE 2.10.** Multifilar plane spiral antennas:
  (a) bifilar
  (b) trifilar
  (c) quadrifilar

The diagram for a quadrifilar flat spiral antenna with electronic beam scanning is shown in Fig. 2.11 [19]. The four elements are driven 90 degrees out of phase, and the distance between the turns is kept constant along the spiral's length. Feed lines enable the spiral to be center-fed. When the feeds are in phase, a sum pattern is generated; when phase shifts are introduced in the feeds, the pattern is deflected and becomes narrower.

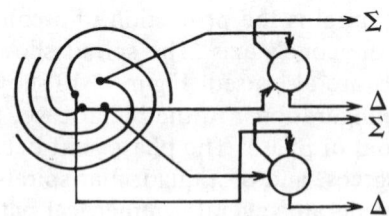

**FIGURE 2.11.** Basic circuit for a plane spiral monopulse antenna:
  sum signal  $(\Sigma)$
  difference signal  $(\Delta)$

The deflection angle may be altered by changing the frequency of operation. During reception, the necessary phase shifts in the receiving channels may be obtained with a phase inverter and matching devices.

## 2.4 PHASED ARRAY ANTENNAS

Wide use is made in monopulse systems of phased array antennas, which are antenna systems consisting of a large number of radiating elements arranged in a prescribed manner. When signals equal in amplitude and phase reach the elements, a beam is formed perpendicular to the plane of the array. The beam may be scanned by introducing the proper phase shifts in the individual signals supplying the array elements.

Three basic types of feeds are distinguished for phased arrays [9]: space feed-through, space-reflecting, and constrained feeds.

Space feed-through arrays (see Fig. 2.12(a)) have two groups of elements, front (2) and rear (1), along with a common feed (3). Between the rear and front radiators are phase shifters ($\phi$). During transmission, the common feed excites the rear radiators, from which the energy is fed through the phase shifters and radiated into space from the front elements. The phase shifter sets the proper phase shifts between the individual elements to steer the beam in the desired direction.

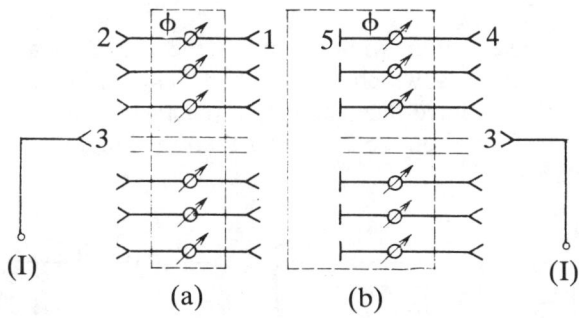

**FIGURE 2.12.** Antenna array with space feeds:
  (a) space feed
  (b) space-reflecting feed
    1   front elements
    2   rear elements
    3   common feed
    4   array element
    5   shorted line segments
    $\phi$   phase shifters
  (I) input

During reception, the arriving beam excites the front elements, from which the energy is channeled through the phase shifters to the rear elements, which, in turn, drive the common feed, now switched to the receiving channel. The received signal will be strongest when the signals from the rear elements arrive at the feed aperture in phase; this is accomplished through shifts introduced by the phase shifters.

In a phased array with a space-reflecting feed system (see Fig. 2.12(b)), there is one common feed (3) and one group of radiating elements (4). During both transmission and reception the phase shifts are produced by the phase shifters (ϕ), which are matched to shorted line segments (5), whose lengths are determined by the operating wavelength. The operation of this antenna is analogous to that of a space feed-through antenna, with the energy passing through the phase shifters twice, having been reflected from the shorted terminals.

In constrained-feed arrays, the output of the transmitter (or input of the receiver) is connected to the array elements through the phase shifters with power dividers and feed lines.

Phased array antennas use diode, gas, and ferrite phase shifters. Three techniques are used to control the phase shifters: continuous, discrete, and binary switching. Binary switching is the most promising method, because it is economical and provides highly stable beam steering. With binary phase-shifting, the process of steering the beam is reduced to on-off switching, and energy is consumed by the shifter control only at the moment of switching.

One such phase shifter widely used is the digital phase shifter with switching resonant irises, which are switched by *pin* diodes (see Fig. 2.13(a, b)) [41]. The diodes introduce capacitive reactance, and the effective iris aperture is therefore about 25% shorter than the length calculated for the iris without diodes.

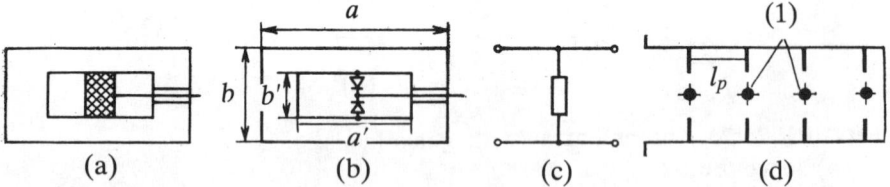

**FIGURE 2.13.** Resonant iris for a discrete phase shifter.
(1) diodes

If no control voltage is applied to the diodes, the electromagnetic waves pass along the waveguide through the aperture with negligible losses, because the resistance of the diodes is high and they have little effect on the circuit. The equivalent circuit for the switched iris (see Fig. 2.13(c)) is a transmission line with a parallel resistance. When a constant voltage is applied to the diodes, their resistance is abruptly reduced by a factor of 250 to 1000, the aperture is shunted and its resonance destroyed, and the propagating waves are almost completely reflected from the iris. If the necessary number of properly switched irises is placed at the shorted end of a waveguide, a switched reflecting phase shifter will be produced (see Fig. 2.13(d)). With uniform spacing, the phase shift is given by

$$\Delta_\phi = (2\pi/\lambda)2l_p \tag{2.12}$$

where $l_p$ is the distance between the irises.

The number of switching irises is

$$n_d = 2\pi/\Delta_\phi - 1 \tag{2.13}$$

An array comprised of $N$ equally spaced elements is shown in the diagram in Fig. 2.14. During reception, the sum voltage $u_0$ is produced at the output of the adder. We will assume that the array elements are isotropic radiators, emitting energy uniformly in all directions. The signal of the first element will be used as a reference with phase $\phi_0$. The phase shifts in neighboring elements are given by

$$\Delta_\phi = (2\pi l_\phi/\lambda) \sin \theta \tag{2.14}$$

Then the expression describing the resulting pattern of an array of istropic elements, called the array factor, has the form [55]:

$$F_a(\theta) = \frac{\sin(N\pi l_\phi \sin \theta/\lambda)}{N \sin(\pi l_\phi \sin \theta/\lambda)} \tag{2.15}$$

The calculations of $F_a(\theta)$ and other array characteristics performed below are approximate and should be used only as a guide.

If the array elements are not isotropic, then the full pattern is described by the expression:

$$F_\phi(\theta) = F_a(\theta)F_e(\theta) = F_e(\theta) \frac{\sin(N\pi l_\phi \sin \theta/\lambda)}{N \sin(\pi l_\phi \sin \theta/\lambda)} \tag{2.16}$$

where $F_e(\theta)$ is the pattern of a single array element.

Equation (2.16) is correct only if the elements have identical patterns. In practice, interaction between the elements makes it impossible to obtain identical patterns, so that (2.16) is only an approximation, which may be unsuitable for array calculations. The exact array pattern may be determined by adding the individual patterns, accounting for both amplitude and phase. In constructing the pattern of each element, the presence of the other elements must be taken into consideration.

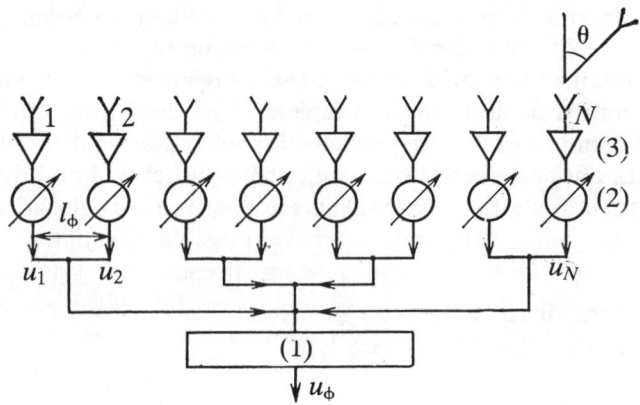

**FIGURE 2.14.** Diagram of a phased array.
(1) adder   (2) phase shifters   (3) amplifiers

For small values of $\theta$, when the aperture width is greater than several wavelengths, (2.15) may be cast in the form of a $(\sin x)/x$-dependence, i.e.,

$$F_a(\theta) = \frac{\sin[N\pi l_\phi \sin(\theta/\lambda)]}{N\pi l_\phi \sin(\theta/\lambda)} \tag{2.17}$$

If $l_\phi = \lambda/2$ in (2.17), then the half-power beamwidth in degrees will be

$$\theta_3 \approx 101.6/\sqrt{N} \tag{2.18}$$

The corresponding maximum gain with the beam axis perpendicular to the aperture plane (i.e., in the axial direction) is given by the approximation:

$$G_\phi \approx \pi N \tag{2.19}$$

With a beam deflection of $\theta$, the directional gain of the array is

$$G_\phi(\theta) \approx \pi N \cos \theta \tag{2.20}$$

The beamwidth increases approximately in proportion to the secant of the deflection angle $\theta$.

The beam may be steered through the angle $\theta_0$ by introducing a discrete linear phase shift from element to element, such that the phase shift between adjacent elements is $2\pi(l_0/\lambda) \sin \theta$. For beam directions in the range of $\pm 90$ degrees, there will be only one main lobe if the spacing between elements is equal to a half wavelength, i.e., $l_0 = \lambda/2$. If $l_0 > \lambda/2$, then grating lobes with amplitude equal to that of the main beam will appear in the pattern as the beam is scanned. It can be seen from (2.17) that the grating lobes will appear when both the numerator and denominator are equal to zero, i.e., when $\pi(l_0/\lambda)\sin\theta = 0$, $\pi$, $2\pi$, and so on. Thus, for example, with $l_0 = \lambda$, grating lobes will appear for $\theta = \pm 90°$, and when $l_0 = 2\lambda$ they appear for $\theta = \pm 30°$ and $\pm 90°$.

In practice, the array-element spacing is selected to obtain the desired pattern width and beam limits. It is also possible to suppress the grating lobes with nonuniform element spacing, because the grating lobes will appear in different directions for different portions of the array, and thus be "smeared." Furthermore, with nonuniform spacing, the number of array elements may be reduced without significantly altering the beamwidth. In an antenna array with nonuniform spacing, the element sites are most dense at the center of the array and become more separated with distance from the center, according to a prescribed arrangement [*see,* for example, *"Nonuniform Arrays"* by Merrill Skolnik, in *Antenna Theory,* Part 1 (R. Collin and Francis Zucker, eds., McGraw-Hill, 1969)—tr.]. The elements are placed symmetrically about the center of the array.

It is possible to generate multiple beams simultaneously with an array antenna. This is accomplished through the use of special multiports, with some inputs connected to the array elements, and the other ports connected to independent channels corresponding to the different beams. Such a multiport is called a beam-forming circuit, and is designed to generate the proper amplitude and phase distributions at the elements as its inputs are driven.

The two most common beam-forming circuits are the serial and parallel matrix circuits [9].

The serial matrix circuit (see Fig. 2.15) consists of two intersecting feed lines (1), connected at the intersections by low-gain directional couplers (2). The horizontal line inputs are the antenna inputs, and the vertical line outputs are connected to the array elements (3). The other end of the horizontal lines are terminated with matched loads (4).

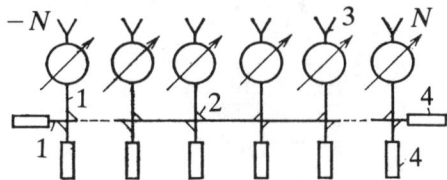

**FIGURE 2.15.** Serial beam forming circuit.

The basic elements of the parallel matrix circuit are eight-ports constructed either from waveguide slot bridges or sum-and-difference ring
bridges. Figure 2.16 shows a greatly simplified diagram of a monopulse
antenna array for single-coordinate angle sensing which employs a parallel
matrix beam-former.

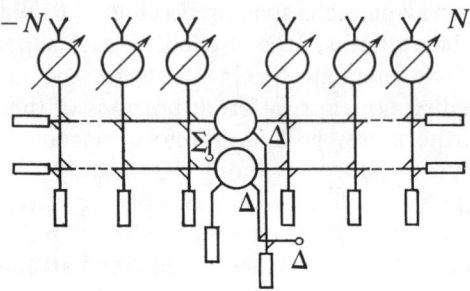

**FIGURE 2.16.** Diagram of a monopulse antenna array for angle sensing
in one coordinate.

In the design of monopulse antenna arrays, the aperture is divided into
subarrays which are combined to produce the necessary sum-and-difference patterns [2, 36, 107]. A method for obtaining these patterns is shown
in Fig. 2.17, for the case where amplification during both transmission and
reception is performed in the subarray (A). The noise factor during reception is determined by the preamplifiers (B), which may then be followed

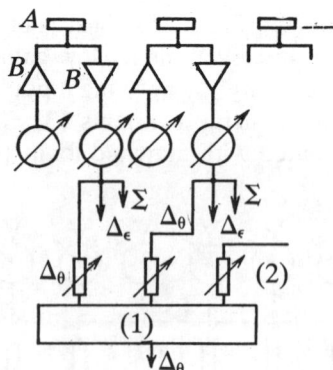

**FIGURE 2.17.** Use of subarrays in forming the sum and difference
patterns.
    (1) adder
    (2) weighted amplitude coefficients

by circuits with large noise factors without seriously degrading system performance. The receiving channel is divided into one sum and two difference channels. The signal is then weighted appropriately and added to the signals from the other subarrays. During transmission, the maximum beam strength is obtained by driving all of the subarray power amplifiers with identical signals. The addition of phase shifters at the input (output during reception) of the subarray amplifiers simplifies beam steering and allows identical control signals to be applied to all of the subarrays.

## 2.5 MONOPULSE RADAR ANTENNA FEEDS

A monopulse antenna feed may consist of a dipole, waveguide-dipole, slot, horn, or other type of device, which will achieve the following:

1. Provide a coherent field distribution at the aperture to generate a sum pattern.
2. Provide antiphase distributions in the upper and lower, as well as left and right, halves of the aperture to form the difference patterns.
3. Radiate no energy in a direction opposite to that of the sum pattern, because this will distort the sum pattern and cause deviations in the nulls of the difference patterns.
4. Create a sum pattern with maximum directional gain, and generate difference patterns with maximum slope for the angle-sensing characteristic.
5. Have minimal transverse dimensions.

Horn feeds are used most often in the centimeter-wave band.

In phase-comparison monopulse systems, each of the four feeds (two for each coordinate) usually has its own reflector. In amplitude-comparison systems, there is usually one antenna with four feeds, arranged as in Fig. 2.18, illuminating one reflector (subreflector) or lens. In systems using four bridges, the feeds are arranged this way for both angle-sensing methods, and the angle measurement is derived from comparison of paired signal sums: the sum of signals 1 + 2 is compared with 3 + 4 for measurement in elevation, while azimuth is derived from the sums 1 + 3 and 2 + 4. Waveguide bridges are used at the marked points. The sum signal is formed at bridge III from the signals 1, 2, 3, and 4, the first pair being added at bridge I and the second at II. The azimuth difference signal is obtained at III by subtracting the sum of the signals (2 + 4) from the signals (1 + 3). The elevation difference signal is obtained at bridge IV by adding the difference signals (1 − 3) and (2 − 4), which are generated at I and II; this gives the necessary signal (1 + 2) − (3 + 4), since the system is linear.

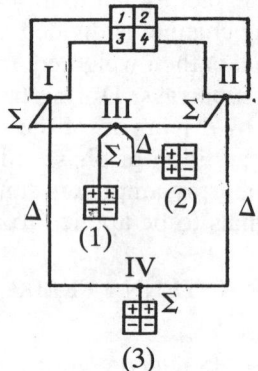

**FIGURE 2.18.** Signals in a four-horn feed system.
(1) sum signal
(2) azimuth error signal
(3) elevation error signal

To generate the sum pattern, the feed circuitry drives all four horns in phase (Fig. 2.19). To produce the maximum radiation, the horns are designed to give the greatest antenna gain when generating the sum pattern.

To form the difference patterns during reception, horn pairs are driven in antiphase (Fig. 2.19). This produces two regions of opposite polarity, creating the two lobes necessary for tracking in the associated angular coordinate.

If the feed-aperture dimensions are chosen to produce the maximum gain for the sum patterns, the difference patterns generated at the feeds are twice as wide as the angular dimension of the reflector. As a result, peaks in the difference illumination are formed at the edge of the reflector, and nearly half of the emitted energy bypasses the reflector (spill-over from the antenna system). This gives the difference patterns lower gain and higher sidelobes.

Figure 2.20 shows the optimal illumination methods, which eliminate spill-over losses [21]. The feed aperture generating the sum pattern retains its original dimensions. All that is necessary to limit the illumination for a given difference pattern to the reflector boundary is an increase in the corresponding aperture dimension of a factor of about 2. Thus, for optimum illumination, the feed dimensions and connecting circuits must be different for each pattern. Two approaches to the design of such a feed system will now be described (sections 2.5.1–2.5.2).

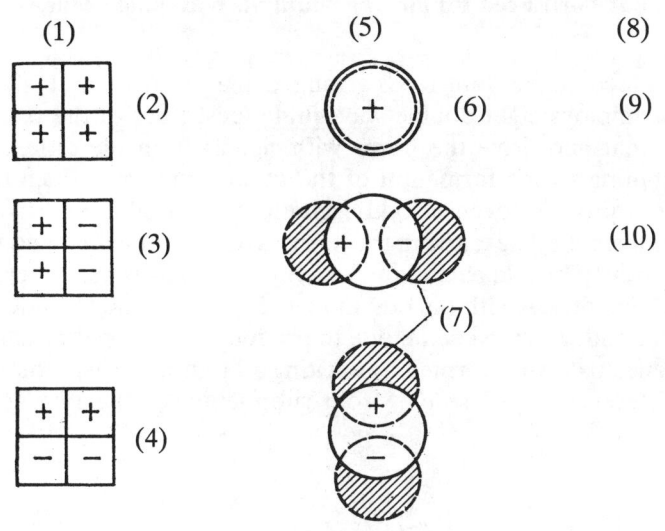

**FIGURE 2.19.** Ordinary monopulse illumination.

(1) feed excitation
(2) sum signal
(3) azimuth difference signal
(4) elevation difference signal
(5) reflector illumination

(6) primary aperture
(7) spill-over
(8) radiation characteristics
(9) good
(10) low gain, high side lobes

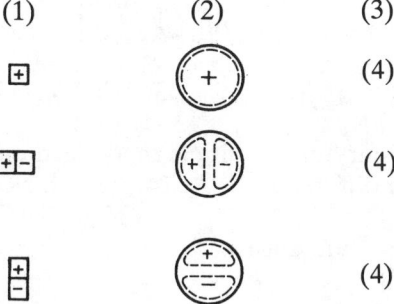

**FIGURE 2.20.** Optimal feed patterns for a monopulse antenna.

(1) feed excitation
(2) reflector illumination
(3) radiation characteristics
(4) good

### 2.5.1 Four-horn Feed Employing Multiple Waveguide Modes

This feed utilizes four feeds arranged in a row (Fig. 2.21) [21]. The feed circuit employs eight double-waveguide tees. Four of the tees are placed in the rear and drive the horns with signals from the different channels contributing to the formation of the various patterns. Each of the other four tees provides even or odd excitation in one of the horns.

The even mode excited in the horns is a sum of the first and third modes (both even); for odd excitation, second-mode waves are generated. Horns 2 and 3 are driven with the odd modes to generate the elevation difference pattern, and with the even mode to produce the sum pattern. The azimuth difference pattern is formed by exciting all four horns with the even mode; the difference signal results from subtraction of the two added signals 1 + 2 and 3 + 4.

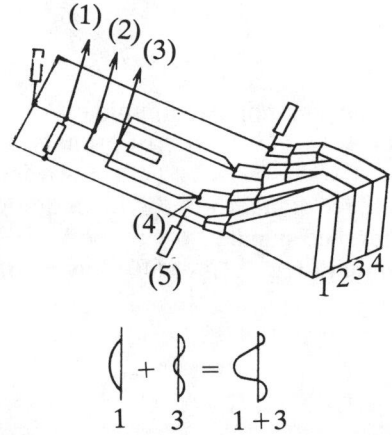

**FIGURE 2.21.** Four-horn, multiple mode feed operation.
(1) azimuth difference          (4) input for odd waves
(2) sum                         (5) higher-order even excitations
(3) elevation difference

The field distributions corresponding to all three patterns are shown in Fig. 2.22. The dotted lines represent the excited regions of the aperture for each case, and the effective aperture dimensions are seen to correspond to those shown in Fig. 2.20.

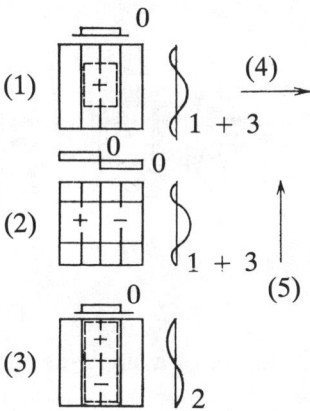

**FIGURE 2.22.** Multiple mode excitation in a four-horn feed.
(1) sum signal                          (4) *E*-plane
(2) azimuth difference signal           (5) *H*-plane
(3) elevation difference signal

### 2.5.2 Single-horn Feed

To examine the operation of a single-horn feed, we will assume that it is to receive signals polarized parallel to the narrow side of its throat [21].

When the signal arrives from a source in the *H*-plane (elevation plane), $TE_{10}$ and $TE_{20}$ modes are excited in the throat, the direction of arrival determining their phase and amplitude. As a result, in both side branches, $TE_{10}$ modes equal in amplitude but 180° out of phase are excited (the phase shift arising from the odd $TE_{20}$ excitation). The angle off axis in the *H*-plane of the received signal may be determined by the difference of the side-branch signals. If the signal arrives from a point on the feed axis, only the $TE_{10}$ wave is excited, and the waves excited in the side branches will be equal in amplitude and phase, their difference being zero.

If the signal arrives from a point in the *E*- (azimuth) plane, then the $TE_{10}$, $TE_{11}$, and $TM_{11}$ modes may be excited. These waves excite $TE_{10}$ waves in the upper and lower branches which are equal in amplitude but 180° out of phase. The difference of the upper- and lower-branch signals gives the off-axis angle in the antenna's *E*-plane of the arrival direction. When the signal arrives from a point on the feed axis, the branch signals are equal and their difference is zero.

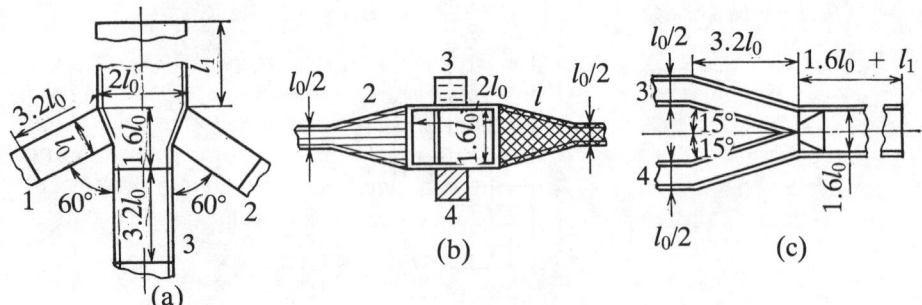

**FIGURE 2.23.** Basic structure of a single-horn feed:
(a) top view
(b) side view
(c) vertical section along symmetry axis
  1  left side branch
  2  right side branch
  3  upper branch
  4  lower branch

A signal arriving from a point on neither the *H*-plane nor the *E*-plane excites the $TE_{21}$ and $TM_{21}$ modes in the feed throat. The interaction of these waves with the four branches depends on the angle of arrival relative to the axes in the *H*- and *E*-planes, and the components of the angle of arrival in either one of these planes may be determined by comparing the signals in the appropriate pair of branches.

The feed should be designed so that the $TE_{10}$, $TE_{20}$, $TE_{11}$, $TM_{11}$, $TE_{21}$, and $TM_{21}$ modes can all propagate through its throat. The dimension $l_0$ must be about $0.7\lambda$; the dimension $l_l$ is not critical.

Table 2.1 gives the cut-off wavelength beyond which the various modes will not propagate through a rectangular waveguide with dimensions $a$ and $b$.

**Table 2.1**

| Wave mode | $TE_{10}$ | $TE_{20}$ | $TE_{11}$, $TM_{11}$ | $TE_{21}$, $TM_{21}$ |
|---|---|---|---|---|
| $\lambda_c$ | $2a_c$ | $a_c$ | $\dfrac{2a_c}{1 + (a_c/b_c)^2}$ | $\dfrac{a_c}{1 + (a_c/2b_c)^2}$ |

## 2.6 SELECTING SQUINT ANGLE AND PHASE-CENTER DISPLACEMENT

An important parameter of an amplitude-comparison monopulse radar system is the squint angle of the beam maximum relative to the axis, and for a phase-comparison system, the distance between antenna phase centers. These quantities have a significant effect on the radar's angle accuracy and range. We will determine the optimal values of these parameters for the case of a monopulse radar with sum-and-difference angle discrimination.

The angle error is inversely proportional to the slope of the angle-sensing response and signal-to-noise ratio $q$

$$\sigma_0 = 1/\mu q \tag{2.21}$$

The slope $\mu$ characterizes the angle-sensing sensitivity:

$$\mu = \left. \frac{dS(\theta)}{d\theta} \right|_{\theta = 0}$$

In an amplitude-sum-and-difference monopulse system, the sum pattern is nearly constant and the difference pattern is linear in a region near the axis; hence,

$$\mu = \left. \frac{d}{d\theta} \left| \frac{F_1(\theta) - F_2(\theta)}{F_1(\theta) + F_2(\theta)} \right| \right|_{\theta = 0} = \frac{F'_d(0)}{F_s(0)} \tag{2.22}$$

The optimal squint angle providing the greatest angle-sensing sensitivity could be found from the equation

$$\frac{d}{d\theta} \left| \frac{F'_d(0)}{F_s(0)} \right| = 0 \tag{2.23}$$

but Fig. 2.24, which shows the dependence of $F_s(0)$, $F'_d(0)$, $F'_d(0)/F_s(0)$, and $F_s(0) \cdot F'_d(0)$ on the squint angle, indicates that the ratio $F'_d(0)/F_s(0)$ is monotonically increasing within the beam limits, and that there is therefore no optimum squint angle which gives maximum angle-sensing sensitivity.

The signal-to-noise ratio $q$ is proportional to the sum pattern $F_s(0)$ and, therefore, the product

$$\mu q = k_\theta F'_d(0) \tag{2.24}$$

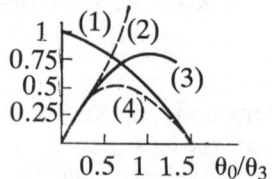

**FIGURE 2.24.** Relative beam-pattern values as a function of the squint angle.

(1)  $F_s(0)$             (3)  $F_d'(0)$

(2)  $F_d'(0)/F_s(0)$      (4)  $F_s(0) \cdot F_d'(0)$

is determined only by the slope of the difference pattern at $\theta = 0$. Substituting (2.24) into (2.21), we see that the angle accuracy in the equisignal direction depends on the slope of the difference pattern.

If we choose the squint angle to maximize the slope of the difference pattern, we can minimize the angle errors. This is a poor criterion, however, because it results in the patterns intersecting at a very low level, and subsequently the received signal power in the sum channel will be much lower than the maximum level possible, significantly decreasing the target detection range.

To achieve a good compromise between the range and accuracy considerations, the squint angle should be chosen so as to maximize the product of the sum-pattern magnitude and difference-pattern slope. In Fig. 2.24, it may be seen that $F_s(0) \cdot F_d(0)$ has a maximum at $\theta_0 = 0.65\theta_3$. Then the two patterns will intersect approximately 3 dB below their maximum, at their common half-power level. The optimum squint angle is thus approximately equal to half of the beamwidth at the half-power level.

The same criterion of achieving a maximum in the product of sum-pattern magnitude and difference-pattern slope is used to determine the optimal displacement between the antenna phase centers in a phase-sum-and-difference monopulse system:

$$\frac{d}{dt} |F_s(0) \cdot F_d'(0)| = 0 \tag{2.25}$$

As can be seen from Fig. 2.25, the aperture of a phase-comparison monopulse antenna is given by

$$d_\phi = 2(d_p - l) \tag{2.26}$$

where $d_p$ is the specified overall antenna dimension.

The amplitude of the sum signal at the receiver input is determined by the product of the transmission and reception patterns and is proportional

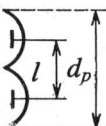

**FIGURE 2.25.** Relation between phase-center displacement and antenna aperture in a phase comparison system.

to the antenna aperture, so we may write

$$F_s^2(0) \equiv C_\Phi(d_p - l) \tag{2.27}$$

where $C_\Phi$ is the proportionality constant.

The difference pattern is determined by the following expression:

$$F_d(0) = F_d(0)\tan(\Delta\phi/2). \tag{2.28}$$

For small target angles, we have, for the phase difference between the signals received by the two patterns:

$$\Delta\phi = 2\pi(l/\lambda)\theta \tag{2.29}$$

Then the slope of the difference pattern at the axis is given by the formula:

$$\frac{d}{d\theta}F_d(\theta)\mid_{\theta=0} = F_d'(0) = \tfrac{1}{2}F_s(0)\Delta\phi' = (\pi l/\lambda)F_s(0) \tag{2.30}$$

Hence, the product of the sum pattern and the slope of the difference pattern is

$$F_s(0){\cdot}F_d'(0) = F_s^2(0)\,(\pi l/\lambda) = C_\Phi(\pi l/\lambda)(d_p - l) \tag{2.31}$$

This product has a maximum when

$$\frac{d}{dl}\left[C_\Phi(\pi l/\lambda)(d_p - l)\right] = 0$$

i.e.,

$$(C_\Phi\pi/\lambda)(d_p - l) - (C_\Phi\pi/\lambda)l = 0$$

from which we have the optimal distance between phase centers:

$$l = d_p/2 \tag{2.32}$$

The optimal phase-center displacement is therefore half the aperture width.

# Chapter 3

# The Basic Functional Components of Monopulse Radars

## 3.1 AUTOMATING RADAR OPERATION WITH COMPUTERS

Modern monopulse radars must satisfy demanding requirements. In general, they must operate over long ranges and have high throughput, resolution, and angular accuracy. They incorporate the most advanced radar technology: phased array antennas, electronic beam steering, complex (wideband) signals, optimal reception, and pulse compression.

Most aspects of radar operation can be automated with computers. They perform beam control, digital processing of the radar signals, target selection and identification, and various other functions. All modern computers incorporate the following basic units: memory, arithmetic, control, input, and output. The structure of a typical computer is shown in Fig. 3.1.

Progress is currently being made in the development of multiprocessor computer systems, which speed up the processing of large amounts of information, and provide improved control of multifunctional devices and systems. The transition to multiprocessor systems is part of an attempt to achieve significant advances in productivity and to perform processing maintenance in the computer itself. The term "central processor" usually refers to that autonomous unit of the computer which performs the elementary (basis) operations written in machine language, i.e., it directly performs transformations, data processing, and implementation of problem-solving software.

Multiprocessor computers are able to run several independent programs, or branches of one program, simultaneously. The stringent reliability requirements for such systems necessitate the incorporation of at least two processors and the existence of two paths between any two components of the computer.

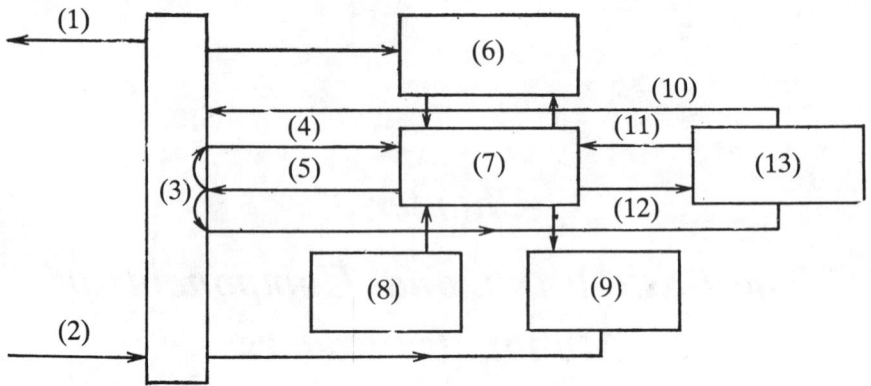

**FIGURE 3.1.** Computer structure.

| | |
|---|---|
| (1) signals to operator | (8) input device |
| (2) input from operator | (9) output device |
| (3) CPU | (10) result flag |
| (4) addresses | (11) data and commands |
| (5) commands | (12) operation codes |
| (6) external memory | (13) arithmetic unit |
| (7) internal memory (RAM) | |

Thus, the properties of multiprocessor systems reflect two current trends in computer system development:

- the principle of modular construction, by which each device is built in several independent modules, which facilitates system modification and elaboration, with module redundancy permitting "graceful degradation" in the event of the failure of a given module;
- simultaneous execution of independent branches of a program, or of several complete programs, reducing the overall problem solution time.

Multiprocessor computer systems have different structures. By structure, we mean the choice of elements (system modules), the nature of their interfaces, and the distribution of tasks among them.

Most multiprocessors in current use have one basic structure (see Fig. 3.2). The system employs several identical central processing units (CPUs), several random access memory (RAM) units, and input/output processors (IOPs). Internal routing can connect any of these units. An external device switch (EDS) connects each IOP with a group of peripherals.

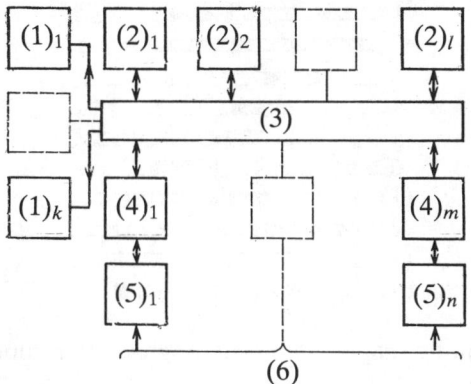

**FIGURE 3.2.** Homogeneous computer structure.

(1) $CPU_{1,k}$          (4) $IOP_{1,...,m}$
(2) $RAM_{1,...,l}$        (5) $EDS_{1,...,n}$
(3) internal data paths  (6) peripherals

In addition to facilitating system unification, modular construction also enhances system reliability. All modules of a given type are interchangeable, and the malfunction of one module does not lead to a catastrophic system failure. Overall control of such a system is effected with just one operating program. Each processor can address the operating system through its own set of interrupts. The process of tasking is significantly simplified in a homogeneous computer system. Incoming and interrupted jobs are placed into a single queue on the basis of priority. Upon completing or interrupting execution of a job, each CPU turns to the next job ready for execution in the queue, eliminating the idle time which results when a CPU must wait for a given task to be read for execution.

The timing diagram for a two-processor system executing six jobs is shown in Fig. 3.3. Program execution is interrupted during peripheral IOP. Through the interrupt, the CPU tells the operating system to invoke the necessary software device controller; the CPU then resumes program execution.

The diagram indicates that $CPU_1$ executes the programs in the order 1, 3, 5, 6, 2, 4, and $CPU_2$ in the order 2, 4, 1, 3, 5. Program execution is necessarily interrupted for short intervals while the peripheral controllers perform their functions. One example of a homogeneous computer system is the multiprocessor *"El'brus"* system.

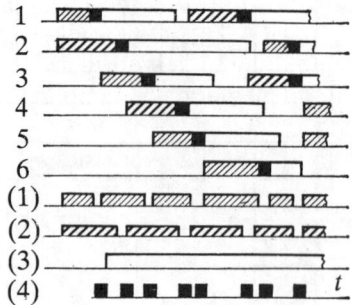

**FIGURE 3.3.** Timing diagram for a homogeneous computer.

(1) CPU$_1$     (3) input/output
(2) CPU$_2$     (4) operating system

## 3.2 DIGITAL SIGNAL PROCESSING

With the improvements in high-speed digital computers, digital signal processing in radar systems has seen much development during the last decade. This is explained by the advantages of digital processing versus analog filtering. Signals are processed in real time with digital methods, removing the difficulties associated with the construction and stability of analog filters.

The primary applications of digital signal processing are digital filtering and spectral analysis. In the case of radar signals, particularly wideband signals, spectral analysis is used the most. In general, spectral analysis makes use of the fast Fourier transform (FFT).

The (forward) Fourier transform, or spectrum, of a signal $f(t)$ is given by the following formula:

$$S(\omega) = \int_{-\infty}^{\infty} f(t)e^{-i\omega t}dt \tag{3.1}$$

The inverse Fourier transform is given by

$$f(t) = \frac{1}{2\pi} \int_{-\infty}^{\infty} S(\omega)e^{i\omega t}d\omega. \tag{3.2}$$

The Fourier transform as implemented on a digital computer makes use of only a finite number of samples in both the time and frequency domains.

We will assume that the function of time $f(t)$ can be represented as a sequence of $N$ samples $f(n, T)$, $0 \leq n \leq N - 1$, where $T$ is the sampling interval in the time domain. Similarly, we will recast $S(\omega)$ as $S(k\Omega)$, where $0 \leq k \leq N - 1$ is the frequency domain sampling interval.

Then the forward and inverse discrete Fourier transforms (DFT) are computed according to the following expressions:

$$S(k\Omega) = \sum_{n=0}^{N-1} f(nT)e^{-i\Omega Tnk} \tag{3.3}$$

$$f(nT) = \frac{1}{N} \sum_{k=0}^{N-1} S(k\Omega)e^{i\Omega Tnk} \tag{3.4}$$

where

$$\Omega = 2\pi/NT \tag{3.5}$$

We can write (3.3) in the simpler form:

$$S_k = \sum_{n=0}^{N-1} f_n W_N^{nk} \tag{3.6}$$

where

$$W_N^{nk} = e^{-i2\pi/N} \tag{3.7}$$

Thus, $W_N^{nk}$ is a periodic sequence with period $N$.

It follows from (3.6) that direct calculation of the $N$-point discrete Fourier transform requires $(N - 1)^2$ complex multiplications and $N(N - 1)$ complex additions. Thus, for sufficiently large $N$ ($N > 1000$), direct computation of the DFT is a very time-consuming operation.

Methods employed to reduce the computation time are based on the fact that the sum of products can be well approximated by the simpler product of sums, and that real operations can be substituted for complex ones. By far the most common approach is that of the fast Fourier transform (FFT) [10]. The fundamental idea behind the FFT is to divide the original sequence into two shorter ones. Let us assume that the $N$ samples of the sequence $f_n$ have been split into two sequences: $f_{n1}$, consisting only of the even-numbered samples, and $f_{n2}$, consisting only of the odd-numbered samples. Then, we may write

$$\begin{aligned} f_{n1} &= f(2n), \quad n = 0,1,...,N/2 - 1 \\ f_{n2} &= f(2n + 1), \quad n = 0,1,...,N/2 - 1 \end{aligned} \tag{3.8}$$

The $N$-point DFT of the sequence $f_n$ can then be placed in the form:

$$
S_k = \sum_{\substack{n=0 \\ n \text{ even}}}^{N-1} f_n W_N^{nk} + \sum_{\substack{n=0 \\ n \text{ odd}}}^{N-1} f_n W_N^{nk}
$$

$$
= \sum_{n=0}^{N/2-1} f(2n) W_N^{2nk} + \sum_{n=0}^{N/2-1} f(2n+1) W_N^{(2n+1)k} \tag{3.9}
$$

Because

$$
W_N^2 = [e^{i2\pi/N}]^2 = e^{i[2\pi/(N/2)]} = W_{N/2} \tag{3.10}
$$

we may rewrite (3.9) as

$$
S_k = \sum_{n=0}^{N/2-1} f_{n1} W_{N/2}^{nk} + W_N^k \sum_{n=0}^{N/2-1} f_{n2} W_{N/2}^{nk} = S_{k1} + W_N^k S_{k2} \tag{3.11}
$$

where $S_{k1}$ and $S_{k2}$ are the $N/2$-point DFTs of the sequences $f_{n1}$ and $f_{n2}$. It can be seen from (3.11) that the $N/2$-point DFTs each require $(N/2)^2$ operations, and that only $N$ more are needed to compute the full DFT $S_k$, resulting in a total of $N + (N/2)^2$ operations. Direct calculation of $S_k$ requires $N^2$ operations, so that when $N^2/2 \gg N$, using (3.11) reduces the number of operations by a factor of 2.

If $N/2$ is divisible by 2, then the $N/2$-point transforms may be reduced to $N/4$-point transforms, with further savings in the computation required to obtain the full DFT; repeated application of this process results in commensurate time savings. This is the essence of the FFT.

Inasmuch as $S_k$ is defined for $0 \leq k \leq N - 1$, but $S_{k1}$ and $S_{k2}$ in (3.11) are defined only for $0 \leq k \leq N/2 - 1$, it is necessary to determine $S_k$ for $k \geq N/2$. Since the DFT is periodic, (3.11) may be expressed as follows:

$$
S_k = \begin{cases} S_{k1}(k) + W_N^k S_{k2}(k), & 0 \leq k \leq N/2 - 1 \\ S_{k1}(k - N/2) + W_N^k S_{k2}(k - N/2), & N/2 \leq k \leq N - 1 \end{cases} \tag{3.12}
$$

The procedure described above is called the time-thinning algorithm, because at each stage the input (time) sequence is divided into two shorter sequences, i.e., the time sequence is thinned at each stage. The base operation in this algorithm consists of transforming the input values $A$ and $B$ into the output quantities $X$ and $Y$ as follows:

$$
X = A + W_N^k B, \qquad Y = A - W_N^k B \tag{3.13}
$$

Figure 3.4(a) shows this operation in the form of a directed graph, or "butterfly diagram." Such a graph is constructed of points (nodes) and arrows (transmission lines). The nodes denote registers containing the input and output files of the DFT, and the arrows represent multiplication by the indicated value. The unshaded circles represent the addition and subtraction operations, with the upper output holding the sum and the lower output holding the difference. These symbols are shown individually in Fig. 3.4(b).

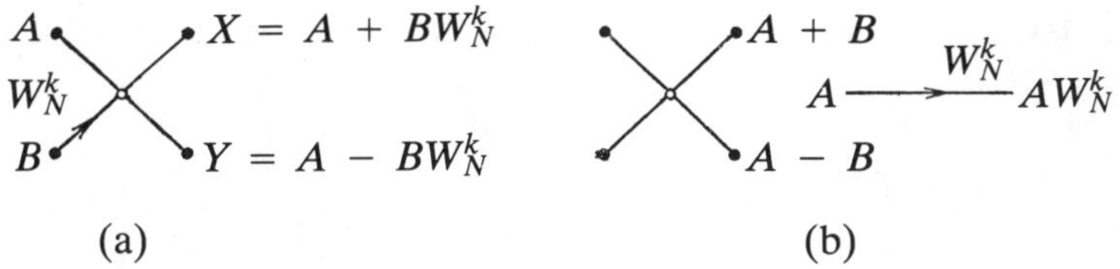

(a)                                                (b)

**FIGURE 3.4.** FFT algorithm base operation.

The butterfly diagram representing an eight-point DFT using two four-point sequences is shown in Fig. 3.5. The input sequence $f_n$ is divided into the two sequences $f_{n1}$ (even terms) and $f_{n2}$ (odd terms), and then the transforms $S_{k1}$ and $S_{k2}$ are taken. $S_k$ is then obtained in accordance with (3.12).

The $N/2$-point transform may be similarly written as the combination of two $N/4$-point DFTs, i.e.,

$$S_{k1} = A_k + W_N^k B_k \tag{3.14}$$

where $0 \leq k \leq N/2 - 1$; and $A_k$ and $B_k$ are the $N/4$-point DFTs.

Figure 3.6 gives the butterfly diagram for the eight-point DFT reduced to two-point DFTs in accordance with (3.14).

The process of reducing the size of the transforms computed can be repeated until only two-point transforms remain, as long as $N$ is a power of 2. A two-point DFT can be calculated without any multiplications by the formulas:

$$
\begin{aligned}
S_k(0) &= f(0) + f(1)W_8^0 \\
S_k(1) &= f(0) + f(1)W_8^4
\end{aligned}
\tag{3.15}
$$

where $f(n)$, $n = 0, 1$, is the two-point sequence to be transformed. Since $W_8^0 = 1$ and $W_8^4 = -1$, no multiplications are needed in (3.15), and the eight-point DFT of Figs. 3.5 and 3.6 can be computed as shown in the butterfly diagram of Fig. 3.7.

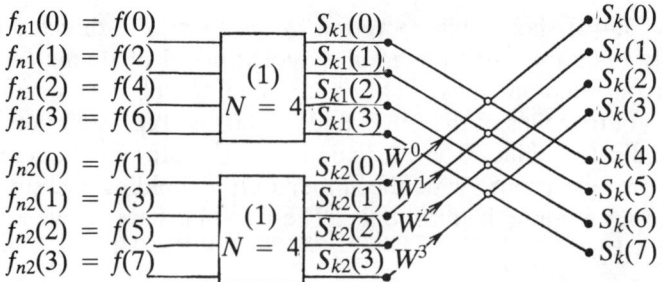

**FIGURE 3.5.** Calculation of eight-point DFT by two four-point DFTs.
(1) DFT

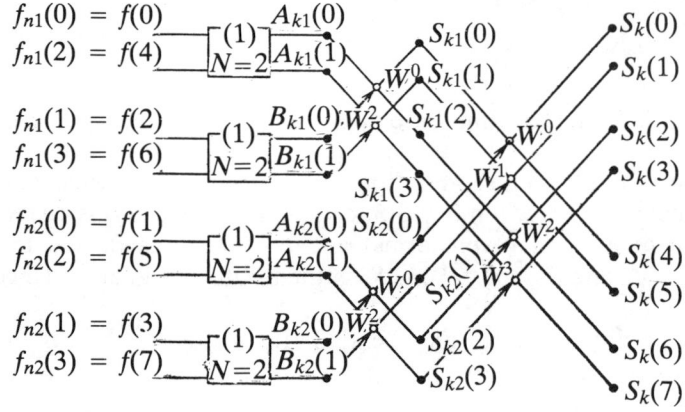

**FIGURE 3.6.** Calculation of eight-point DFT by 4 two-point DFTs.
(1) DFT

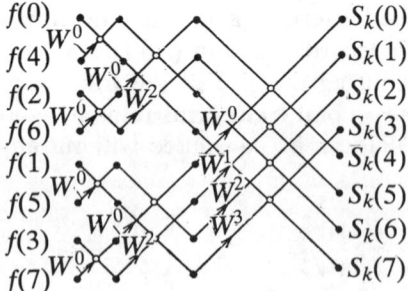

**FIGURE 3.7.** Calculation of eight-point DFT: final decomposition.

Analysis of the transform reduction process shows that $N/2$ complex multiplications are required for each stage of the FFT. Because there are $\log_2 N$ stages, the complete $N$-point FFT requires approximately

$(N/2)\log_2 N$ complex multiplications; this result is inexact since multiplication by $W_N^0$, $W_N^{N/2}$, $W_N^{N/4}$, and $W_N^{3N/4}$ is equivalent to adding complex constants. However, the approximation is very good for large $N$.

Figure 3.7 illustrates several practical aspects of performing the FFT. It is apparent from the diagram that only complex additions and subtractions are needed in the first two stages; it is only in the third stage that two complex multiplications are required. Hence, instead of the 12 ($4 \log_2 8$) multiplications given by the approximation, only two are performed.

Figure 3.7 also shows that $N/2$ base operations are performed in each stage. Each node corresponds to two memory registers (one for each component of a complex number). We will assume that the input sequence is stored in memory in the order:

$$f_0, \ f_4, \ f_2, \ f_6, \ f_1, \ f_5, \ f_3, \ f_7$$

as in Fig. 3.7. Only one more memory location will then be necessary to perform the FFT. The results of the first stage of operations are written over the original data, and further results are written over the output of the previous stage until the final transform sequence is stored in these same memory locations. Such an algorithm is called a replacement FFT, because the output sequence replaces the input sequence in memory.

The eight left-most nodes in Fig. 3.7 are the input registers. During the first stage the value of the next set of nodes to the right are calculated. The values in each pair of nodes are determined only by the values in the corresponding pair of input nodes, which will not be needed after the first stage. If each input node is accessed simultaneously, the pair is then made available for storage of the result. The second stage consists of similar computation by pairs, although the node pairs are now interleaved; this does not affect the replacement operation, however, because the values at any pair of nodes are needed only to compute the values of the nodes to the immediate right. Thus, the new results may again be written over the old values, which are no longer needed.

There are many different FFT algorithms, corresponding to the number of factors of $N$ and their ordering. The concept of a *base* is used to characterize the decomposition. A *mixed base* factoring is one in which the factors are not all equal. Take $N = 60$, for example. This number may be expressed as the product of smaller factors in several ways: $60 = 3 \times 4 \times 5 = 5 \times 4 \times 3 = 12 \times 5 = 2 \times 2 \times 5 \times 3$, and so on. If $N = 64$, we obtain the base-2 algorithms used above. Fixed base algorithms (bases 2, 4, 8) are far more common today than mixed base algorithms, because they simplify analysis, programming, and the design of specialized devices.

### 3.3 SUM-AND-DIFFERENCE DEVICES

It was shown in Section 1.2 that in addition to $n/2$ phase shifters, which multiply by $\pm i$, signal converters include devices which form the sum and difference of the input signals. No signal conversion is necessary to obtain multiplicative ratios of the received signals. Formation of additive ratios requires the use of sum-and-difference devices, such as the hybrid ring junction and the magic *tee*.

The sum-and-difference ring junction (see Fig. 3.8), has four branches spaced 60° apart along a semicircle of length $3\lambda/4$, with the spacing between branches equal to $\lambda/4$. The sum signal results at the port marked $\Sigma$, and the difference signal is produced at branch $\Delta$. When high-frequency in-phase signals are introduced at branches 1 and 2, they travel the same path length to the $\Sigma$ port, and are added in phase; they travel paths differing by $\lambda/2$ to the $\Delta$ branch and are added 180° out of phase. The $\Delta$ output signal has the phase of the input signal with greater amplitude.

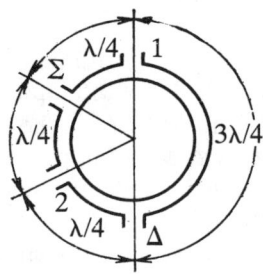

**FIGURE 3.8.** Ring junction.

When the amplitude of signal 1 is greater than that of signal 2, the phase of the difference signal is determined by the phase of signal 1, and is shifted from the input phase at port 1 by the path length $3\lambda/4$. The sum signal is offset from the phase at 1 and 2 by the path length $\lambda/4$. The sum-and-difference signals are therefore 180° out of phase.

When the amplitude of signal 1 is smaller than that of signal 2, the phase of the difference signal is determined by the phase of signal 2, which is shifted relative to port 2 by the path length $\lambda/4$. The sum signal is still shifted by the same amount, and the sum and difference signals are in phase.

When two signals are introduced in phase into the branches 1 and 2 of a magic *tee* (see Fig. 3.9), the sum signal is produced at branch $\Sigma$ (*XOY*-plane), and the difference signal results at branch $\Delta$ (*XOZ*-plane) [19].

This results from the fact that if the electric field lines run up and down (in the z-direction), then they enter the Δ branch lying in opposite directions, but maintain their direction in the Σ branch. The difference signal will have the phase of the stronger input signal; when the input signals are equal, the difference signal is zero.

In addition to these hybrid junctions, directional couplers (balanced slot bridges) can be used as sum-and-difference devices.

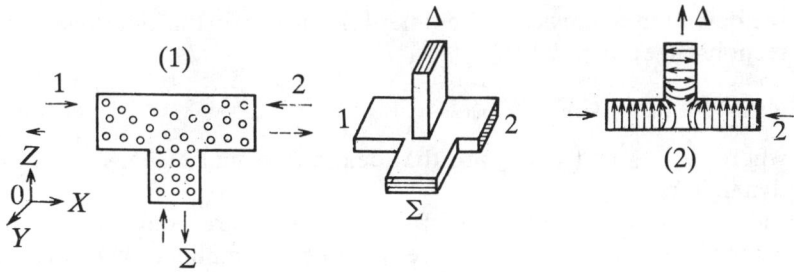

**FIGURE 3.9.** Magic *tee*.

(1) *XOY*-plane      (2) *XOZ*-plane

## 3.4 LOGARITHMIC AMPLIFIERS

A logarithmic amplifier is a nonlinear device for which the output voltage varies logarithmically with the input voltage. The linear portion ($U_{in} \leq U_{in\ t1}$) of the amplitude response of a logarithmic amplifier (see Fig. 3.10) is described by the equation:

$$U_{out} = k_0 U_{in} \tag{3.16}$$

where $k_0$ is the gain over the linear region and $U_{in\ t1}$ is the input voltage at which the amplifier makes the transition to logarithmic operation.

In the logarithmic mode ($U_{in\ t1} < U_{in} < U_{in\ t2}$), the logarithmic amplitude response (LAR) has the form [6]:

$$U_{out} = k_0 U_{in\ t1}[a_l \ln(U_{in}/U_{in\ t1}) + 1] \tag{3.17}$$

where $a_l$ characterizes the slope of the logarithmic response, and $U_{in\ t2}$ is the input voltage at which the transition out of logarithmic operation occurs.

The performance of logarithmic amplifiers is also described by the following parameters:

(1) $U_{\text{out } t1}$ and $U_{\text{out } t2}$—the output voltages corresponding to the transitions in and out of the logarithmic portion of the amplitude response;

(2) the dynamic range of the input and output voltages:

$$D_{\text{in}} = U_{\text{in } t2}/U_{\text{in } t1}, \qquad D_{\text{out}} = U_{\text{out } t2}/U_{\text{out } t1} \tag{3.18}$$

(3) the compression of the amplified voltage:

$$K_c = D_{\text{in}}/D_{\text{out}} \tag{3.19}$$

(4) the deviation of the actual amplifier response from the ideal logarithmic response (see Fig. 3.10):

$$\delta_l = (U_{\text{out } a} - U_{\text{out } id})/U_{\text{out } id} \tag{3.20}$$

where $U_{\text{out } id}$ and $U_{\text{out } a}$ are the ideal and actual responses over the dynamic range;

(5) the stability or repeatability of the logarithmic response, which characterizes the ability to produce a series of amplifiers with identical parameters.

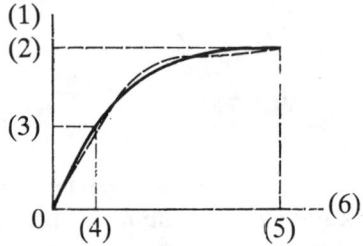

**FIGURE 3.10.** Amplitude response of a logarithmic amplifier: calculated and experimental.

| | | |
|---|---|---|
| (1) $U_{\text{out}}$ | (3) $U_{\text{out } t1}$ | (5) $U_{\text{in } t2}$ |
| (2) $U_{\text{out } t2}$ | (4) $U_{\text{in } t1}$ | (6) $U_{\text{in}}$ |

In Section 1.3, it was shown that, in a monopulse system employing amplitude angular discrimination and a logarithmic amplifier, the angle information is obtained by subtracting the logarithms of two signals, which is equivalent to taking the logarithm of their ratio. In fact, if the logarithmic amplifiers are identical, the voltage at the output of the subtracting device (see Fig. 1.3(a)) will equal

$$u_{sd} = u_{\text{out } 1} - u_{\text{out } 2} = k_{sd}k_0 u_{\text{in } t1} a_l \ln(u_{\text{in } 1}/u_{\text{in } 2}) \tag{3.21}$$

where $k_{sd}$ is the gain of the subtracting device.

Consistent with (1.4) and (1.5), we may write

$$u_{\text{in } 1} = E(t)F_1(\theta) = E(t)F(\theta_0 - \theta) \qquad (3.22)$$
$$u_{\text{in } 2} = E(t)F_2(\theta) = E(t)F(\theta_0 + \theta)$$

Then (3.21) takes the form:

$$u_{sd} = k_{sd}k_0 U_{\text{in } t1}a_l \ln[F(\theta_0 - \theta)/F(\theta_0 + \theta)] \qquad (3.23)$$

Equation (3.23) describes the angle-sensing response, and shows that for such a monopulse system, this response depends on the properties of the logarithmic amplifiers and not on the signal values. Thus, any deviation from true logarithmic response in the amplifier leads to distortion in the angle-sensing response and, therefore, gives rise to errors in target angle measurement.

In determining the ratio $m_u = u_{\text{in } 1}/u_{\text{in } 2}$, for a given acceptable error $\delta_u = \Delta m_u/m_u$, the corresponding acceptable values of absolute and relative deviations of the actual amplifier response from logarithmic response at any point are [6]

$$|\pm \Delta U_{\text{out}}| = 0.7k_0 U_{\text{in } t1}a_l \ln(1 + \delta_u) \qquad (3.24)$$

$$\delta_{\text{out}} = \left| \frac{\pm \Delta U_{\text{out}}}{U_{\text{out}}} \right| = \frac{0.7a_l(1 + \delta_u)}{a_l \ln(u_{\text{in}}/u_{\text{in } t1}) + 1}$$

where 0.7 is the gain of the subtraction circuit.

Equation (3.24) shows that the acceptable absolute deviation from logarithmic response increases with the maximum gain and with the threshold input voltage, and is constant over the logarithmic portion of the response. The relative deviation varies, and decreases with increasing signal levels. In other words, the logarithmic response in monopulse system amplifiers needs to be more precise at the upper end of the range than at lower voltages. This accuracy requirement becomes more stringent with increasing dynamic range.

When the responses of the two amplifiers are not identical (or when the deviation of each response from logarithmic exceeds the acceptable level), the output voltage of the amplitude angular discriminator does not satisfy the requirements for the angle-sensing response. Therefore, special measures must be taken to ensure that the amplifier responses are, to the extent possible, precisely logarithmic and identical to one another.

A logarithmic response cannot be obtained in the range of 80–100 dB, with just one stage. It is, therefore, necessary to employ a multistage cascade in which one stage at a time exhibits a logarithmic response, the others remaining linear. The amplifier is designed so that just as one stage goes out of logarithmic operation, the preceding stage switches in.

The amplitude response of a single nonlinear stage in a cascade is shown in Fig. 3.11, and consists of three regions corresponding to linear (1), logarithmic (2), and quasilinear (3) operation [6]. In the following analysis, the subscripts 1, 2, and 3 refer to these three modes; $t1$ and $t2$ refer to the transitions from one response to the next; input and output voltages for the entire amplifier will be denoted by $U$, and the input and output voltages for the $i$th stage by $u(i)$. If the input voltage $U_{in} < U_{in\ t1}$, then all stages operate as linear amplifiers with gain $k_1$. When the input voltage reaches $U_{in\ t1}(n)$, the last ($n$th) stage begins to operate in the logarithmic mode, and its input voltage at this transition is

$$u_{t1} = u(n)_{in\ t1} = U_{in\ t1(n)}k_1^{n-1} \tag{3.25}$$

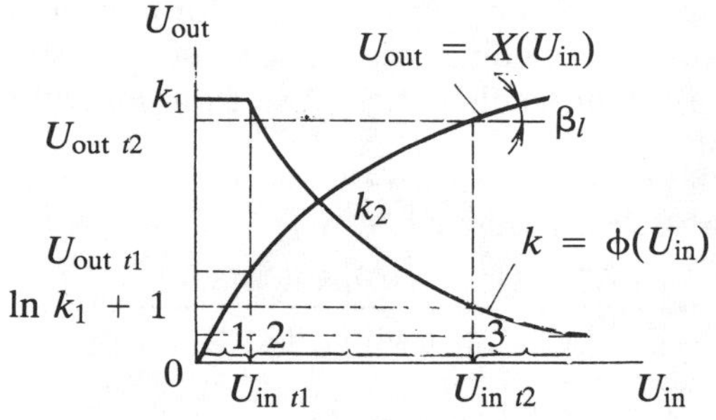

FIGURE 3.11. Amplitude response of nonlinear stage.

This last stage operates with logarithmic response as its input voltage rises to the value $u_{t2} = u(n)_{in\ t2}$, and the other stages all have linear response. As a result, the overall amplifier response is logarithmic. The output voltage at the last ($n$th) stage while in the logarithmic mode is

$$u(n)_{out\ 2} = k_1u_{t1}[a_l \ln(u/u_{t1}) + 1] \tag{3.26}$$

where $u$ is the input voltage at the last ($n$th) stage.

To simplify the analysis, we will now assume that $a_l = 1$. With voltage at the input to the last stage in the range $u_{t1}$ to $u_{t2}$, the cascade's input voltage lies in the range $(u_{t1} \rightarrow u_{t2})/k_1^{n-1}$, and the amplifier gain with the $n$th stage in logarithmic operation is

$$k_0(n) = k_1^{n-1}(k_1u_{t1}/u_{in})[\ln(u/u_{t1}) + 1] \tag{3.27}$$

where $u = u(n)_{in} = U_{in}k_1^{n-1}$ is the voltage at the input of the $n$th stage, and $U_{in}$ is the amplifier input voltage.

The amplifier's logarithmic response is then given by the expression:

$$U_{out}(n) = U_{in}k_0(n) = k_1u_{t1}[\ln(u/u_{t1}) + 1]$$
$$= k_1u_{t1}[\ln(u/u_{t1}) + 1] \tag{3.28}$$

When the voltage at the input of the $n$th stage reaches $u_{t2}$, the stage goes into quasilinear operation. In order for the stages to go into logarithmic operation one at a time, the $(n - 1)$th stage must go into logarithmic operation just as the $n$th stage goes into quasilinear operation. In other words, $u(n) = u_{t2}$ when $u(n - 1) = u_{t1}$, because the two transition voltages are the same for all of the stages. Because the $(n - 1)$th stage is linear up to this point, the voltages must be related by $u_{t2} = u(n) = k_1u(n - 1) = u_{t1}$, which is clearly satisfied if and only if

$$k_1 = u_{t2}/u_{t1} = d_{in} \tag{3.29}$$

In other words, the gain is given by the range of logarithmic operation $d_{in}$ for a single stage.

With the amplifier input in the range $(u_{t1} \to u_{t2})/k_1^{n-2}$, the $(n - 1)$th stage is in its logarithmic region, all preceding stages have linear response, and the last stage is in the quasilinear mode. In this case, the amplifier gain is given by the expression:

$$k_0(n - 1) = k_1^{n-2}(k_1u_{t1}/u)[\ln(u/u_{t1}) + 1]k_3(n) \tag{3.30}$$

where $u = u_{in}(n - 1) = U_{in}k_1^{n-2}$ is now the input voltage at the $(n - 1)$th stage, and $k_3(n)$ is the (quasilinear) gain of the $n$th stage.

All the preceding stages, exhibiting a linear response, have no effect on the form of the amplifier's amplitude response. The last stage, operating in its quasilinear region, causes no distortion in the logarithmic response as long as its gain is constant and equal to unity. It may be variable and greater than unity if the differential gain $k_3(n)$ is unity.

We will now find this differential gain. The gain of the stage at the threshold between logarithmic and quasilinear operation equals

$$k_2 = (k_1u_{t1}/u_{t2})[\ln(u_{t2}/u_{t1} + 1] = \ln k_1 + 1 \tag{3.31}$$

Because $k_1 > 1$ and $\ln k_1 > 0$, $k_2 > 1$ and, as a result, as the stage goes from logarithmic to quasilinear operation, its gain $k_3 = k_2 > 1$.

We find the differential gain of a stage in its quasilinear region by equating the first derivatives and ordinates at the point of transition from region 2 (logarithmic) to region 3 (quasilinear):

$$b_l = du_{\text{out } 2}/du_{\text{in } 2} = du_{\text{out } 3}/du_{\text{in } 3}$$

Using (3.29), we may write

$$\frac{d\{k_1 U_{t1}[\ln(u_{\text{in } 2}/u_{t1}) + 1]\}}{du_{\text{in } 2}} = \frac{d[k_1 u_{t1}(\ln k_1 + 1) + b_1(u_{\text{in } 3} - u_{t2})]}{du_{\text{in } 3}}$$

Differentiating, we obtain

$$k_1 u_{t1}/u_{t2} = b_l$$

Because $u_{t2} = u_{t1}k_1$, the differential gain $b_l = 1$.

The output voltage of a stage in quasilinear operation is

$$u_{\text{out } 3} = u_{\text{out } t2} + b_l(u_{\text{in } 3} - u_{\text{in } t2}) \tag{3.32}$$
$$= k_1 u_{t1}(\ln k_1 + u_{\text{in } 3}/u_{t2})$$

where $u_{\text{out } t2}$ is the output voltage of the stage at the transition point. Then the gain of the stage over the quasilinear portion of its amplitude response is

$$k_3 = (k_1 u_{t1}/u_{\text{in } 3})(\ln k_1 + u_{\text{in } 3}/u_{t2}) \tag{3.33}$$
$$= (k_1 u_{t1} \ln k_1)/u_{\text{in } 3} + 1$$

From (3.33) it can be seen that $k_3$ is a variable quantity, and tends toward unity as the input voltage $u_{\text{in } 3}$ grows.

Using this value for $k_3$ in (3.30), for the amplifier gain with the $(n - 1)$th stage in the logarithmic mode, we obtain

$$k_0(n - 1) = k_1^{n-2}(k_1 u_{t1}/u)[\ln(u/u_{t1}) + 1] \tag{3.34}$$
$$\times \{[(k_1 u_{t1} \ln k_1)/u_{\text{in}}(n)] + 1\}$$

where $u = u_{\text{in}}(n - 1)$ is the input at the $(n - 1)$th stage. Consequently, the amplifier output voltage is equal to

$$U_{\text{out}}(n - 1) = U_{\text{in}}k_0(n - 1) \tag{3.35}$$
$$= k_1 u_{t1} \ln k_1 + u_{t1}k_1[\ln(u/u_{t1}) + 1]$$

With further increase in the amplifier input voltage, the $(n - 1)$th cascade goes into quasilinear operation, the $(n - 2)$th stage enters logarithmic operation, and the gain and output voltage of the amplifier are accordingly given by

$$k_0(n - 2) = k_1^{n-3}(k_1 u_{t1}/u)[\ln(u/u_{t1}) + 1]$$
$$\times \{[(k_1 u_{t1} \ln k_1)/u_{\text{in}}(n - 1)] + 1\} \qquad (3.36)$$
$$\times \{[(k_1 u_{t1} \ln k_1)/u_{\text{in}}(n)] + 1\}$$

$$U_{\text{out}}(n - 2) = 2k_1 u_{t1} \ln k_1 + u_{t1}k_1[\ln(u/u_{t1}) + 1] \qquad (3.37)$$

where $u$ now equals the input voltage at the $(n - 2)$th stage.

Continuing in this manner, we obtain expressions for the gain and output voltage for an $n$-stage amplifier with the first stage operating in its logarithmic region:

$$k_0(1) = (k_1 u_{t1}/u)[\ln(u/u_{t1}) + 1]$$
$$\times \{[(k_1 u_{t1} \ln k_1)/u_{\text{in}}(2)] + 1\} \times \cdots \qquad (3.38)$$
$$\times \{[(k_1 u_{t1} \ln k_1)/u_{\text{in}}(n - 1)] + 1\}$$
$$\times \{[(k_1 u_{t1} \ln k_1)/u_{\text{in}}(n)] + 1\}$$

$$U_{\text{out}}(1) = (n - 1)k_1 u_{t1} \ln k_1 + u_{t1}k_1[\ln(u/u_{t1}) + 1] \qquad (3.39)$$

where $u$ is now the voltage at the input to the first stage, i.e., the amplifier input voltage.

On the basis of the foregoing, we can write the general expression for the gain of an $n$-stage amplifier with the stages switching into logarithmic operation one at a time:

$$k_0 = k_1^{n-m}[u_{t1}k_1/u_{\text{in}}(n - m + 1)]\{\ln[u_{\text{in}}(n - m + 1)/u_{t1}] + 1\}$$
$$\times \Pi\{[(k_1 u_{t1} \ln k_1)/u_{\text{in}}(n - m + 1)] + 1\} \qquad (3.40)$$

where $n - m$ is the number of stages operating linearly, $n - m + 1$ is the index of the stage operating in its logarithmic region, and $m - 1$ is the number of stages in quasilinear operation.

The amplifier output is simply the input multiplied by this gain:

$$U_{\text{out}} = k_0 U_{\text{in}} \qquad (3.41)$$

It is now a simple matter to express the parameters of an $n$-stage logarithmic amplifier for the general case $a_l \neq 1$, through the parameters of the individual stages.

The beginning of the logarithmic amplitude response for the cascade is that input voltage at which the last stage goes into logarithmic operation:

$$U_{\text{in } t1} = u_{t1}/k_1^{n-1} \qquad (3.42)$$

where $u_{t1}$ is the $n$th stage input voltage when it goes into logarithmic operation. In this case, the amplifier's output voltage equals

$$U_{\text{out } t1} = u_{\text{in } t1}k_1^n \tag{3.43}$$

The upper bound on the logarithmic response corresponds to that input voltage at which the first nonlinear stage goes into quasilinear operation:

$$U_{\text{in } t2} = U_{\text{in } t1}k_1^n \tag{3.44}$$

Then the amplifier output voltage is

$$U_{\text{out } t2} = nk_1u_{t1}a_l \ln k_1 + k_1u_{t1} = U_{\text{in } t2}(na_l \ln k_1 + 1) \tag{3.45}$$

The dynamic ranges for the amplifier input and output voltages over the logarithmic response are then given by

$$D_{\text{in}} = U_{\text{in } t2}/U_{\text{in } t1} = k_1^n = d_{\text{in}}^n \tag{3.46}$$

$$D_{\text{out}} = U_{\text{out } t2}/U_{\text{out } t1} = na_l \ln d_{\text{in}} + 1 \tag{3.47}$$

Consequently, the amplifier compression coefficient may be written in the form:

$$K_c = D_{\text{in}}/D_{\text{out}} = d_{\text{in}}^n/(na_l \ln d_{\text{in}} + 1) \tag{3.48}$$

There are several methods for constructing an amplifier with a logarithmic amplitude response. The most common methods at present are shunting the amplifier loads by nonlinear elements and sequential addition of stage outputs in a cascaded amplifier.

*Logarithmic Amplifier with Nonlinear Shunting Elements*

A simplified circuit for a common emitter amplifier stage with collector load shunted by a nonlinear element is shown in Fig. 3.12. A semiconductor diode is employed as the shunting element, and is represented in the simplified schematic as an active nonlinear resistance $R_{nl}$.

With the load shunted by a simple nonlinear element, it is possible to build a cascaded logarithmic amplifier with a series of *pnp* or *npn* transistors. The *pnp* and *npn* transistors must have negative and positive input signals, respectively. In analyzing a multistage logarithmic amplifier, consideration must be given to the fact that the input impedance of a transistor stage depends on the load impedance. Because the load impedance changes least in a common-emitter configuration, it is best to use common-emitter circuits in the design of multistage logarithmic amplifiers.

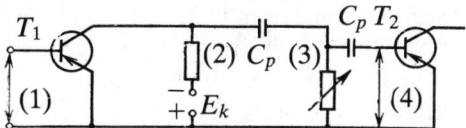

**FIGURE 3.12.** Simplified stage circuit with collector load shunted by nonlinear element.

(1) $U_{in}$   (2) $R_t$   (3) $R_{n1}$   (4) $U_{out}$

The equivalent circuit of a common emitter stage for midfrequency operation is shown in Fig. 3.13. The input dynamics of the stage are linearized with a feedback resistance, and the transistor may be considered to be a linear amplifying element [7].

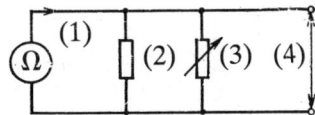

**FIGURE 3.13.** Equivalent circuit of transistor stage at IF.

(1)   $I_{n1}$   (2)   $R_e$   (3)   $R_{n1}$   (4)   $U_{out}$

The voltage gain of this stage is

$$k_v = (\beta_e/R_{in\ 0})[R_e R_{nl}/(R_e + R_{nl})] = \beta_e R_l/R_{in\ 0} \qquad (3.49)$$

where $\beta_e$ is the current gain of the common-emitter circuit; $R_{in\ 0}$ is the input impedance of the cascade including the feedback loop in the emitter circuit;

$$1/R_e = 1/R_{out} + 1/R_t + 1/R_n + 1/R_{in}$$

where $R_{out}$ is the output impedance of the transistor; $R_t$ is the impedance of the tank circuit; $R_n$ is the equivalent resistance of the base supply circuit in the following stage; $R_{in}$ is the input impedance of the following transistor; $R_l = R_e R_{nl}/(R_e + R_{nl})$ is the total load impedance.

The output voltage of the stage is given by

$$U_{out} = U_{in}k_v = U_{in}\beta_e(R_l/R_{in\ 0}) \qquad (3.50)$$

For linear operation of the stage, the resistance of the nonlinear element must be large, constant, and must not shunt the collector load, i.e.,

$R_{nl\ 1} \gg R_e$ (as before, the subscripts 1, 2, and 3 refer to linear, logarithmic, and quasilinear operation). In this case, $k_1 = \beta_e R_e$.

An expression for the variation of the nonlinear resistance $R_{nl\ 2}$ as a function of the output voltage in the logarithmic mode is found by equating the right-hand sides of (3.26) and (3.50):

$$k_1 U_{in}\left(a_l \ln \frac{U_{in\ 2}}{U_{in\ l1}} + 1\right) = \beta_e U_{in\ 2}\left(\frac{R_e R_{nl\ 2}}{R_e + R_{nl\ 2}}\right) \tag{3.51}$$

Introducing the notation $p_l = a_1 \ln(U_{in\ 2}/U_{in\ l1}) + 1$ and transforming the equations, we obtain

$$R_{nl\ 2} = R_e R_l / \exp[(p_l - 1)/a_l - p_l] \tag{3.52}$$

Analogous manipulations using (3.32) and (3.50) give an expression for $R_{nl\ 3}$, the nonlinear resistance in quasilinear operation, as a function of the stage's output voltage:

$$R_{nl\ 3} = R_e/[(k_l/a_l p_l)(p_l - a_l \ln k_1 - 1 + a_l) - 1] \tag{3.53}$$

Equation (3.53) is valid when the input impedance $R_{in}$ of the following transistor is constant and independent of the input signal, which is true if the following stage has a common-collector configuration. Thus, it is necessary to insert linear emitter-followers with unity gain between the nonlinear stages forming the logarithmic response of the entire cascade.

In practice, the nonlinear element input resistance $R_{in} = R_{nl}$ is understood to be the ratio of the video pulse amplitude to the amplitude of the current flowing through the nonlinear element:

$$R_{nl} = U_{nl}/I_{nl} \tag{3.54}$$

The nonlinear element impedance for low-input signal levels (tenths of a volt) should be small, in the range of hundreds or tens of ohms, depending on the specific case.

In choosing the nonlinear element, it must be kept in mind that exact logarithmic response of a multistage amplifier with sequential nonlinear stage operation is obtained only when the maximum input voltage for the $n$th stage $U_{in\ 3}$, corresponding to the end of the quasilinear portion of its response, does not exceed the voltage $U_{in\ lim}$ at the upper end of the transistor's linear operation. More simply, the transistor must not be saturated. Thus, the following inequality must be satisfied:

$$U_{in\ 3} = k_1 U_{in\ 1}[(n - 1)a_l \ln D_{in\ 1} + 1] \le U_{in\ lim} \tag{3.55}$$

It follows that the input voltage corresponding to the beginning of the entire amplifier's logarithmic response is given by

$$U_{\text{in } 1} \lesseqgtr U_{\text{in lim}}/[k_1(n - 1)a_l \ln D_{\text{in } 1} + 1] \tag{3.56}$$

*Logarithmic Amplifier Adding Stages' Output Voltages*

Sequential addition of output voltages from the stages can be used to obtain a logarithmic response in $n$-stage selective and aperiodic amplifiers. In general, such an amplifier consists of $n$ stages in series, with their outputs connected to a voltage adder (see Fig. 3.14(a)); the output of the adder is the sum voltage of all the outputs of the stages. The adder for the case of a selective amplifier of sinusoidal oscillations can be a delay line (DL) or active resistance. In aperiodic pulsed signal amplifiers, the adder can take the form of ordinary stages having linear response over a wide range and unity gain, connected in parallel with the amplifier stages (see Fig. 3.14(b)).

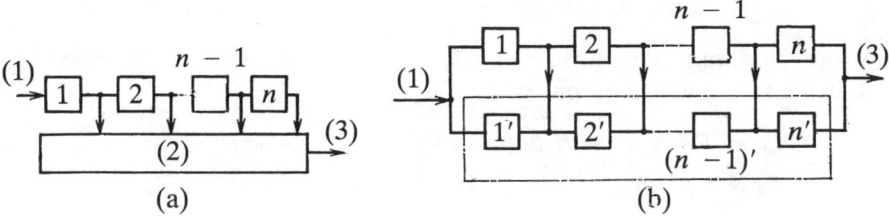

**FIGURE 3.14.** Logarithmic amplifier with stage output summation. (1) $U_{\text{in}}$  (2) adder  (3) $U_{\text{out}}$

As an example of this method for obtaining logarithmic response, we will use the amplifier whose block diagram is shown in Fig. 3.14(a). We will assume that all of the amplifier stages are identical and have gain $k_1$ in the linear mode, and constant output voltage in saturation, independent of the input voltage. We will also assume that the signals are added linearly in the adder and that the adder's gain is unity.

Then, with small input signals, all of the stages operate over the linear region of their responses and the adder's output voltage is

$$U'_{\text{out}} = U'_{\text{in}}(k_1^n + k_1^{n-1} + \cdots + k_1^2 + k_1) \tag{3.57}$$

Denoting the gain of the linear mode for the cascade by $k_r$, (3.57) may be written as

$$U'_{\text{out}} = U'_{\text{in}} k_r \tag{3.58}$$

When the input voltage reaches the value $U'_{\text{in}} = U_{\text{in } t1}$, the last ($n$th) stage saturates and the voltage at the output of the detector of the last stage is

$$U_{\text{out } t1} = U_{\text{in } t1} k_1^n$$

where, as before, the subscript $t1$ indicates the point at which the amplifier makes the transition into logarithmic operation. The output voltage at the adder is

$$U_{\text{out } t1} = U_{\text{in } t1}(k_1^n + k_1^{n-1} + \cdots + k_1) = U_{\text{in } t1} k_r \tag{3.59}$$

When the input voltage rises to the value $U''_{\text{in}} = k_1 U_{\text{in } t1}$, the $(n-1)$th stage saturates and the adder's output voltage (for $n \gg 1$) is

$$U''_{\text{out}} = k_1^n U_{\text{in } t1} + U_{\text{in } t1}(k_1^n + k_1^{n-1} + k_1^2)$$
$$- k_1^n U_{\text{in } t1} + U_{\text{in } t1} k_r$$
$$= U_{\text{out } t1} + k_1^n U_{\text{in } t1}$$

When the input voltage reaches $U'''_{\text{in}} = k_1^2 U_{\text{in } t1}$, the $(n-2)$th stage saturates and the adder output voltage is

$$U'''_{\text{out}} = 2k_1^n U_{\text{in } t1} + U_{\text{in } t1}(k_1^n + k_1^{n-1} + \cdots + k_1^3)$$
$$- 2k_1^n U_{\text{in } t1} + U_{\text{in } t1} k_r$$
$$= U_{\text{out } t1} + 2k_1^n U_{\text{in } t1}$$

Analogously, with saturation of the first stage, for the input voltage we may write

$$U^{(n)}_{\text{in } t1} = U_{\text{in } t2} = k_1^{n-1} U_{\text{in } t1} \tag{3.60}$$

where $U_{\text{in } t2}$ is the voltage at which the amplifier as a whole saturates and as a consequence goes out of logarithmic operation, and the corresponding adder output voltage is

$$U^{(n)}_{\text{out}} = U_{\text{out } t2} = (n-1)k_1^n U_{\text{in } t1} + U_{\text{in } t1} k_1^n \tag{3.61}$$
$$= nk_1^n U_{\text{in } t1}$$

The above analysis shows that as the stages saturate one at a time, the input voltage varies exponentially and the output varies linearly. As a result, there is a logarithmic dependence between the input and output voltages.

We may write (3.60) in the form:

$$\ln U_{\text{in } t2} = (n - 1)\ln k_1 + \ln U_{\text{in } t1}$$

and then find the value of $n$:

$$n = (1/\ln k_1)\ln(U_{\text{in } t2}/U_{\text{in } t1}) + 1 \qquad (3.62)$$

Substituting this value of $n$ into (3.61), we find

$$U_{\text{out}} = [k_1^n U_{\text{in } t1}/(\ln k_1)]\ln(U_{\text{in}}/U_{\text{in } t1}) + k_1^n U_{\text{in } t1} \qquad (3.63)$$
$$= k_1^n U_{\text{in } t1}\{1/[(\ln k_1)\ln(U_{\text{in}}/U_{\text{in } t1})] + 1\}$$

Comparing (3.17) and (3.63), we obtain

$$a_l = 1/(\ln k_1)$$

The dynamic ranges of the input and output of such an amplifier are given by

$$D_{\text{in}} = U_{\text{in } t2}/U_{\text{in } t1} = k_1^{n-1} \qquad (3.64)$$

$$D_{\text{out}} = U_{\text{out } t2}/U_{\text{in } t2} = nk_1^n/k_r \qquad (3.65)$$

With $k_1 \gg 1$ and $k_r \approx k_1^n$, $D_{\text{out}} = n$, and the amplified voltage compression factor is

$$K_c = D_{\text{in}}/D_{\text{out}} = k_1^{n-1}k_r/nk_1^n \qquad (3.66)$$

With the assumption that the response of each stage is linear through saturation, the response of the amplifier will not be exactly logarithmic. An exact logarithmic response can be obtained only with precise tailoring of the individual responses.

An $n$-stage amplifier will produce an exact logarithmic response when the combined response of the stages and the adder is described by (3.16), (3.26), and (3.32) when $a_l = 1$, and by (3.16), (3.43), and (3.45) when $a_l \neq 1$. Then the output of the $i$th amplifier stage may be written in the general form:

$$u_{\text{out}}(i) = u_{\text{out } r} - u_{\text{out}}(i - 1) = u_{\text{out } r} - u_{\text{in}}(i) \qquad (3.67)$$

where $U_{\text{out } r}$ is the required voltage, varying according to (3.16), (3.26) or (3.32), and (3.43); $u_{\text{in}}(i) = U_{\Sigma i}$ is the adder voltage, equal to the input voltage of the stage as long as the adder has unity gain.

For the $i$th stage, we have

$$U_{\Sigma i} = u_{\text{in}}(i) = U_{\text{in}}\sum_{1}^{i} k_1^{i-1} \qquad (3.68)$$

where $U_{\text{in}}$ is the amplifier input voltage.

If the amplifier goes into logarithmic operation at the input voltage $U_{\text{in } t1}$, a logarithmic response will be obtained, according to (3.67), when the amplitude responses of the stages satisfy the following conditions.

With the input voltage of the $i$th stage lying in the range 0 to $u_{t1} = U_{\text{in } t1}k_1^{n-1}$ all of the cascades other than the first should have a linear response described by the expression:

$$u_{\text{out } 1} = u_{\text{in } 1}k_1 - u_{\text{in}}\sum_{m=1}^{i-2}(1/k_1^{m-1}) = u_{\text{in } 1}k_1' \qquad (3.69)$$

where $u_{\text{in } 1}$ is the input voltage at the $i$th stage; $m$ is the number of the preceding stage whose response is considered in calculating the amplitude responses; $k_1'$ is the linear gain of the stage:

$$k_1' = k_1 - \sum_{m=1}^{i-2}(1/k_1^{m-1}) \qquad (3.70)$$

If $k_1 \gg 1$, then the adder response to only the preceding stage need be considered, and (3.70) takes the form:

$$k_1' = k_1 - 1 \qquad (3.71)$$

If $k_1 \geq 10$, then the error in the calculation of the amplitude response, introduced by using (3.71) instead of (3.70), is less than 1%.

If the $i$th stage input voltage lies in the range $u_{t1}$ to $u_{\text{in } s} = k_1 U_{\text{in } t1}$, its output voltage will vary according to the expression:

$$u_{\text{out } 2} = k_1 u_{\text{in } t1}[a_l \ln (u_{\text{in } 2}/u_{\text{in } t1}) + 1] - u_{\text{in } 2} \qquad (3.72)$$

and (3.29) is satisfied. If the $i$th-stage input voltage $u_{\text{in } 3} > u_{\text{in } s}$, the stage will be saturated and its output voltage will be constant and equal to

$$u_{\text{out } 3} = k_1 u_{\text{in } t1}(a_l \ln d_{\text{in}} + u_{\text{in } 3}/u_{\text{in } t1}) - u_{\text{in } 3} \qquad (3.73)$$
$$= k_1 u_{\text{in } t1}a_l \ln d_{\text{in}}$$

An adder is necessary to obtain a logarithmic response from an $n$-stage amplifier by sequential addition of the stage outputs. In aperiodic pulse amplifiers, ordinary amplifier stages of gain less than or equal to unity can be used as the adder [7]. These stages are called followers to distinguish them from the amplifying stages.

In Fig. 3.14(b), the amplifier stages are denoted by the numbers $1, 2, \ldots, n - 1, n$, and the followers by $1', 2', \ldots, (n - 1)', n'$. A follower is connected in parallel with each amplifier stage and should have linear responses over a wide dynamic range. This linear range must increase at each stage. As a follower, it is best to use a common-emitter stage with a strong negative-current feedback loop.

A single nonlinear stage consisting of an amplifier stage and follower in parallel should have an amplitude response described by (3.16), (3.43), and (3.45), i.e., a response such that neighboring stages switch operating modes simultaneously. This requires the stages to have the responses described by (3.72) and (3.73), while the adder should have a strictly linear response and unity gain.

## 3.5 PHASE DETECTORS

An important element in monopulse systems employing phase and sum-and-difference angle discriminators is the phase detector, with which the target angle is determined. The circuit of a vector-measurement phase detector is shown in Fig. 3.15, in which $R_1 = R_2$ and $C_1 = C_2$.

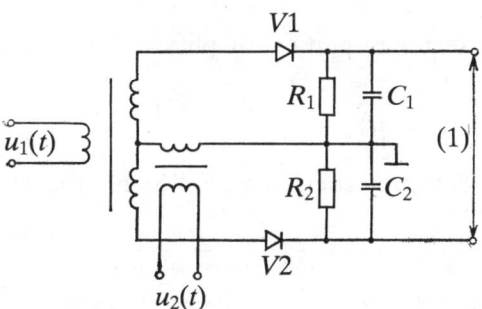

**FIGURE 3.15.** Balanced vector-measurement phase detector.
(1) $u_{pd}$

The input of amplitude detector $V1$ is the sum of the voltages $u_1(t)$ and $u_2(t)$, and the input of detector $V2$ is the difference of those voltages. For the case of a sum-and-difference angle discriminator, $u_1(t)$ and $u_2(t)$ correspond to the normalized difference and sum voltages, and, for a phase discriminator, they are the linear IF amplifier outputs. The detector voltages are subtracted by means of special loads.

If the amplitude detector extracts the envelope or the square of the envelope of the input random process, it is possible to derive a mathematical description of the phase-detector operation. The phase-detector input voltages are written in the form:

$$u_1(t) = U_1 \sin(\omega_c t + \phi_1)$$

$$u_2(t) = U_2 \sin(\omega_c t + \phi_2)$$
(3.74)

where $\omega_c$ is the down-converted frequency.

The input vector voltages are shown in Fig. 3.16, which clearly shows the voltages detected at $V1$ and $V2$. These voltages correspond to the vector sums:

$$U_+ = U_1 + U_2, \qquad U_- = -U_1 + U_2$$
(3.75)

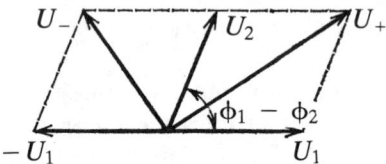

**FIGURE 3.16.** Vector diagram of phase-detector voltages.

With linear detector responses the phase detector output voltage will be

$$u_{pd} = K_{pd}[|U_+| - |U_-|]$$
(3.76)

where $K_{pd}$ is the detection coefficient. Following from the vector diagram (Fig. 3.16):

$$|U_+| = \sqrt{U_1^2 + U_2^2 + 2U_1U_2 \cos(\phi_1 - \phi_2)}$$
(3.77)

$$|U_-| = \sqrt{U_1^2 + U_2^2 + 2U_1U_2 \cos[180° - (\phi_1 - \phi_2)]}$$
(3.78)

$$= \sqrt{U_1^2 + U_2^2 - 2U_1U_2 \cos(\phi_1 - \phi_2)}$$

Consequently,

$$u_{pd} = K_{pd}[\sqrt{U_1^2 + U_2^2 + 2U_1U_2 \cos(\phi_1 - \phi_2)} - \sqrt{U_1^2 + U_2^2 - 2U_1U_2 \cos(\phi_1 - \phi_2)}]$$
(3.79)

If $U_1 \ll U_2$, which is true for monopulse systems using a sum-and-difference angle discriminator, then by expanding each term of (3.79) as a power series and retaining the terms of first and second order, we obtain

$$u_{pd} = 2K_{pd}\frac{U_1 U_2}{\sqrt{U_1^2 + U_2^2}}\cos(\phi_1 - \phi_2)$$

$$= 2K_{pd}U_1\cos(\phi_1 - \phi_2) \tag{3.80}$$

When the signal amplitudes are equal, $U_1 = U_2 = U$ (which can occur with a phase angle discriminator) the phase detector voltage will have the form:

$$u_{pd} = K_{pd}\{\sqrt{2U^2[1 + \cos(\phi_1 - \phi_2)]} - \sqrt{2U^2[1 - \cos(\phi_1 - \phi_2)]}\}$$

$$= 2K_{pd}U\left[\sqrt{\frac{1 + \cos(\phi_1 - \phi_2)}{2}} - \sqrt{\frac{1 - \cos(\phi_1 - \phi_2)}{2}}\right]$$

$$= 2K_{pd}U\left(\cos\frac{\phi_1 - \phi_2}{2} - \sin\frac{\phi_1 - \phi_2}{2}\right) \tag{3.81}$$

Figure 3.17 shows the variation of output voltage with phase shift for the two cases. With $U_1 \ll U_2$ (or $U_2 \ll U_1$), the output exhibits a cosine dependence on the input phase shift, and, for $U_1 = U_2$ with the phase shift in the range 0 to $n$, the dependence is linear.

With square-law amplitude detectors, the phase-detector voltage is proportional to the average value of the difference of the currents flowing through the diodes:

$$u_{pd} = K_{pd}\overline{\{[u_1(t) + u_2(t)]^2 + [u_1(t) - u_2(t)]^2\}}$$

$$= K_{pd}\overline{u_1(t)u_2(t)} \tag{3.82}$$

Using the values of $u_1(t)$ and $u_2(t)$ from (3.74), we obtain

$$u_{pd} = K_{pd}U_1 U_2 \overline{\sin(\omega_c t + \phi_1)\sin(\omega_c t + \phi_2)} \tag{3.83}$$

$$= \tfrac{1}{2}K_{pd}U_1 U_2 \cos(\phi_1 - \phi_2)$$

It is apparent from (3.83) that a phase detector employing diodes with quadratic responses is equivalent to a simple multiplier (neglecting the higher-order harmonics $\omega_c$), and the output voltage exhibits a cosine dependence on the phase shift.

The shortcoming of this phase detector is that it limits the input phase shift to a 180°-range, while the actual shift provided by an unambiguous angle sensor may often lie outside this range.

The Kirkpatrick phase detector shown in Fig. 3.18 permits a wider range of input values. The angular range is increased by decreasing the sensitivity along the axis, and because the signal level is highest on axis, performance is not significantly degraded by this reduction in sensitivity [32].

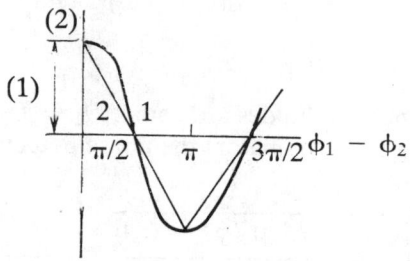

**FIGURE 3.17.** Phase-detector response.

1: for $U_1 \ll U_2$
2: for $U_1 = U_2$
(1) $2K_{pd}U_1$    (2) $u_{pd}$

A comparison of Figs. 3.15 and 3.18 shows that the detection portion of the Kirkpatrick phase detector is the same as for the simpler circuit, consisting of two diodes and differential loads. In both detectors, the output voltage is generated by the difference in diode voltages. The difference lies in the diode input voltages; in the simple detector they are the sum-and-difference voltages, in the Kirkpatrick detector the diode voltages are superpositions of the input voltage components.

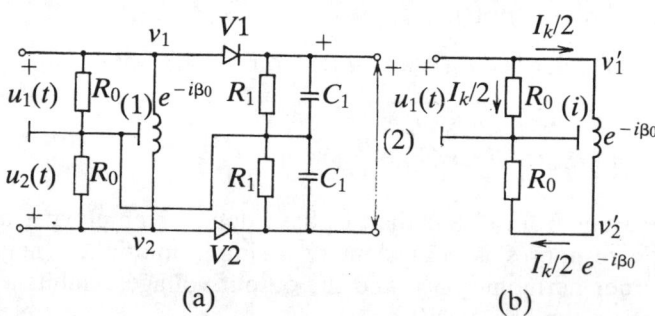

**FIGURE 3.18.** Kirkpatrick phase detector:
(a) simplified detector circuit
(b) part of circuit forming diode voltages
(1) DL    (2) $u_{pd}^k$

We will denote the voltages occurring at the diode inputs as $v_1$ and $v_2$. Then the voltage $u_1(t)$ forms components $v_1'$ and $v_2'$ which influence only the components shown in Fig. 3.18(b). The current $I_k$ driven by $u_1(t)$ flows through two parallel impedances, one of which is $R_0$ and the other a delay line with impedance $R_0$. As a result, the current $I_k$ is evenly divided between the two paths; hence, the two components generated by $u_1(t)$ are

$$v_1' = u_1(t) \quad \text{and} \quad v_2' = u_1(t)e^{-i\beta_0}$$

Analogously, but in the opposite order, $u_2(t)$ forms corresponding components. Consequently, the output voltage of a Kirkpatrick phase detector with linear diodes will equal

$$
\begin{aligned}
u_{pd}^k &= K_{pd}(|u_1(t) + u_2(t)e^{-i\beta_0}| - |u_1(t)e^{-i\beta_0} + u_2(t)|) \\
&= K_{pd}\left[\sqrt{U_1^2 + U_2^2 + 2U_1U_2\cos(\phi_1 - \phi_2 - \beta_0)}\right. \\
&\quad \left. - \sqrt{U_1^2 + U_2^2 + 2U_1U_2\cos(\phi_1 - \phi_2 + \beta_0)}\right] \\
&= K_{pd}\left[\sqrt{U_1^2 + U_2^2}\sqrt{1 + \frac{2U_1U_2}{U_1^2 + U_2^2}\cos(\phi_1 - \phi_2 - \beta_0)}\right. \\
&\quad \left. - \sqrt{U_1^2 + U_2^2}\sqrt{1 + \frac{2U_1U_2}{U_1^2 + U_2^2}\cos(\phi_1 - \phi_2 + \beta_0)}\right]
\end{aligned}
\tag{3.84}
$$

For a system employing a sum-and-difference angle discriminator ($U_1 \ll U_2$), expanding (3.84) and retaining terms of first and second order gives

$$
\begin{aligned}
u_{pd}^k &= K_{pd}\left[\sqrt{U_1^2 + U_2^2} + \frac{U_1U_2}{\sqrt{U_1^2 + U_2^2}}\cos(\phi_1 - \phi_2 - \beta_0)\right. \\
&\quad \left. - \sqrt{U_1^2 - U_2^2} - \frac{U_1U_2}{\sqrt{U_1^2 + U_2^2}}\cos(\phi_1 - \phi_2 + \beta_0)\right] \\
&\approx 2K_{pd}U_1\sin(\phi_1 - \phi_2)\sin\beta_0
\end{aligned}
\tag{3.85}
$$

In the case of a monopulse system with a phase angle discriminator ($U_1 = U_2 = U$), the output of the Kirkpatrick phase detector is

$$
\begin{aligned}
u_{pd}' &= 2K_{pd}U\left[\sqrt{\frac{1 + \cos(\phi_1 - \phi_2 - \beta_0)}{2}} - \sqrt{\frac{1 + \cos(\phi_1 - \phi_2 + \beta_0)}{2}}\right] \\
&= 4K_{pd}U\sin\frac{\phi_1 - \phi_2}{2}\sin\frac{\beta_0}{2}
\end{aligned}
\tag{3.86}
$$

Comparison of (3.80), (3.85), (3.81), and (3.86), along with the response curves of Fig. 3.19, shows that the output voltages of the simple and Kirkpatrick phase detectors differ only by $\pi/2$ for $\beta_0 = \pi/2$, and the angular coverage is limited to the phase shift interval $-\pi/2 \leq \phi_1 - \phi_2 \leq \pi/2$. However, this range extends to 360° as $\beta_0$ goes to zero. Figure 3.19 shows the output voltage and the widened range for $\beta_0 = 0.1$. It is also apparent that the slope of the output decreases substantially with the increase in angular range.

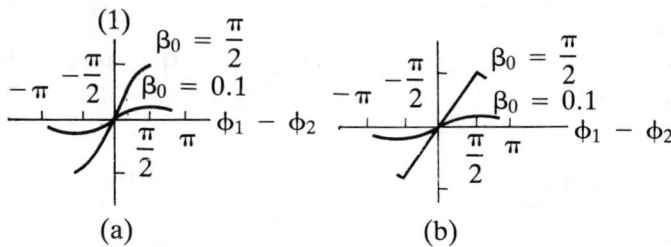

(a)                                                    (b)

**FIGURE 3.19.** Response of Kirkpatrick phase detector.
  (a) for $U_1 \ll U_2$
  (b) for $U_1 = U_2$
  (1) $u_{pd}^k$

## 3.6 AUTOMATIC GAIN CONTROL

Figure 3.20 shows block diagrams for pulse automatic gain control (AGC) systems employed in most monopulse radars. Usually, the input to the AGC detector is the delay voltage $U_d$, so that voltage control occurs above $U_d$. The delay voltage is often obtained in AGC circuits with silicon diodes, which are known to switch off sharply in the range 0.4 to 0.6 $V$. The AGC system may be amplified or unamplified, depending on whether or not an amplifier is present in the circuit.

Amplification in AGC systems can occur before the AGC detector with ac voltage (see Fig. 3.20(b)), or after the detector with dc voltage (see Fig. 3.20(c)). Modern monopulse systems also often employ multiloop AGC units with parallel operating loops (see Fig. 3.21). The controlled amplifiers in these AGC systems are usually the first stages of the IF amplifier, thus ensuring little nonlinear distortion in the IF circuit. The basic characteristics of an AGC system include the amplitude responses of the AGC circuit and controlled amplifier, with and without gain control, and the gain control response.

The amplitude response of an AGC circuit is the dependence of the control voltage on the output voltage of the controlled amplifier: $U_c = f(U_{out})$. Figure 3.22 shows the form of this dependence for an AGC system using a delay voltage. When the amplifier exceeds this delay voltage, the control voltage will be nonzero. With proper selection of the circuit parameters, the operating amplitude response should remain linear with further increase in $U_{\text{out}}$. The slope of the response determines the feedback circuit gain:

$$\tan \alpha_c = k_f$$

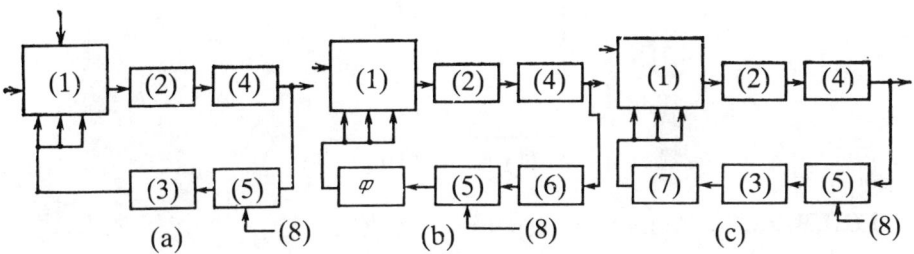

(a)    (b)    (c)

**FIGURE 3.20.** Block diagrams for pulse AGC systems:
  (a) "unamplified"
  (b) "amplified" with ac
  (c) "amplified" with dc

|  |  |
|---|---|
| (1) controlled IF amplifier | (5) AGC detector |
| (2) amplitude detector | (6) AGC amplifier |
| (3) filter | (7) IF amplifier |
| (4) video amplifier | (8) $U_d$ |

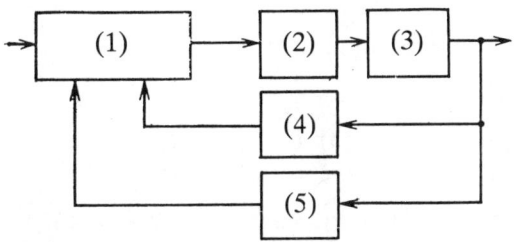

**FIGURE 3.21.** Block diagram for a multiloop AGC system.

|  |  |
|---|---|
| (1) controlled IF amplifier | (4) AGC-1 |
| (2) amplitude detector | (5) AGC-2 |
| (3) video amplifier |  |

$$k_c = f(U_c)$$

Figure 3.23 shows the amplitude response of the controlled IF amplifier, which determines the dependence of the amplifier output voltage ($U_{\text{out}}$) on the input voltage ($U_{\text{in}}$), both with and without AGC. The control characteristic of the amplifier (see Fig. 3.24) determines the relation between the controlled amplifier gain and the control voltage:

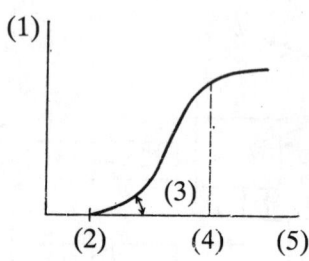

**FIGURE 3.22.** Amplitude response of AGC circuit.

(1) $U_c$     (2) $U_d$     (3) $\alpha_c$     (4) $U_{out\ max}$     (5) $U_{out}$

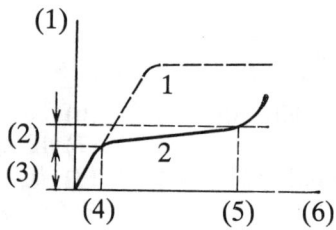

**FIGURE 3.23.** Amplifier amplitude response.
   1: without AGC
   2: with AGC

(1) $U_{out}$      (3) $U_t$       (5) $U_{in\ max}$
(2) $\Delta U_{out}$      (4) $U_{in\ min}$      (6) $U_{in}$

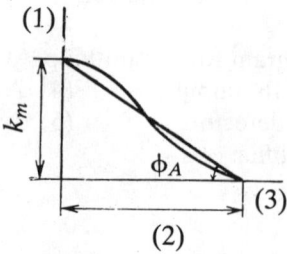

**FIGURE 3.24.** Controlled amplifier response.
   (1) $k_c$        (2) $U_{c\ max}$        (3) $U_c$

The form of this dependence varies, depending on the type and number of controlled stages. In analyzing actual systems it is necessary to know the AGC control characteristic, which is nonlinear. This nonlinearity increases strongly with the number of controlled stages, and various approximations can be used for analysis; linear, exponential, polynomial, hyperbolic, and other functions can be used. The best approximation must be selected separately for each situation, depending on the methods utilized. In general, the linear approximation greatly simplifies the analysis without introducing significant errors.

AGC systems have been analyzed in a great number of studies, the most comprehensive of which is [37]. Here we will examine only the basic dynamic properties of AGC systems. We will make use of simplifying assumptions, which do not distort the fundamental phenomena, while affording the simplest derivations of the necessary relations.

We will assume that there is no nonlinear distortion in the receiver, and that the receiver passband is much wider than that of the AGC circuit. We will further assume that the AGC circuit incorporates a single-section RC filter and a wideband detector. The behavior of an AGC system with a single-section filter is described by the following system of equations:

$$U_{\text{out}} = \begin{cases} k_m U_{\text{in}} & \text{for } U_{\text{out}} < U_d \\ k_c U_{\text{in}} & \text{for } U_{\text{out}} > U_d \end{cases}$$
$$U_c = k_f F_A(p)(U_{\text{out}} - U_t) \quad \text{for } U_{\text{out}} > U_d \tag{3.87}$$

where $k_m$ is the maximum gain for $U_c = 0$; $U_d$ is the AGC delay voltage; $k_f = k_D k_A$ is the gain of the feedback loop, equal to the product of the detector gain and the AGC gain; $F_A(p) = 1/(pT_A + 1)$ is the single-section filter response; $T_A$ is the time constant of the AGC circuit.

With the linear approximation to the control characteristic:

$$k_c = k_m - b_A U_c \tag{3.88}$$

where $b_A = \tan \alpha_A = k_m/U_{bm}$ is the angular coefficient of the control characteristic, and $U_{bm} = k_m/b_A$ is the voltage at which the receiver gain falls to zero.

Making use of the approximation (3.88), the system of equations in (3.87) becomes

$$U_{\text{out}} = \begin{cases} k_m U_{\text{in}} & \text{for } U_{\text{out}} < U_d \\ (k_m - b_A U_c)U_{\text{in}} & \text{for } U_{\text{out}} > U_d \end{cases}$$
$$(pT_A + 1)U_c = k_f(U_{\text{out}} - U_d) \quad \text{for } U_{\text{out}} > U_d \tag{3.89}$$

We will find the response of the system to a voltage jump exceeding $U_{in\ min}$. With $U_{in} = U_A = \text{const}$, (3.89) becomes an inhomogeneous differential equation of first order with constant coefficients:

$$T_A\, dU_{out}/dt + U_{out}(1 + b_A k_f U_A) = b_A k_f U_A U_d + k_m U_A \qquad (3.90)$$

The solution of this equation has the form:

$$U_{out} = B_A + C_A e^{-t/\tau_A} \qquad (3.91)$$

where the partial solution of the inhomogeneous equation is

$$B_A = (k_m U_A + b_A k_f U_d)/(1 + k_A) \qquad (3.92)$$

the duration of the voltage spike is

$$\tau_A = T_A/(1 + k_e) \qquad (3.93)$$

and

$$k_e = b_A k_f U_A \qquad (3.94)$$

is the equivalent gain of the AGC system.

Setting the initial conditions, we will find the constant of integration $C_A$. We will assume that at $t = 0$ the voltage across the filter capacitor is zero; in this case $U_c = 0$ and $U_{out} = k_m U_A$. Then (3.91) can be written in the form:

$$U_{out} = U_A\left[\frac{b_A k_f U_d + k_m}{1 + k_e} + \left(k_m - \frac{b_A k_f U_d + k_m}{1 + k_e}\right)e^{-t/\tau_A}\right] \qquad (3.95)$$

Setting $U_d = 0$, (3.95) and (3.89) become, respectively,

$$U_{out} = k_m U_A\,(1 + k_e e^{-t/\tau_A})/(1 + k_e) \qquad (3.96)$$

$$U_c = [k_m k_f U_A/(1 + k_e)]\,(1 - e^{-t/\tau_A}) \qquad (3.97)$$

$$k_c = k_m\,(1 + k_e e^{-t/\tau_A})/(1 + k_e) \qquad (3.98)$$

Figure 3.25 shows plots of the curves for the output and control voltages as calculated from (3.96) and (3.97). The output voltage (see Fig. 3.25(a)) abruptly increases at $t = 0$ to $k_m U_A$ due to the spike, and then decays exponentially to the steady-state value $k_m U_A/(1 + k_e)$. The control voltage, on the other hand, grows exponentially to its steady-state value $k_m k_f U_A/(1 + k_e)$. The duration of the transient is the same in both cases and depends, not only on the time constant and circuit parameters of the AGC system, but also on the amplitude $U_A$ of the spike. As follows from (3.93) and (3.94), the time duration of the transient decreases with increasing $U_A$, and the control voltage increases more rapidly.

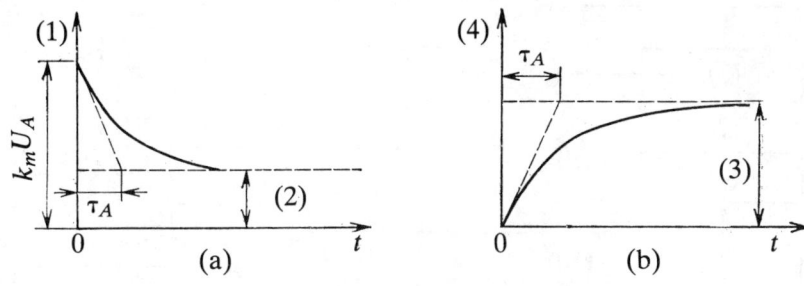

**FIGURE 3.25.** AGC response to an input voltage transient.

(1) $U_{\text{out}}$    (2) $\dfrac{k_m U_A}{1 + k_e}$    (3) $\dfrac{k_m k_f U_A}{1 + k_e}$    (4) $U_c$

In conclusion, we will consider the operation of an AGC system in an amplitude-sum-and-difference monopulse system. Normalization of the sum and difference signal amplitudes by the sum signal renders the error signal independent of the received signal amplitudes. The structure of a possible AGC circuit is shown in Fig. 3.26 [21].

The sum and difference signals generated by one of the methods discussed in Section 3.3 are fed to the IF amplifier. Each channel has its own AGC circuit. The gain in each IF amplifier is controlled by feeding it with short IF control pulses from a pulse generator, the pulses being proportional to the required gain. These control pulses lead the transmitted pulses slightly.

The blocking oscillator produces signals which enable the pulse generator. These same signals reach the transmitter modulator through a delay line, providing the lag time between the control and transmitted pulses. The control pulses are direct inputs of the two-stage variable IF amplifier, and are attenuated before being fed to the IF amplifiers in the sum and difference channels. The control signals are then detected and fed to the video amplifier, and then to the AGC detector. These detectors also receive blanking pulses from the blocking generator. The delay voltage is applied to the detectors, which, therefore, only detect voltages exceeding that value.

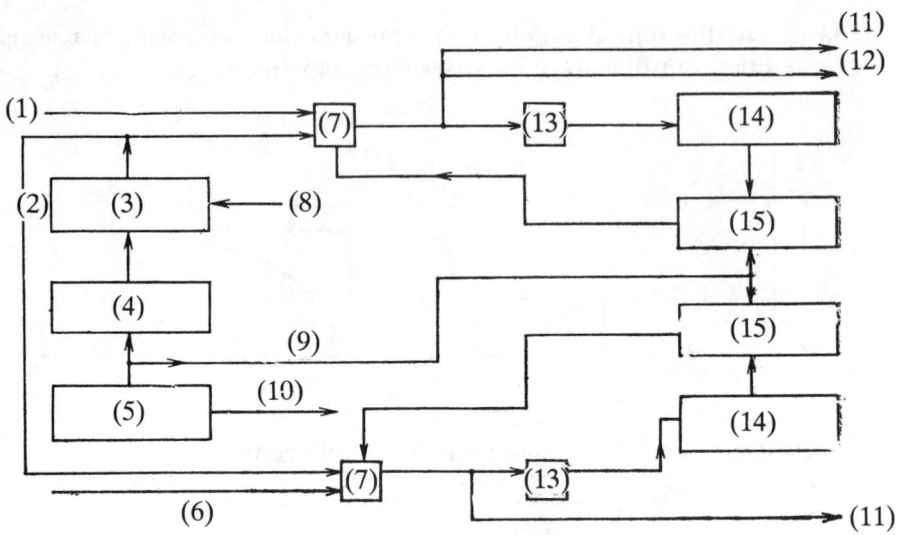

**FIGURE 3.26.** Block diagram for an AGC system in an amplitude-sum-and-difference radar.

(1)  sum signal
(2)  control impulses
(3)  2-stage IF amplifier
(4)  pulse generator
(5)  blocking generator
(6)  difference signal
(7)  IF amplifier
(8)  blanking pulses

(9)  voltage proportional to sum signal
(10) to transmitter modulator
(11) to phase detector
(12) to amplitude detector
(13) amplitude detector
(14) video amplifier
(15) AGC detector

When the target is detected and the system locks on in range, the two-stage IF amplifier input is a positive voltage proportional to the sum signal, which causes the control pulses to be amplified. As a result, the AGC circuit accomplishes normalization of the sum and difference signals by the sum signal, and ensures that the control pulses fed to the sum and difference IF amplifiers have equal amplitude. This ensures that the amplification of the echo signal will be the same when the echo signal is equal to or greater than the noise level.

# Chapter 4

# General Design Principles for Modern Monopulse Radar Systems

## 4.1 MONOPULSE TRACKING RADARS

### 4.1.1 Amplitude-Amplitude Systems

As was shown in Section 1.1, angle sensing is achieved in amplitude-amplitude systems with an antenna that generates two beams in each coordinate plane, the beams squinted from the axis by the angles $\pm \theta_0$ (see Fig. 1.1). The amplitude imbalance in the receiver channels associated with each beam is directly related to the tracking error, a greater error causing a greater imbalance. This amplitude difference, and hence tracking error, is zero when the received signals are equal. Thus, tracking is performed by rotating the antenna system (or steering the beam) until the reflected signals have equal amplitudes.

The block diagram of an amplitude-amplitude system for tracking in one coordinate is shown in Fig. 4.1. Normalization of the error signal in this system is achieved with logarithmic amplifiers. If the target echo at the antenna is $E(t) = \underline{E}_m e^{i\omega t}$, and the target angle relative to the axis is $\theta$, then the signals in the two receiver channels will be given by

$$\underline{E}_1(t, \theta) = E_m F_1(\theta)\exp i\omega t = E_m F(\theta_0 - \theta)\exp i\omega t \qquad (4.1)$$
$$\underline{E}_2(t, \theta) = E_m F_2(\theta)\exp i\omega t = E_m F(\theta_0 + \theta)\exp i\omega t$$

After frequency conversion, *IF* amplification, and linear detection, the signals at the input of the subtraction device will be

$$u_1(\theta) = \ln[k_1 E_m F(\theta_0 - \theta)] \qquad (4.2)$$
$$u_2(\theta) = \ln[k_2 E_m F(\theta_0 + \theta)]$$

where $k_1$ and $k_2$ are the channel signal gains.

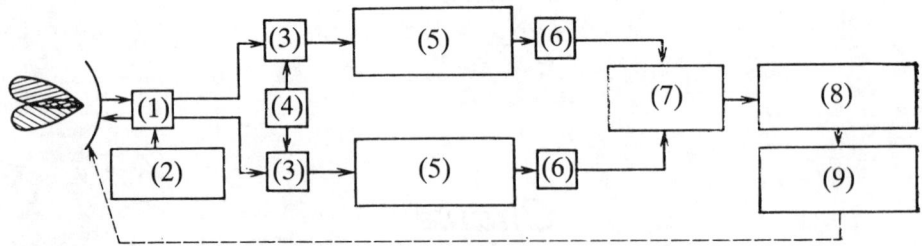

**FIGURE 4.1.** Block diagram of an amplitude-amplitude monopulse system for target tracking in one coordinate.

(1) receiving/transmitting switch  (6) amplitude detector
(2) transmitter                    (7) subtraction circuit
(3) mixer                          (8) error signal amplifier
(4) local oscillator               (9) antenna control system
(5) logarithmic IF amplifier

At the output of the subtraction circuit, we obtain

$$S(\theta) = \ln[k_1 F(\theta_0 - \theta)/k_2 F(\theta_0 + \theta)] \tag{4.3}$$

which is the angle-sensing response.

This error signal is amplified and then fed to the antenna control system. Equation (4.3) shows that the angle-sensing response depends upon the antenna patterns, along with the parameters and similarity of the logarithmic amplifiers. Instability and imbalance in the amplitude responses of the amplifiers, therefore, lead to distortion in the angle-sensing response, and, as a result, generate tracking errors.

With identical receiver channels ($k_1 = k_2 = k$) and small angle errors, (4.3) may be written as follows:

$$S(\theta) = \ln\frac{F(\theta_0 - \theta)}{F(\theta_0 + \theta)} = \ln\frac{F(\theta_0)(1 + \mu\theta)}{F(\theta_0)(1 - \mu\theta)} = \ln\frac{1 + \mu\theta}{1 - \mu\theta}$$

$$\approx 2[\mu\theta + (\mu^2\theta^2)/2 + \cdots] \approx 2\mu\theta, \tag{4.4}$$

where $F(\theta_0)$ is the antenna gain on the axis, and $\mu$ is the slope of either squinted beam at the axis (which, in the linear approximation above, is one-half the slope of the resultant difference pattern at the axis).

The main drawback to using an amplitude angle discriminator is the requirement for precise and identical logarithmic amplifier responses.

### 4.1.2. Phase-Phase Monopulse Systems

A simplified block diagram of a phase-phase monopulse system for angle sensing in one coordinate is shown in Fig. 4.2. The antenna system in this approach generates two parallel beams in each coordinate plane.

With identical beams, the signals received at the antenna are

$$E_1(t, \theta) = E_m F(\theta) \exp[i(\omega t + \Delta\phi/2)]$$
$$E_2(t, \theta) = E_m F(\theta) \exp[i(\omega t + \gamma_\phi - \Delta\phi/2)]$$

(4.5)

where $\Delta\phi$ is the phase shift arising from the difference in path length as given by (1.1), and $\gamma_\phi = \pi/2$ is a biasing phase shift which transforms the cosine dependence of the response on the phase difference into a sine dependence, thus producing a null error signal when the target is on the axis ($\theta = 0$).

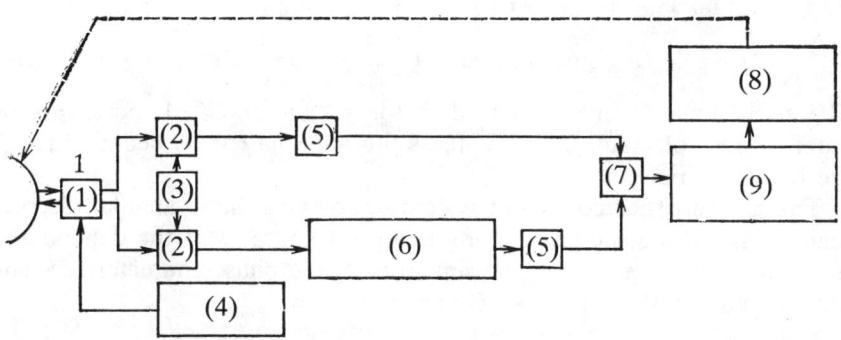

**FIGURE 4.2.** Block diagram of a phase-phase monopulse system for target tracking in one coordinate.

(1) receiving/transmitting switch  (6) 90° phase shifter
(2) mixer                          (7) phase detector
(3) local oscillator               (8) antenna control system
(4) transmitter                    (9) error signal amplifier
(5) IF amplifier

At the output of the IF amplifier the down-converted signals are

$$\underline{u}_1(t, \theta) = k_1 E_m F(\theta) \exp[i(\omega t_c + \Delta\phi/2)]$$
$$\underline{u}_2(t, \theta) = k_2 E_m F(\theta) \exp[i(\omega_c t + \pi/2 - \Delta\phi/2)]$$

(4.6)

If normalization is achieved through limiting, the signals at the phase-detector input may be expressed as

$$\underline{u}'_1(t, \theta) = U_{\lim} \exp[i(\omega_c t + \Delta\phi/2)] \qquad (4.7)$$
$$\underline{u}'_2(t, \theta) = U_{\lim} \exp[i(\omega_c t + \pi/2 - \Delta\phi/2)]$$

where $U_{\lim}$ is the limiter threshold amplitude.

If there are product detectors at the front of the phase detector, i.e., if the phase detector multiplies and averages the input signals, then the phase-detector output will take the form:

$$S(\theta) = K_{pd} \, \text{Re}[\underline{u}'_1(t, \theta)\underline{u}'_2(t, \theta)] \qquad (4.8)$$

where $K_{pd}$ is the gain of the phase detector.

Consequently,

$$S(\theta) = K_{pd} U^2_{\lim} \sin \Delta\phi \qquad (4.9)$$

Replacing $\Delta\phi$ with its value from (1.1), we obtain

$$S(\theta) = K_{pd} U^2_{\lim} \sin[(2nl/\lambda)\sin \theta] \qquad (4.10)$$

The error signal from the phase detector is then amplified and fed to the antenna control system, which steers the antenna (or its beam) through the required angle.

The greatest shortcoming of systems employing phase angular discriminators, as will be shown in more detail in Chapter 7, is the dependence of the tracking accuracy on the similarity of the phase characteristics and on the stability of the receiver channels.

### 4.1.3 Amplitude-Sum-and-Difference Monopulse Systems

The requirement for identical receiver channels is not as stringent in systems that use sum-and-difference angular discriminators, which are, therefore, employed most widely in modern monopulse radars. In such a system, the target echo signals presented at the antenna output are passed to a sum-and-difference device (see Section 3.3), where they are added and subtracted. The resulting output signals are then fed to the sum and difference channels of the receiver to be down-converted to IF and amplified as necessary. The amplitude of the difference signal indicates the magnitude of the angular error, and the phase difference between the sum and difference signals gives the direction of the off-axis error. The block diagram of such a system for tracking in one coordinate is shown in Fig. 4.3.

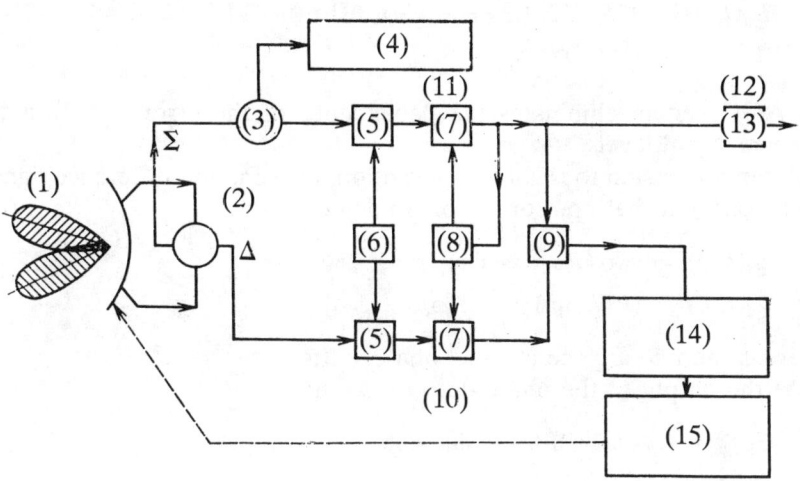

**FIGURE 4.3.** Block diagram of an amplitude-sum-and-difference monopulse system for target tracking in one coordinate.

(1) receiving/transmitting antenna
(2) hybrid ring junction
(3) receiving/transmitting switch
(4) transmitter
(5) mixer
(6) local oscillator
(7) IF amplifier
(8) AGC

(9) phase detector
(10) difference channel
(11) sum channel
(12) ranging signal
(13) amplitude detector
(14) error signal amplifier
(15) antenna control system

With small target angles relative to the axis, the signals at the output of the antenna system in the first and second channels have the form:

$$\begin{aligned}
\underline{E}(t, \theta) &= E_m F_1(\theta)\exp i\omega t = E_m F(\theta_0 - \theta)\exp i\omega t \\
&= E_m F(\theta_0)(1 + \mu\theta)\exp i\omega t \\
\underline{E}_2(t, \theta) &= E_m F_2(\theta)\exp i\omega t = E_m F(\theta_0 + \theta)\exp i\omega t \\
&= E_m F(\theta_0)(1 - \mu\theta)\exp i\omega t
\end{aligned} \tag{4.11}$$

The signals at the output of the (balanced) sum-and-difference device will be

$$\begin{aligned}
\underline{E}_s(t, \theta) &= (1/\sqrt{2})[\underline{E}_1(t, \theta) + \underline{E}_2(t, \theta)] \\
&= \sqrt{2}E_m F(\theta_0)\exp i\omega t
\end{aligned} \tag{4.12}$$

$$\underline{E}_d(t, \theta) = (1/\sqrt{2})[\underline{E}_1(t, \theta) - \underline{E}_2(t, \theta)]$$
$$= \sqrt{2}E_m F(\theta_0)\mu\theta \exp i\omega t$$

An AGC system eliminates the dependence of the error signal on the received signal levels.

After conversion to IF and amplification, the sum-and-difference signals (with AGC) at the input of the phase detector are

$$\underline{u}_s(t, \theta) = \exp[i(\omega_c t + \phi_1)]$$
$$\underline{u}_d(t, \theta) = (k_2/k_1)\mu\theta \exp[i(\omega_c t + \phi_2)]$$

(4.13)

where $\phi_1$ and $\phi_2$ are the channel phase shifts.

At the output of the phase detector we have

$$S(\theta) = K_{pd}(k_2/k_1)\mu\theta \cos(\phi_1 - \phi_2)$$

(4.14)

The sum and difference patterns for a sum-and-difference monopulse system are shown in Fig. 4.4, using plus and minus signs to denote the phase relations. It is apparent from the curves that the phase of the difference signal at the antenna output depends on the target angle relative to the axis, and can only be in phase, or directly out of phase, with the sum signal. When there is no imbalance in the channels, that is, when the target lies on the axis, the target echoes have equal amplitudes in the receiving beams, and the difference signal is zero.

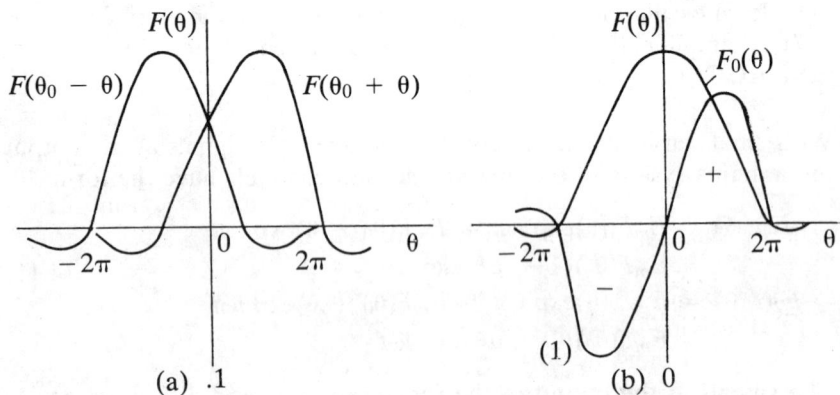

**FIGURE 4.4.** Antenna patterns for an amplitude-sum-and-difference monopulse system.
(a) partial
(b) sum and difference
(1) $F_d(\theta)$

The difference signal is used directly by the antenna control system for target direction-finding or automatic tracking. In addition to its function as a reference, the sum signal is used for target detection, ranging, and velocity measurement.

### 4.1.4 Phase-Sum-and-Difference Monopulse Systems

We will now consider the signal processing for a phase-sum-and-difference monopulse system (see Fig. 4.5). As with the phase-phase system examined in 4.1.2, the signals at the output of the antenna are given by

$$E_1(t, \theta) = E_m F(\theta)\exp[i(\omega t + \Delta\phi/2)]$$
$$E_2(t, \theta) = E_m F(\theta)\exp[i(\omega t - \Delta\phi/2)]$$ (4.15)

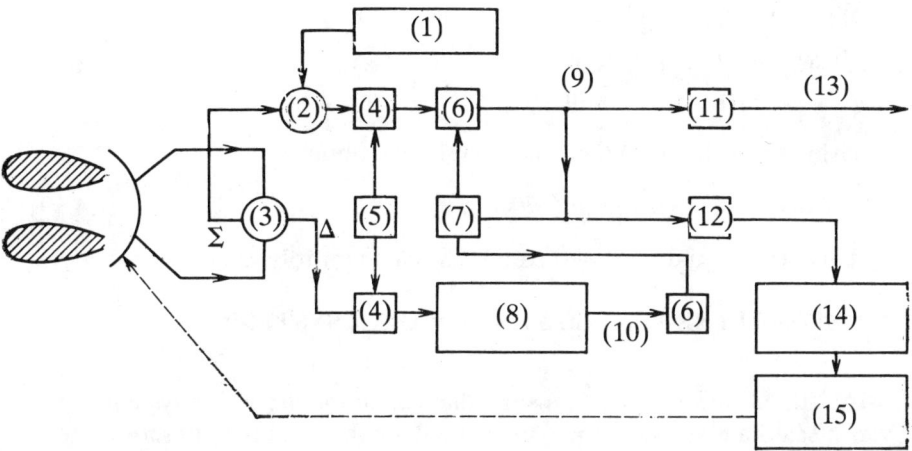

**FIGURE 4.5.** Block diagram of a phase-sum-and-difference monopulse system for target tracking in one coordinate.

(1) transmitter
(2) receiving/transmitting switch
(3) hybrid ring junction
(4) mixer
(5) local oscillator
(6) IF amplifier
(7) AGC
(8) 90° phase shifter

(9) sum channel
(10) difference channel
(11) amplitude detector
(12) phase detector
(13) ranging signal
(14) error signal amplifier
(15) antenna control system

At the output of the sum-and-difference device we have the signals:

$$\underline{E}_s(t, \theta) = (1/\sqrt{2})E_mF(\theta)\{\exp[i(\omega t + \Delta\phi/2)]$$
$$+ \exp[(i(\omega t - \Delta\phi/2)]\} \qquad (4.16)$$
$$\underline{E}_d(t, \theta) = (1/\sqrt{2})E_mF(\theta)\{\exp[i(\omega t + \Delta\phi/2)]$$
$$- \exp[i(\omega t - \Delta\phi/2)]\}$$

After conversion to IF, amplification with AGC, and the addition of a $\pi/2$ phase shift in the difference channel, at the output of the phase detector, we have

$$S(\theta) = \text{Re}[\underline{u}_s(t, \theta)\underline{u}_d^*(t, \theta)]/\underline{u}_s(t, \theta)\underline{u}_s^*(t, \theta) \qquad (4.17)$$

where $\underline{u}_s^*$ and $\underline{u}_d^*$ are the complex conjugates of the signals at the output of the sum-and-difference channels.

With (4.16), relation (4.17) takes the form:

$$S(\theta) = (k_2/k_1)K_{pd}(\sin \Delta\phi)/(1 + \cos \Delta\phi) \qquad (4.18)$$
$$= (k_2/k_1)K_{pd} \tan(\Delta\phi/2)$$

Using the value of $\Delta\phi$ given by (1.1), we obtain

$$S(\theta) = (k_2/k_1)K_{pd} \tan[(\pi l/\lambda)\sin \theta)] \qquad (4.19)$$

This error signal is passed to the antenna control system.

## 4.2 MONOPULSE SURVEILLANCE RADAR SYSTEMS

As stated in Section 1.1, a surveillance monopulse radar system is one which scans a given sector in a prescribed manner as it searches for targets, determining the coordinates of all resolved targets within the beam for each beam position. As an example, we will consider a surveillance monopulse system utilizing amplitude-comparison angle sensing.

The system incorporates a phased antenna array with rapid electronic beam steering. Wideband pulses are transmitted and compressed upon reception. A computer automates the radar operations, performing beam control and signal processing in addition to other functions.

Single target elements or several targets may be acquired either on the basis of external target indicators or autonomously. With both detection schemes, the echo signals at the output of the receiver sum channel are compared with a threshold $h_t$. If the signal amplitude exceeds the threshold within a certain gate, a detection is assumed and target measurements are made. These measurements are fitted to a trajectory for the tracked target, and any measurements not fitting tracked trajectories are used as the basis of new trajectories.

Estimates of the measured parameters in such radar systems are calculated on the basis of functions set up *a priori* from estimates of the amplitudes and delays of the measured signals. We will examine a method for obtaining these estimates for the case where the signal envelope and attendent noise are sampled discretely in time, which is characteristic of most modern monopulse systems with computerized signal processing [31, 99, 107].

Each signal at the output of the measurement channel can be expressed as the realization of some random process:

$$u(t) = L[y(t)] \tag{4.20}$$

formed by the envelope of the signal and white Gaussian noise emerging from a linear filter matched to the signal.

The statistics of the process $y(t)$ are given by

$$L[\bar{y}(t)] = A_m \mu_t(t - t_d) \tag{4.21}$$
$$L\{[y(t_1) - \bar{y}(t_1)][y(t_2) - \bar{y}(t_2)]\} = \sigma_{nf}^2 \mu_t(t_1 - t_2)$$

where $\mu_t$ is the known ambiguity function of the transmitted signal; $A_m$ and $t_d$ are the unknown amplitude and time delay of the measured signal; $\sigma_{nf}^2$ is the variance of the noise at the output of the filter.

There is a known relation in surveillance radars between the amplitude $A_m$ and the signal intensity and target angle off-axis.

We will assume that $2n + 1$ discrete samples of the random process $u(t)$ are observed:

$$u_i = u(t_\phi + i\Delta_t), \qquad i = 0, \pm 1, \pm 2,..., \pm n \tag{4.22}$$

where $\Delta_t$ is the sampling interval, and $t_\phi$ is a fixed time corresponding to the center of the sample, usually determined by the results of previous measurements or with a detector.
Then,

$$L(\bar{y}_i) = A_m \mu_t(t_\phi - t_d + i\Delta_t) = A_m \mu_t(t_x + i\Delta_t) \tag{4.23}$$
$$L(\overline{y_i y_j} - \bar{y}_i \bar{y}_j) = \sigma_{nf}^2 \mu_t[(i - j)\Delta_t]$$

where

$$t_x = t_\phi - t_d$$

An estimate of the amplitude $A_m$ and time delay $t_d$ are determined by the method of maximum likelihood with the system of equations:

$$\partial\Lambda(A_m,\, t_x)/\partial A_m = 0, \qquad \partial\Lambda(A_m,\, t_x)/\partial t_x = 0 \tag{4.24}$$

where $\Lambda(A_m,\, t_x) = \ln W(u_{-n}, u_{-n+1}, \ldots, u_n/A_m,\, t_x)$ and $W$ is the $(2n + 1)$-dimensional angular density function.

With the assumption that $u(t)$ is a normal random process, exact expressions for $\hat{A}_m$ and $\hat{t}_x$ may be obtained. However, comparison of the estimates obtained with the exact formulas and these approximate formulas:

$$\hat{A}_m = \bar{u}_0/\mu_0(t), \qquad \hat{t}_x = (1/2\mu_1)(u_{-1} - u_1)\mu_0 \tag{4.25}$$

where $\mu_1 = \mu_t(t_x + t\Delta_t)$, leads to the conclusion that it is more expedient, in practice, to use the approximations (4.25) to calculate the estimate of the amplitude and time delay with sample sizes of 1 and 3, respectively.

### 4.2.1 Amplitude-Amplitude Monopulse Systems

As was shown earlier, amplitude angle discriminators with logarithmic amplifiers suffer from the requirement for stable and identical logarithmic responses in the amplifiers. This constraint greatly complicates their use in amplitude-amplitude monopulse systems.

We will consider an approach to the construction of such systems that avoids the use of logarithmic amplifiers. Normalization in this system (see Fig. 4.6) is accomplished, instead, with respect to the sum signal at video frequency. We will now describe the formation of the angle-sensing response for an amplitude-amplitude system with this method of normalization.

For an off-axis target angle $\theta$, the signal at the output of the antenna is given by (4.11). At the output of the detectors, the signals take the form:

$$u_1(\theta) = k_1 E_m F(\theta_0)(1 + \mu\theta) \tag{4.26}$$
$$u_2(\theta) = k_2 E_m F(\theta_0)(1 - \mu\theta)$$

If the receiver channels are identical $(k_1 = k_2 = k)$, then the adder output signal will be

$$u_s = u_1(\theta) + u_2(\theta) = 2k E_m F(\theta_0) \tag{4.27}$$

and the subtraction circuit output will be

$$u_d = u_1(\theta) - u_2(\theta) = 2k E_m F(\theta)_0 \mu\theta \tag{4.28}$$

The output of the division circuit will then be

$$S(\theta) = u_d/u_s = 2kE_mF(\theta_0)\mu\theta/2kE_mF(\theta_0) = \mu\theta \qquad (4.29)$$

Comparison of (4.4) and (4.29) shows that they are identical except for a factor of 2, i.e., the output of the system with sum signal normalization is proportional to the angle error, and, therefore, it may be used to determine the target's angular coordinates.

In practice, it is not possible to use linear IF amplifiers in the design of amplitude-amplitude monopulse surveillance radars, which have a dynamic range of 80–100 dB. In these cases logarithmic IF amplifiers can be used, securing the necessary balance between channels by normalizing with the help of control signals; the rest of the processing circuit is like that shown in Fig. 4.6.

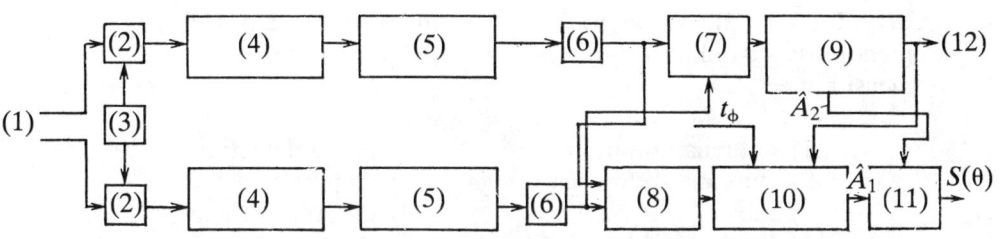

**FIGURE 4.6.** Block diagram of an amplitude-amplitude monopulse surveillance system for a single-coordinate angle measurement.

<table>
<tr><td>(1) antenna output</td><td>(8) subtraction circuit</td></tr>
<tr><td>(2) mixer</td><td>(9) $\hat{A}_2$ and $t$ calculation</td></tr>
<tr><td>(3) local oscillator</td><td>(10) $\hat{A}_1$ calculation</td></tr>
<tr><td>(4) optimum filter</td><td>(11) division circuit</td></tr>
<tr><td>(5) linear IF amplifier</td><td>(12) ranging signal</td></tr>
<tr><td>(6) amplitude detector</td><td></td></tr>
<tr><td>(7) adder</td><td></td></tr>
</table>

### 4.2.2 Amplitude-Sum-and-Difference Monopulse Systems

The diagram of an amplitude-sum-and-difference surveillance monopulse radar is shown in Fig. 4.7. As is evident from a comparison of Figs. 4.3 and 4.7, normalization in both systems is with respect to the sum signal. In surveillance systems, however, the normalization is not performed at IF with AGC, but at video frequency, dividing the phase detector output by the sum signal. This is a consequence of the fact that surveillance radars must often track targets which are close together, which cannot be accomplished with AGC circuits and their long time constraints.

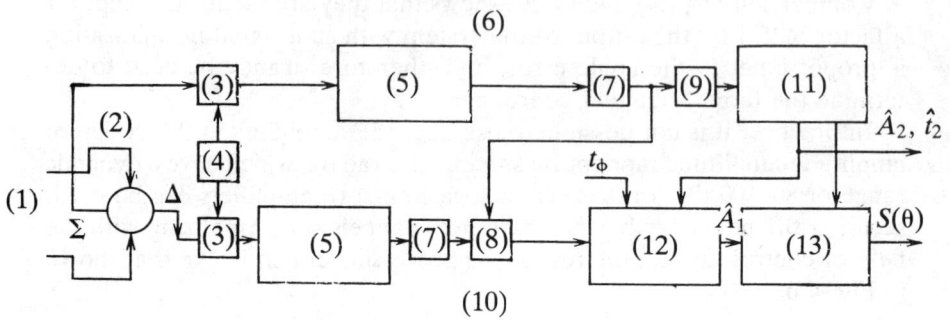

**FIGURE 4.7.** Block diagram of an amplitude-sum-and-difference monopulse surveillance system for a single-coordinate angle measurement.

| | |
|---|---|
| (1) antenna output | (8) phase detector |
| (2) sum-and-difference bridge | (9) amplitude detector |
| (3) mixer | (10) difference channel |
| (4) local oscillator | (11) $\hat{A}_2$ and $\hat{t}_2$ calculation |
| (5) optimum filter | (12) $\hat{A}_1$ calculation |
| (6) sum channel | (13) division circuit |
| (7) IF amplifier | |

We will examine the formation of the angle-sensing response in an amplitude-sum-and-difference surveillance radar. The signals presented by the sum-and-difference device are given by (4.12). At the output of the phase detector, they may then be written as

$$u_s(t, \theta) = \sqrt{2}k_1 E_m F(\theta_0)\exp[i(\omega_c t + \phi_1)] \tag{4.30}$$
$$u_d(t, \theta) = \sqrt{2}k_2 E_m F(\theta_0)\mu\theta \exp[i(\omega_c t + \phi_2)]$$

The sum signal at the output of the amplitude detector, assuming it has unity gain, will equal

$$u_s = \sqrt{2}k_1 E_m F(\theta_0) \tag{4.31}$$

and the voltage at the output of the square-law phase detector will be

$$u_{pd} = 2K_{pd}k_1 k_2 [E_m F(\theta_0)]^2\mu\theta \cos(\phi_1 - \phi_2) \tag{4.32}$$

If the square-law phase detector is used in the block diagram of Fig. 4.7, it is necessary to square the output of the linear amplitude detector. Then, at the output of the division circuit, we obtain

$$S(\theta) = u_{pd}/u_s^2 = K_{pd}(k_2/k_1)\mu\theta \cos(\phi_1 - \phi_2) \tag{4.33}$$

Comparison of (4.33) and (4.14) shows that they give the same response, i.e., normalization by the sum signal at video frequency has the same effect as normalization at IF with AGC.

## 4.3 DESIGN PRINCIPLES OF MONOPULSE SYSTEMS FOR DIRECTION FINDING IN TWO PLANES

The monopulse systems discussed so far are designed for angle measurement in one plane. Two-coordinate direction-finding systems are somewhat more complicated.

### 4.3.1 Monopulse Systems for Two-Coordinate Target Angle Measurement with Amplitude and Phase Angle Discriminators

Such systems may be designed around the combination of two monopulse systems, one designed for angle sensing in azimuth and the other in elevation. As shown in the block diagram of Fig. 4.8, such an amplitude-amplitude system for two-coordinate angle measurement must use four antenna patterns and four amplifier channels. However, it is possible to use just one amplifier for both the azimuth and elevation planes in systems not using sum-and-difference angle discriminators, which simplifies construction of both the antenna and the system as a whole.

An example of the basic structure of one of the first phase-phase monopulse systems is shown in Fig. 4.9. The antenna consists of four tightly connected parabolic reflectors and feed assemblies. One of the antennas transmits and the other three receive, from which it may be deduced that one of the receiving antennas serves for both azimuth and elevation processing.

The main problem with this system is the inefficient use of the antenna aperture, inasmuch as part of it is used solely for angle sensing in azimuth, another part just for elevation, and a third devoted to illumination. With four receiving patterns, this shortcoming is eliminated, because the entire aperture may be used for both transmission and reception.

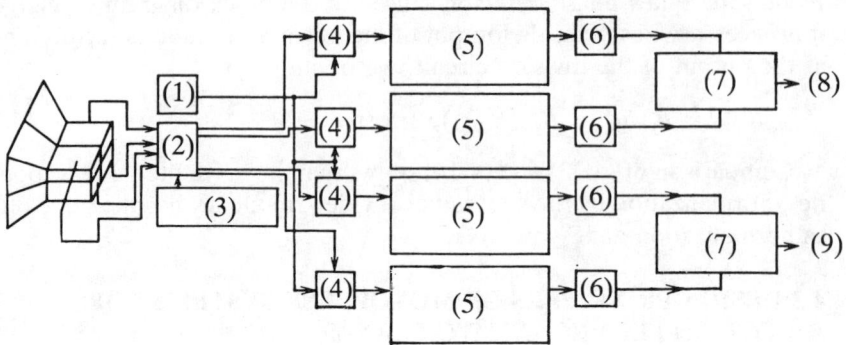

**FIGURE 4.8.** Block diagram of an amplitude-amplitude monopulse system for two-coordinate angle measurement.

(1) local oscillator             (6) amplitude detector
(2) receiving/transmitting switch (7) subtraction circuit
(3) transmitter                  (8) azimuth error signal
(4) mixer                        (9) elevation error signal
(5) logarithmic IF amplifier

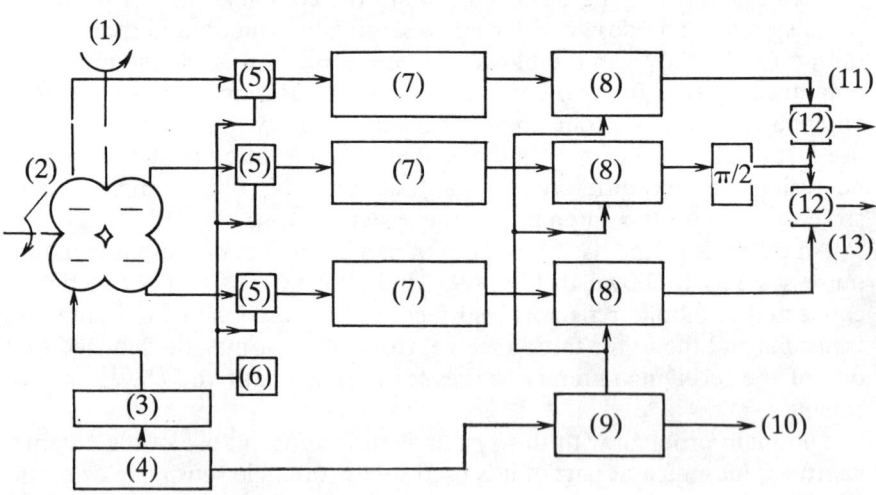

**FIGURE 4.9.** Block diagram of a phase-phase monopulse system for two-coordinate angle measurement.

(1) azimuth control     (8) IF amplifier and AGC
(2) elevation control     (9) range gate
(3) magnetron     (10) ranging signal
(4) modulator     (11) azimuth error signal
(5) mixer     (12) phase detector
(6) local oscillator     (13) elevation error signal
(7) IF preamplifier

### 4.3.2 Monopulse Systems with Sum-and-Difference Angle Discriminators

Monopulse systems employing sum-and-difference discriminators for angle measurement in two coordinates differ from single-coordinate systems in the use of four patterns (instead of two), three or four sum and difference devices (not one), an additional difference channel, and several other elements. Such a system, incorporating three sum-and-difference devices, is laid out in the block diagram of Fig. 4.10 [21], and its signal processing is outlined in Fig. 4.11. The signal processing for a system utilizing four bridges is shown in Fig. 4.12.

The tracking monopulse system of Fig. 4.10 employs a Cassegrain antenna. The sum and difference signals emerging from the sum-and-difference bridges are fed to the receiver, which amplifies and normalizes them through the use of AGC circuits. The sum channel output is used for target detection and ranging. The magnitude of the difference channel output determines the magnitude of the off-axis target angle, and the phase difference between the sum and difference signals gives the target direction. Target detection is effected with target indicator data and an angular search in the given sector; in the search mode, the accumulation of $k$ of $m$ pulses constitutes a detection.

In monopulse systems using four sum-and-difference bridges, the sum signal is formed by initial pairwise summation of signals 1 and 2 at bridge I, and of signals 3 and 4 at bridge II, and then completed at bridge III. The RF transmitted signal is distributed in power and phase through the bridges to the four antennas, which form the sum pattern. The elevation difference signal is formed by subtracting the sum of signals 3 and 4 from the sum of 1 and 2 at bridge III, while the azimuth difference signal is similarly formed from $(1 - 2) + (3 - 4)$ at bridge IV.

These signals may be written as follows:

• the sum signal emerging from the sum branch of bridge III is

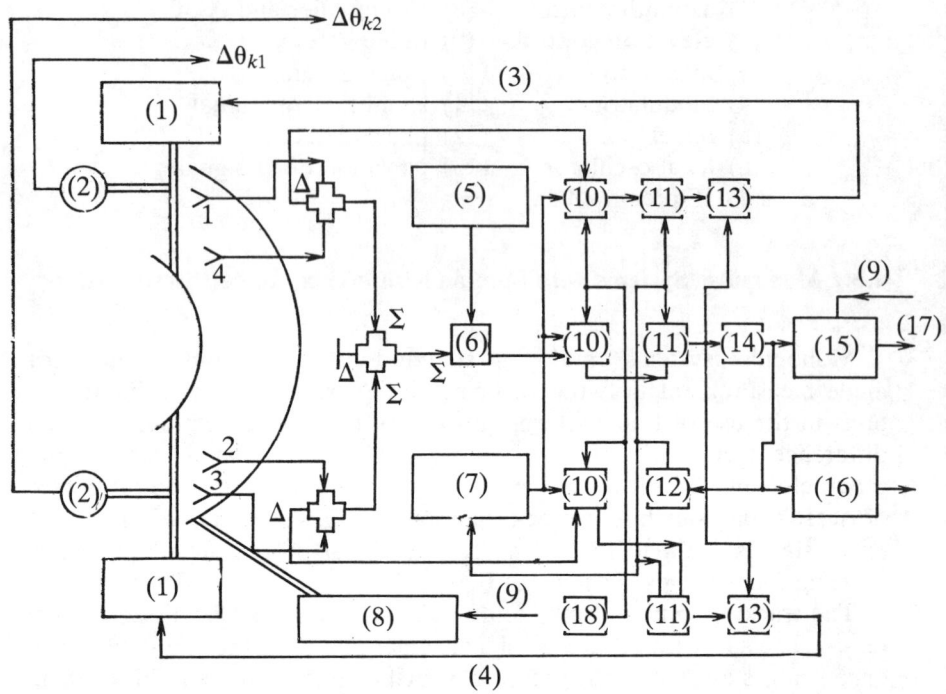

**FIGURE 4.10.** Block diagram for an amplitude-sum-and-difference
monopulse system for two-coordinate angle measurement.

| | |
|---|---|
| (1) subreflector control | (10) mixer |
| (2) sensor | (11) IF amplifier |
| (3) error signal $\Delta\theta_{d1}$ | (12) AGC |
| (4) error signal $\Delta\theta_{d2}$ | (13) phase detector |
| (5) transmitter | (14) amplitude detector |
| (6) receiving/transmitting switch | (15) range finder |
| (7) control signal generator | (16) target detector |
| (8) main reflector control | (17) $\Delta R$ |
| (9) target indicator data | (18) local oscillator |

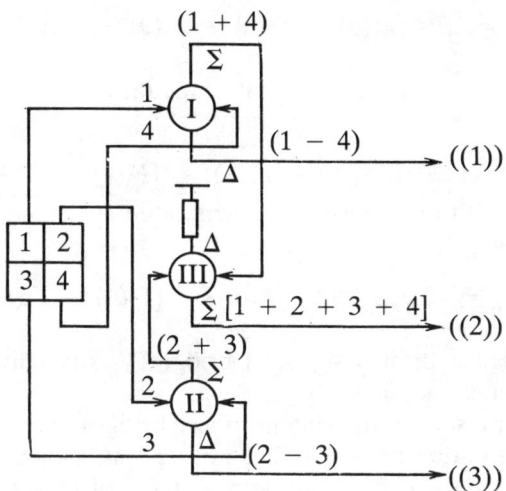

**FIGURE 4.11.** Formation of sum-and-difference signals with three bridges.

((1)) elevation difference channel
((2)) sum channel
((3)) azimuth difference channel

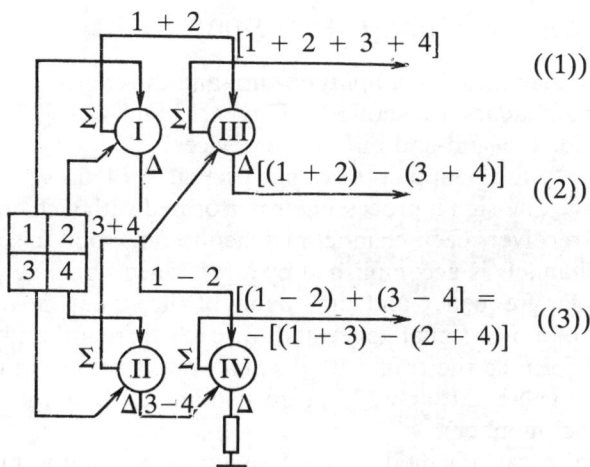

**FIGURE 4.12.** Formation of sum-and-difference signals with four bridges.
((1)) sum channel
((2)) elevation difference channel
((3)) azimuth difference channel

$$E_s(t, \theta) = \tfrac{1}{2}E(t)[F_1(\theta) + F_2(\theta) + F_3(\theta) + F_4(\theta)] \qquad (4.34)$$

- the elevation difference signal at the difference branch of bridge III is

$$E_{d\epsilon}(t, \theta) = \tfrac{1}{2}E(t)\{[F_1(\theta) + F_2(\theta)] - [F_3(\theta) + F_4(\theta)]\}, \qquad (4.35)$$

- the azimuth difference signal generated at the sum branch of bridge IV is

$$E_{d\theta}(t, \theta) = \tfrac{1}{2}E(t)[F_1(\theta) + F_3(\theta)] - [F_2(\theta) + F_4(\theta)] \qquad (4.36)$$

The difference branch signal of bridge IV is usually unused, and terminated in a matched load.

In a monopulse system using just three bridges, the sum signal is formed by pairwise addition of signals 2, 3 and 1, 4 at bridges I and II, and final addition at bridge III. The difference signal of bridge III is unused and terminated with a matched load. One of the two difference signals is formed by the difference of signals 1 and 4, and the other from the difference of signals 2 and 3.

With three bridges, the sum signal is still given by (4.34), and the difference signals are

$$E_{d\epsilon} = (1/\sqrt{2})E(t)[F_1(\theta) - F_4(\theta)] \qquad (4.37)$$

$$E_{d\theta} = (1/\sqrt{2})E(t)[F_2(\theta) - F_3(\theta)] \qquad (4.38)$$

Block diagrams for amplitude-sum-and-difference surveillance and pulse Doppler radars are shown in Figs. 4.13 and 4.14 [25]. Three bridges are used for the sum-and-difference devices.

The pulse Doppler example of Fig. 4.14 uses wideband phase-shift keying. The signal processing is performed with a channelized correlation-filter receiver, each channel matched to one point in space. This tuning of the channels is accomplished by a reference signal generator fed with the Doppler frequency and time delay of the signals in the detection mode. The reference signal generated for each correlation filter channel has the same form as the transmitted signal, but is shifted in time and frequency by the values $\Delta t_k$ and $\Delta f_k$, corresponding to the target indicators ($k$ is the channel number).

In the tracking mode, the reference signal generator is fed with estimates of the Doppler frequency $\hat{f}_D$ and the time delay $\hat{f}_d$, obtained from previous measurements of these quantities. To form the tracking discriminator responses, pairs of detuned channels in the multichannel correlation filter device are used. To form the tracking discrimination response for the Doppler frequency (radial target velocity), signals from two channels, offset by $\pm \Delta f_D$ from the frequency $\hat{f}_D$, are used. To form the tracking ranging response, signals are taken from two channels offset by $\pm \Delta t$ relative to the delay $\hat{t}_d$.

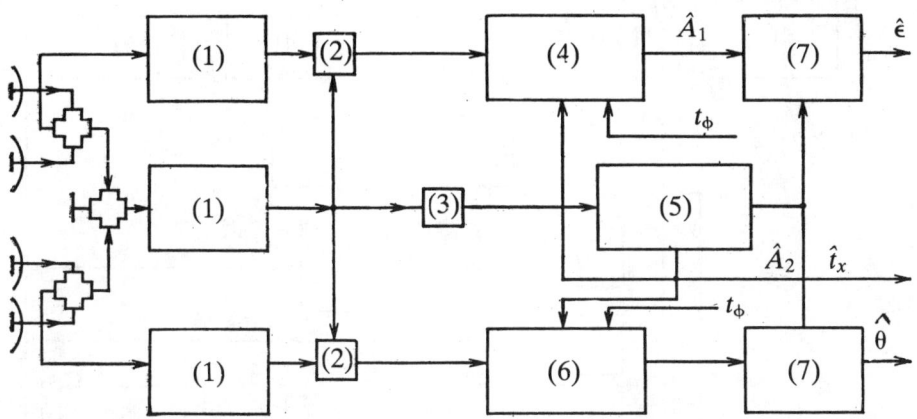

**FIGURE 4.13.** Block diagram of an amplitude-sum-and-difference surveillance radar.

| | |
|---|---|
| (1) optimum filter | (5) $\hat{A}_2$ and $\hat{t}_x$ calculation |
| (2) phase detector | (6) $\hat{A}_3$ calculation |
| (3) amplitude detector | (7) division circuit |
| (4) $\hat{A}_1$ calculation | |

A computer is used to calculate the tracking error and estimate the slopes of the tracked target range and radial velocity according to the output signals of the corresponding channels. The sum signal is passed directly from the receiver filter to the phase detector; the difference channel output is first phase-shifted by 90°. The mixing voltage has the same form as the transmitted signal, and is shifted in time and frequency (to within IF bandwidth) by the estimates $\hat{f}_D$ and $\hat{t}_d$.

For exact tracking in range and Doppler frequency, demodulated signals are fed to the inputs of the IF amplifiers in the sum-and-difference channels. The receiver filters are matched to the demodulated signal. A phase shifter introduces a 180° phase shift between the sum and difference signals at the input to the phase detector. From a data extraction device, the computer receives the information necessary to estimate the angular coordinates of the tracked target and form the antenna control signals. The computer also normalizes the signals. The information required for these purposes can be obtained from the output signals of the correlation-filter system.

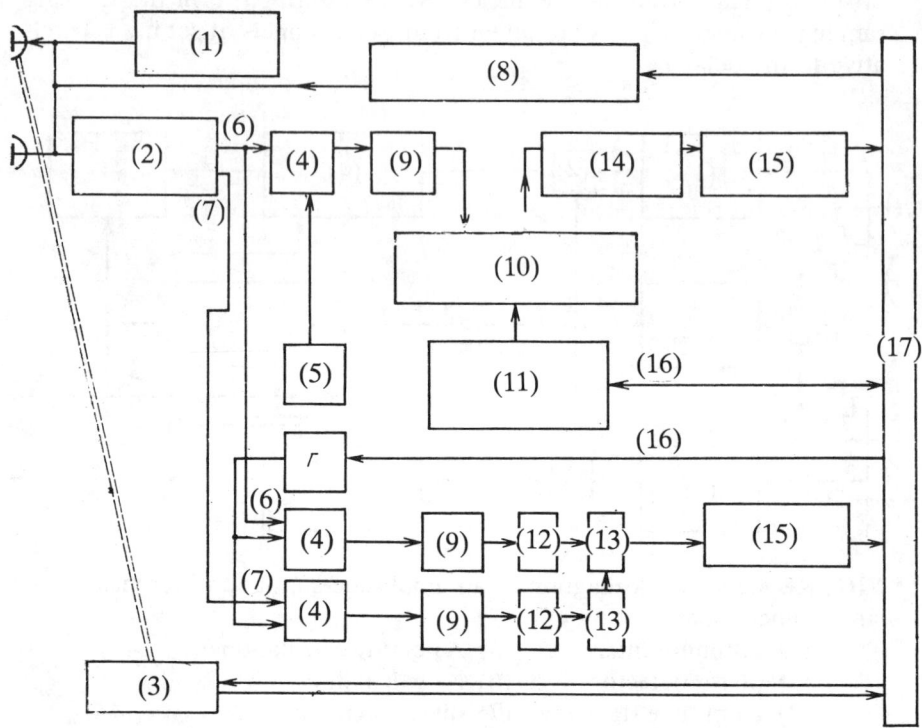

**FIGURE 4.14.** Block diagram of a pulsed Doppler amplitude-sum-and-difference radar.

(1) transmitter
(2) sum-and-difference device
(3) power drives
(4) mixer
(5) local oscillator
(6) $u_s$
(7) $u_d$
(8) beam control
(9) IF amplifier

(10) correlation filter channels
(11) reference signal generator
(12) filter
(13) phase detector
(14) channel switch
(15) data extractor
(16) $t_d$ and $f_D$
(17) computer

Amplitude- and phase-sum-and-difference monopulse systems are identical in their angular processing, except for the antenna patterns. The target angle information in an amplitude-comparison system is derived from the ratio of the signal amplitudes, and, in a phase-comparison system, from the phase shift caused by the differing distances from the target to the corresponding antennas.

### 4.3.3 Monopulse Systems with Digital Signal Processing

It was already stated in Section 3.2 that the primary applications of digital signal processing in monopulse systems are digital filtering and spectral analysis. In modern radar systems utilizing digital techniques, the greatest use is made of digital spectral analysis, especially in wideband systems. For this purpose, the fast Fourier transform (FFT) is usually used (see Section 3.2).

A hypothetical amplitude-sum-and-difference monopulse system using digital signal processing and FFTs is shown in Fig. 4.15. Separate transmitting and receiving antennas are used in this system. This system differs from the usual surveillance amplitude-sum-and-difference system (see Fig. 4.7) in the incorporation of analog-to-digital (A/D) converters, a device for storing the reference signal spectra, a device for multiplying spectra, and circuits for performing forward and reverse Fourier transforms. These forward and reverse transforms are performed in a computer, as discussed in Section 3.2.

The computer is also responsible for signal normalization, dividing the difference signal by the sum signal. The normalized signals are used to measure the target coordinates in each coordinate plane.

### 4.4 COMBINATION MONOPULSE SYSTEMS

There are several design advantages to a combination monopulse system. Such a system uses antenna patterns providing simultaneous independent amplitude and phase information about the target from the received signals. This makes it possible to use just two interconnected channels for two-coordinate angle measurement with only one bridge connected to their inputs (see Fig. 4.16).

The vector diagram of Fig. 4.17 illustrates the method of simultaneous formation of the two-dimensional angle error. The signals in the first and second channels are denoted in vector form by $U_1$ and $U_2$, and differ in both phase and amplitude. It may be seen in the diagram that the difference signal $U_2 - U_1$ may be decomposed into two components, one of which is in quadrature with the sum signal, as in a phase-comparison system, and the other which is either in phase or 180° out of phase with the sum signal, as in an amplitude-comparison system. As will be shown below, the first component can be used as the azimuth error signal, and the second as the elevation error signal.

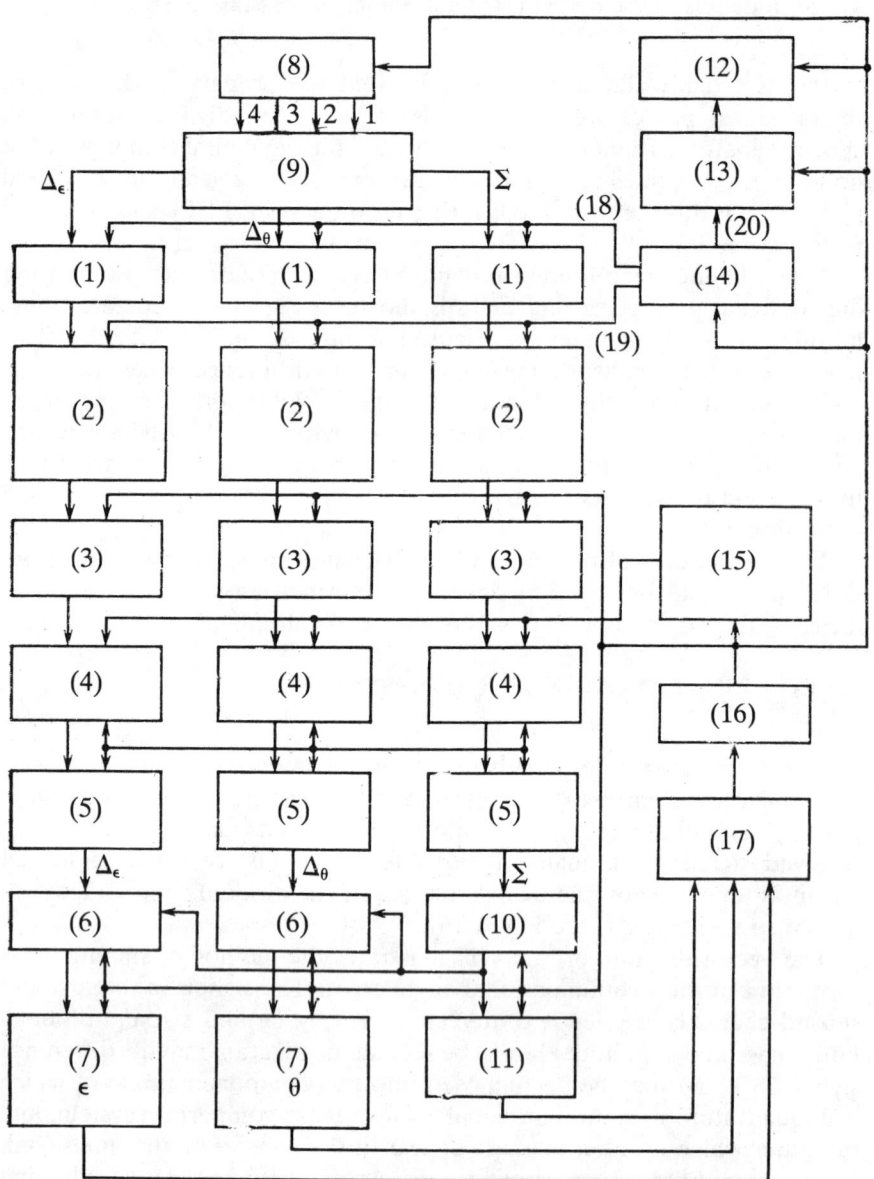

**FIGURE 4.15.** Block diagram of an amplitude-sum-and-difference monopulse system with digital signal processing.

(1) mixer and IF amplifier
(2) synchronous detection and A/D conversion
(3) forward FFT
(4) spectra multiplier
(5) reverse FFT
(6) normalization
(7) angle measurement ($\epsilon$ or $\theta$)
(8) receiving phased array
(9) sum-and-difference signal formation
(10) target detection

(11) range and radial velocity measurement ($R$ and $v_r$)
(12) transmitting phased array
(13) transmitter
(14) signal generator
(15) reference signal spectra storage
(16) system control
(17) computer
(18) $f_{\text{mix}}$
(19) $f_{\text{ref}}$
(20) $f_0$

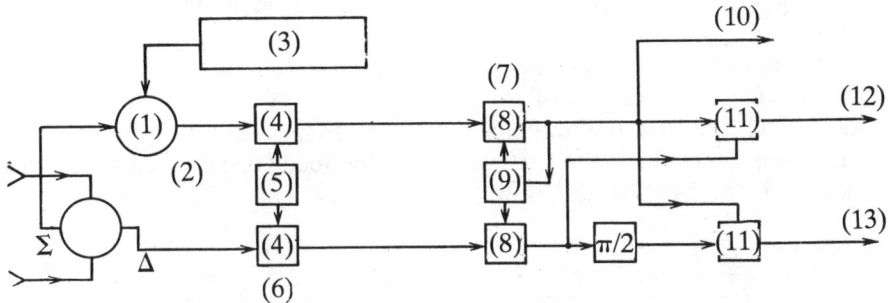

**FIGURE 4.16.** Structure of a combination monopulse system.

(1) receiving/transmitting switch
(2) hybrid ring junction
(3) transmitter
(4) mixer
(5) local oscillator
(6) difference channel
(7) sum channel

(8) IF amplifier
(9) AGC
(10) ranging signal
(11) phase detector
(12) elevation difference signal
(13) azimuth difference signal

There are just two antenna patterns. In the vertical plane they are displaced from one another by the angle $2\theta_0$, and in the horizontal plane they are parallel to one another and displaced at their phase centers by the distance $l$. Thus, angle sensing is accomplished in the vertical plane by amplitude comparison, and in the horizontal plane by phase comparison.

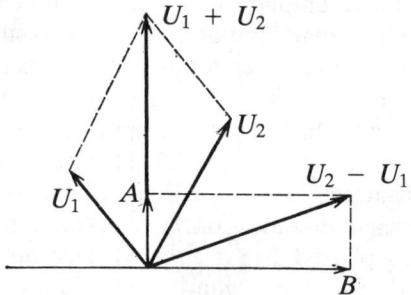

**FIGURE 4.17.** Vector diagram for combination angle sensing.
   A: in-phase component of difference signal (elevation error)
   B: quadrature component of difference signal (azimuth error)

In order to form beams that are parallel in the azimuth plane, the linear antenna feeds are displaced from one another, and illuminate the right and left halves of a parabolic cylinder (see Fig. 4.18). In order to have the beams deflected in the vertical plane, one feed is positioned above the focal plane and the other below.

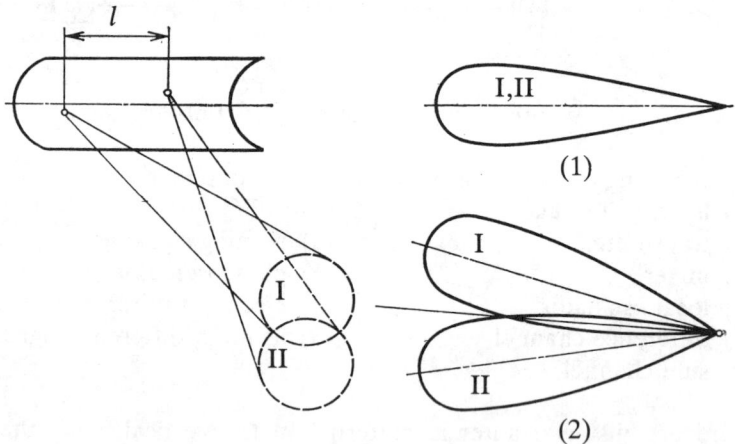

**FIGURE 4.18.** Beam formation in combination monopulse systems.
   (1) horizontal plane   (2) vertical plane

In order to obtain the angle information in each coordinate, two phase detectors are used at the output of the receiver channels, one forming the elevation error signal, and the other forming the azimuth error signal. The difference signal reaches the azimuth phase detector 90° out of phase with the sum signal, which is used as a reference. The remaining elements of a combination-sum-and-difference system are analogous to the sum-and-difference systems considered earlier.

We will now examine the formation of the angle-sensing response in the combination system. Let the target lie in a direction off-axis in azimuth by the angle $\theta$ and in elevation by the angle $\epsilon$. Then, the phase difference of the signals presented by the antennas will depend on the azimuth angle, and the amplitude difference will depend on the elevation angle; the signals may be expressed as

$$\underline{E}_1(t, \theta) = E_m F(\theta) F(\theta_0 - \epsilon) \exp[i(\omega t - \Delta\phi/2)] \tag{4.39}$$
$$\underline{E}_2(t, \theta) = E_m F(\theta) F(\theta_0 + \epsilon) \exp[i(\omega t + \Delta\phi/2)]$$

At the output of the bridge, the sum and difference signals (dropping the constant power division factors) may be written in the form:

$$\underline{E}_s(t, \theta) = E_k\{F(\theta_0 - \epsilon)\exp[i(\omega t + \Delta\phi/2]$$
$$+ F(\theta_0 + \epsilon)\exp[i(\omega t - \Delta\phi/2)]\} \tag{4.40}$$
$$\underline{E}_d(t, \theta) = E_k\{F(\theta_0 - \epsilon)\exp[i(\omega t + \Delta\phi/2)]$$
$$- F(\theta_0 + \epsilon)\exp[i(\omega t - \Delta\phi/2)]\}$$

where $E_k = (1/\sqrt{2})E_m F(\theta)$.

After frequency conversion and IF amplification, not considering the associated phase shifts and signal normalization, we obtain

$$\underline{u}_s(t, \theta) = E_k k_s\{F(\theta_0 - \epsilon)\exp[i(\omega_c t + \Delta\phi/2]$$
$$+ F(\theta_0 + \epsilon)\exp[i(\omega_c t - \Delta\phi/2)]\} \tag{4.41}$$
$$\underline{u}_d(t, \theta) = E_k k_d\{F(\theta_0 - \epsilon)\exp[i(\omega_c t + \Delta\phi/2)]$$
$$- F(\theta_0 + \epsilon)\exp[i(\omega_c t - \Delta\phi/2)]\}$$

At the output of the product phase detector, taking into account the normalization in the receiver, the error signal will be

$$S(\theta) = K_{pd} \, \text{Re}[\underline{u}_s(t, \theta)\underline{u}_d^*(t, \theta)]/\underline{u}_s(t, \theta)\underline{u}_s^*(t, \theta) \tag{4.42}$$

After elementary transformations, for the response at the output of the elevation phase detector, we obtain

$$S(\epsilon) = K_{pd}\left(\frac{k_d}{k_s}\right)$$
$$\times \frac{F^2(\theta_0 - \epsilon) - F^2(\theta_0 + \epsilon)}{F^2(\theta_0 - \epsilon) + F^2(\theta_0 + \epsilon) + 2F(\theta_0 - \epsilon)F(\theta_0 + \epsilon)\cos\Delta\phi} \tag{4.43}$$

Note that this elevation error response is dependent on the azimuth error through the term $\cos \Delta\phi$.

For small angle errors, a linear approximation to the antenna patterns is applicable:

$$F(\theta_0 \mp \epsilon) = F(\theta_0)(1 \pm \mu\epsilon) \tag{4.44}$$

$$F^2(\theta_0 \mp \epsilon) \approx F^2(\theta_0)(1 \pm 2\mu\epsilon) \tag{4.45}$$

Then, (4.43) with equal channel gains may be put into the form:

$$S(\epsilon) = K_{pd}(\mu\theta)/\cos^2(\Delta\phi/2) \tag{4.46}$$

When there is no azimuth error ($\Delta\phi = 0$), we obtain

$$S(\epsilon) = K_{pd}\mu\theta \tag{4.47}$$

which is analogous to (4.14), calculated for an amplitude-sum-and-difference system, as long as the receiver channels have identical amplitude-phase characteristics.

At the azimuth phase-detector output, including the additional phase shift of 90°, for $k_s = k_d$ we obtain

$$\begin{aligned}
S(\theta) &= K_{pd}\frac{\text{Re}[\underline{u}_s(t, \theta)\underline{u}_d^*(t, \theta)\exp(i\pi/2)]}{\underline{u}_s(t, \theta)\underline{u}_s^*(t, \theta)} \\
&= K_{pd}\frac{2F(\theta_0 - \epsilon)F(\theta_0 + \epsilon)\sin \Delta\phi}{F^2(\theta_0 - \epsilon) + F^2(\theta_0 + \epsilon) + 2F(\theta_0 - \epsilon)F(\theta_0 + \epsilon)\cos \Delta\phi} \\
&\approx K_{pd}(\sin \Delta\phi)/(1 + \cos \Delta\phi) = K_{pd} \tan(\Delta\phi/2)
\end{aligned} \tag{4.48}$$

Equation (4.48) is the same as (4.18), which was derived for a phase-sum-and-difference system. These error signals are used by the antenna control system.

Thus, a combination sum-and-difference system can generate the error signals for both coordinates with just two channels. This is the main advantage of a combination system over pure amplitude- or phase-comparison systems, which require four antenna patterns, three or four bridges, and three receiver channels. These benefits, however, are obtained at the cost of higher sidelobes, lower sum pattern gain, and an angle-sensing sensitivity about 3 dB lower than is obtained with an optimal design. Therefore, the choice of a combination monopulse system should be based on a detailed trade-off analysis of the performance requirements. A further consideration is the enhanced angular resolution possible with a combination system using the same number of elements as other systems, instead of just the lower number necessary for basic operation. The use of four or more independent feeds can provide the extra data necessary to generate the additional relations determining the spatial position of the target, and can also allow several targets to be resolved at the same range within the main beam. This will be discussed in more detail in Chapter 5.

The advantages of combination systems make them especially attractive for use on ships and aircraft, where size and weight considerations are of primary importance. One of the first combination monopulse systems is the AN/APG-25 experimental airborne radar, designed for fire control in the tail protection of US Navy patrol aircraft.

In addition to sum-and-difference discriminators, amplitude and phase discriminators can also be employed in these designs combining amplitude- and phase-comparison techniques; the resulting systems are combination-amplitude and combination-phase systems.

## 4.5 DESIGN PRINCIPLES FOR CW AND PULSED DOPPLER MONOPULSE SYSTEMS

Monopulse systems using CW and pulsed Doppler transmission usually perform target selection on the basis of velocity, a feature which leads to a distinctive structure. A typical block diagram of a receiver for such a monopulse system is shown in Fig. 4.19 [21].

The structure of each channel is very similar to that of receivers in single-channel CW radars. After frequency conversion and IF amplification, the signals are fed to the mixers 4 and 5; the mixing voltages here are the IF transmitter reference signals. The Doppler frequency signals emerging from the mixers are further amplified in Doppler frequency (DF) amplifiers and selected by frequency.

The Doppler frequency selection, which corresponds to selection by target velocity, is performed with the help of mixers 6 and 7, which are followed by narrowband Doppler filters $\Phi_1$ and $\Phi_2$, the bandwidth of which is chosen on the basis of the target velocity resolution requirements. The frequency conversion is effected with the signals from the tracking oscillator, controlled by signals from the frequency discriminator.

The signals from the Doppler filters, that contain the target angle information, are fed to a phase detector, where the error signal is formed. This error signal is then used in the usual fashion. The passbands of velocity selection angle-sensing systems are fairly narrow, which leads to a greater requirement for balanced receiver channels. Differences in the phase or amplitude characteristics of the channels, especially in the Doppler filters, will lead to errors in target angle measurement. Inasmuch as the slope of the phase-frequency response in narrowband filters is rather steep, and the target velocity and associated Doppler frequency vary within wide limits during tracking, even a slight difference in the Doppler filter responses will cause a large phase-shift error in the phase-detector input signals, and large angle errors will result.

Systems employing quasi-CW transmission, which is characterized by pulses of short duration, can also perform target selection by velocity. The design of such a monopulse system is analogous to that just described. The only difference is the incorporation of IF range gating in the quasi-CW receiver. The requirements for balanced channel response are as stringent as for the CW system.

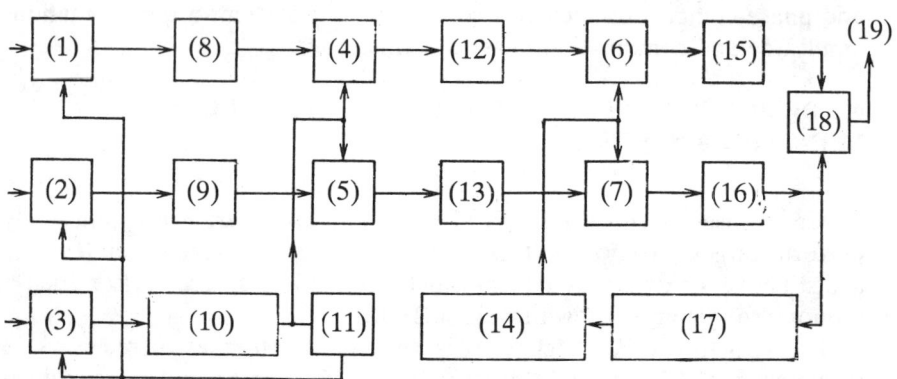

**FIGURE 4.19.** Block diagram for a monopulse CW-receiver.

| | |
|---|---|
| (1) mixer 1 | (11) reference signal |
| (2) mixer 2 | IF amplifier |
| (3) mixer 3 | (12) Doppler frequency amplifier 1 |
| (4) mixer 4 | (13) Doppler frequency amplifier 2 |
| (5) mixer 5 | (14) tracking oscillator |
| (6) mixer 6 | (15) filter 1 |
| (7) mixer 7 | (16) filter 2 |
| (8) IF amplifier 1 | (17) frequency discriminator |
| (9) IF amplifier 2 | (18) phase detector |
| (10) local oscillator | (19) error signal |

## 4.6 MONOPULSE SYSTEMS USING CONICAL SCANNING (CONOPULSE SYSTEMS)

The system we will now examine combines the methods of both monopulse and conical scanning. The basic advantages of such a "conopulse" system are that echo fluctuations do not affect the angle measurement accuracy (as with regular monopulse systems), and also that it requires only two channels, and accordingly fewer components, to achieve the same accuracy as a regular monopulse system. The main drawbacks are the lower data rate associated with such systems, and the difficulties associated with the design of an antenna which must scan two beams.

The essence of conopulse [100] (see Fig. 4.20) is the use of two beams, as opposed to one for ordinary conical scanning. The target echoes $E_1$ and $E_2$ in the corresponding beams are modulated to equal degrees both by target fluctuations and the beam scanning. The modulation coefficients of $E_1$ and $E_2$ due to scanning are similar to those that result when the single beam is observed in an ordinary conical scanning system. If the two beams are identical except for their half-scan displacement, then the two scanning modulation coefficients will also be identical, except for the shift introduced by the half-scan period.

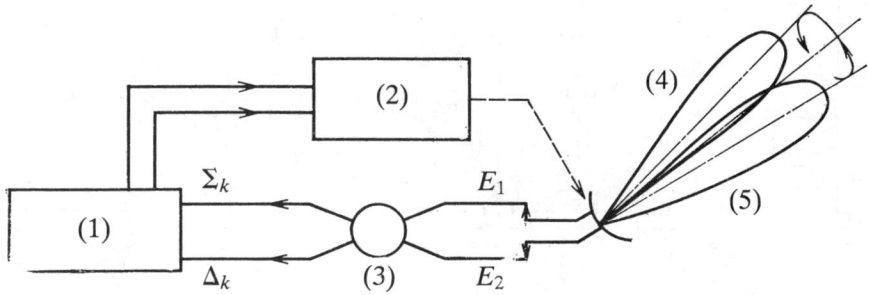

**FIGURE 4.20.** Simplified structure of a conopulse system.
             (1) signal processor    (4) beam 1
             (2) antenna control     (5) beam 2
             (3) bridge

Each signal $E_1$ and $E_2$ can be processed to measure the target angle as in a single-beam conical scanning system, but the use of both beams eliminates the effects of amplitude fluctuations. The idea is to obtain a ratio of two signals, each of which is itself a function of $E_1$ and $E_2$, such that the factor due to amplitude fluctuation is canceled out, just as in monopulse processing. Thus, the receiver combines the processing approaches of both conical scanning and monopulse systems. The error signals obtained as a result of this processing are proportional to the off-axis target angle. As will be shown below, these error signals are formed using the techniques of conical scanning, which require the processing of a sequence of pulse echoes. It is, therefore, impossible to obtain target angle information with just one pulse in conopulse; the contribution of the monopulse concept is simply the elimination of amplitude fluctuation effects.

The diagram of a possible conopulse system is shown in Fig. 4.21 [100]. The monopulse processing required to excise the effects of fluctuations is performed first, and then target angle measurement is accomplished with conical scanning methods. The issue of generating the two independent scanned beams has not been satisfactorily resolved and is not addressed. The sum $\Sigma_k$ and difference $\Delta_k$ of the two signals $\underline{E}_1$ and $\underline{E}_2$ are initially obtained in a sum-and-difference bridge. Then the ratio $\bar{S}(\theta)$ of the difference-to-sum signal is formed, just as in ordinary monopulse systems (see Chapter 1). The basic operations performed in the system are shown in Fig. 4.22 [100].

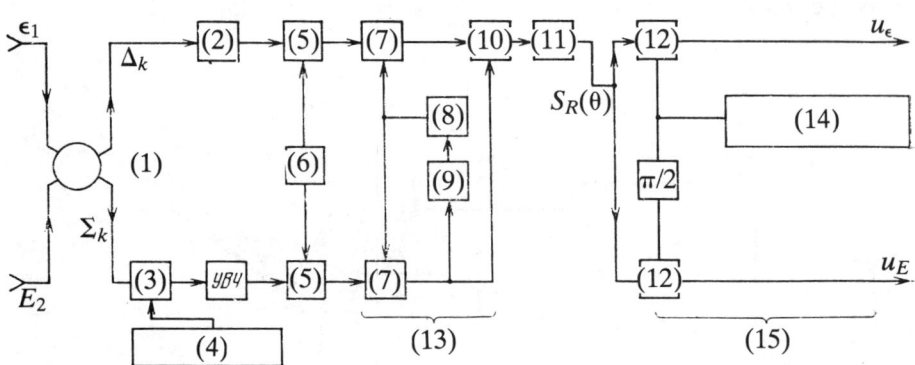

**FIGURE 4.21.** Block diagram for a possible conopulse system.

| | |
|---|---|
| (1) bridge | (9) amplitude detector |
| (2) video amplifier | (10) phase detector |
| (3) receiving/transmitting switch | (11) low-pass filter |
| (4) transmitter | (12) phase detector |
| (5) mixer | (13) monopulse signal |
| (6) local oscillator | normalization |
| (7) IF amplifier | (14) scanning generator |
| (8) AGC | (15) conical scanning detectors |

The transmitter output signal is

$$u_t = A_t(t)\exp[i(\omega t + \phi_t)] \tag{4.49}$$

where $\omega$ is the angular frequency of operation, and $\phi_t$ is an arbitrary phase shift. $A_t(t)$ represents the group of video pulses of duration $\tau_c$; the pulse repetition frequency is $T_t$.

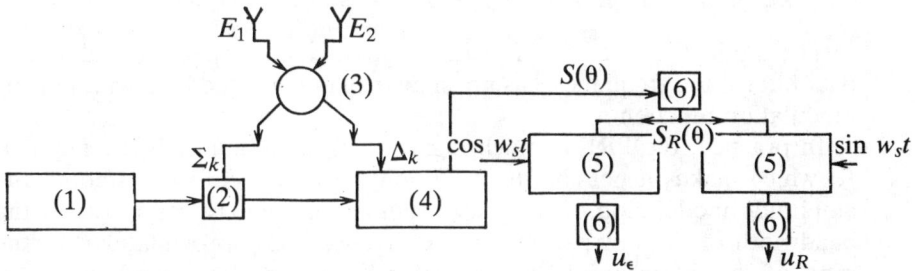

**FIGURE 4.22.** Simplified diagram showing basic operations performed in a conopulse system.

| | |
|---|---|
| (1) transmitter | (4) division circuit |
| (2) receiving/transmitting switch | (5) multiplier |
| (3) bridge | (6) low-pass filter |

We will assume that the beams are axisymmetric, so that they are functions only of the off-axis angle $\theta$. If the antenna patterns are $F_1(\theta)$ and $F_2(\theta)$, we may write the sum-and-difference signals as

$$u_{sk} = k_p[F_1(\theta) + F_2(\theta)]^2 A_t(t - t_d)\exp\{i[\omega t + \phi_t + \psi(t)]\}$$
$$u_{dk} = k_p[F_1(\theta) + F_2(\theta)][F_1(\theta) - F_2(\theta)] \qquad (4.50)$$
$$\times A_t(t - t_d)\exp\{i[\omega t + \phi_t + \psi(t)]\}$$

where $\psi(t)$ is the phase shift induced by the target and its associated phase fluctuations; $t_d$ is the time delay; $k_p$ is a proportionality constant determined both by the target amplitude fluctuations and the coefficients of the radar range equation.

Then the ratio of the difference-to-sum signal will be

$$S(\theta) = \frac{u_{dk}}{u_{sk}} = \frac{F_1(\theta) - F_2(\theta)}{F_1(\theta) + F_2(\theta)} \sum_{k=-\infty}^{\infty} \text{rect}\left[\frac{(t - t_d - k_p T_t)}{\tau_c}\right] \qquad (4.51)$$

where

$$\text{rect}(t/\tau_c) = \begin{cases} 1 & \text{if } |t| < \tau_c/2 \\ 0 & \text{otherwise} \end{cases}$$

The second (bracketed) factor in (4.51) is a periodic sequence of rectangular pulses, repeated at the transmission PRF with period $T_t$, and the first factor is the modulation induced by the scanning.

The Fourier transform of $S(\theta)$ will consist of the modulation spectra at the frequencies:

$$kf_t = k/T_t, \qquad k = 0, \pm 1, \pm 2, \ldots$$

The modulation function for these signals can be separated with a low-pass filter if the frequency $f_t$ is no smaller than twice the bandwidth of the modulation function.

In practice, the PRF is much higher than the scanning frequency $f_t \gg f_S$, which makes it possible to use a low-pass filter (LPF) to smooth the amplitude modulation of the pulse train and obtain the ratio $S(\theta)$. If the bandwidth of the low-pass filter (see Fig. 4.22) is approximately $f_t/2$, the ratio of the difference to sum signal will be given by the following approximation:

$$S_R(\theta) = (\tau_c/T_t) \, [F_1(\theta) - F_2(\theta)]/[F_1(\theta) + F_2(\theta)] \qquad (4.52)$$

where $S_R(\theta)$ is approximately a periodic function with angular frequency $\omega_S = 2\pi f_S$. Then the multiplier and low-pass filter (see Fig. 4.22) will act as synchronous detectors, operating on the Fourier term with frequency $\omega_S$. Consequently, $u_\epsilon$ and $u_\theta$ will be proportional to the coefficients of the Fourier cosine and sine series for frequency $\omega_S$:

$$\left. \begin{matrix} u_\epsilon \\ u_\theta \end{matrix} \right\} = \frac{1}{T_S} \int_{-T_S/2}^{T_S/2} S_R(\theta) \left\{ \begin{matrix} \cos \omega_S t \\ \sin \omega_S t \end{matrix} \right\} dt \qquad (4.53)$$

where $T_S = 1/f_S$.

The error signals of (4.53) are usually calculated in a computer after selection of a suitable approximation to the antenna patterns.

Thus, it is possible, in principle, to eliminate the influence of amplitude fluctuations with this system, determining the error signals by (4.53). However, the difficulty of designing an antenna system for scanning the two beams simultaneously still prevents advanced development of conopulse systems.

# Chapter 5

# The Angular Resolution and Angle-Sensing Sensitivity of Monopulse Radars

## 5.1 THE CONCEPT OF ANGULAR COORDINATE RESOLUTION

The angular resolution of a radar is the ability of the system to measure the angular coordinates of two proximite but separate signal sources, which are not selected by other means (time, frequency, polarization, *etcetera*). The resolution is the minimum angular displacement between signal sources, whose coordinates can be measured separately and with the required accuracy. Although the issue of angular resolution has been treated in many works, there is, at present, no generally accepted resolution index or criterion. As a result, individual systems are attributed with different capabilities, and the judgements of various specialists are contradictory [45, 101].

The Rayleigh criterion, as originally applied to optical systems, states that two point sources are resolved if the peak intensity of the diffraction pattern of the first signal lies on or beyond the circle of the first null of the other signal's diffraction pattern. This assumes that the two signals are of equal intensity (amplitude) and there are no interfering signals.

Two signal sources are thus resolved according to the Rayleigh criterion if they are separated by an angle greater than or equal to

$$\delta\theta = 1.22\lambda/d$$

where $\lambda$ is the wavelength of the light, and $d$ is the optics aperture.

The Rayleigh resolution criterion is so simple and sufficiently objective that it has been applied to one- and two-dimensional radar problems, as well as optics. The form of the criterion suggests that an estimate of the resolution be made from the width of the autocorrelation function of the input signals:

$$\Psi(\Delta\theta) = \mathrm{Re}\left[\int\limits_{-\pi}^{\pi}\int\limits_{0}^{T}\underline{E_0}(t,\,\theta)\underline{E_0^*}(t,\,\theta\,+\,\Delta\theta)\,dt\,d\theta\right]$$

where $\underline{E_0}(t,\,\theta)$ is the first target's input signal, $\underline{E_0}(t,\,\theta\,+\,\Delta\theta)$ is the second target's input signal, $\theta$ and $\theta\,+\,\Delta\theta$ are the target angles, and $T$ is the duration of observation.

When measuring the angles of a pair of targets, at the receiver input there will be at least two signals, the resolution of which depends on the relative signal energies and on the extent to which the resolved quantities (angle, in this case) overlap. It is clear that the less the overlap in this quantity, the greater will be the ability to track the targets individually, and the greater will be the radar's resolution. This makes it possible to estimate the resolution by the width of the autocorrelation function, inasmuch as this function represents the extent to which the two signals are correlated in the given parameter.

The analysis in [21] shows that, with a Gaussian antenna pattern, the correlation function is

$$\Psi(\Delta\theta) = 2E_{0s}\exp(-\gamma_0^2\Delta\theta^2/2)$$

where $E_{0s}$ is the energy of the high frequency signal; $\gamma_0$ is a coefficient corresponding to the beamwidth, equal to $1.66/\theta_3$; and $\Delta\theta$ is the angular target separation. The form of $\Psi(\Delta\theta)$ is shown in Fig. 5.1.

If the system target resolution is taken to be the width of the input signal autocorrelation function at the level of one half of the peak value, then

$$\delta\theta = 1.4\theta_3$$

There results a direct variation of the width of the autocorrelation function with the radar beamwidth.

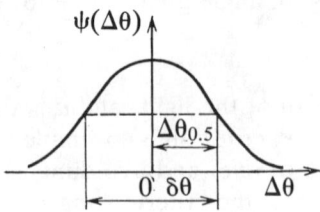

**FIGURE 5.1.** Form of the signal envelope autocorrelation function.

The half-peak width of the input signal autocorrelation function is often used for the potential resolution. This quantity is simple, and therefore attractive, but, as will be shown below, is an inaccurate measure of the resolution in the majority of cases.

We will estimate the resolution of a monopulse radar and compare it with the resolution of a single-channel conical-scanning radar.

## 5.2 THE ANGULAR RESOLUTION OF MONOPULSE RADARS

We will analyze the angular resolution initially without considering input signal fluctuations or receiver noise, for two of the most common types of single-coordinate monopulse systems.

### 5.2.1 Single-Coordinate Amplitude-Sum-and-Difference Monopulse Systems

The complex target echoes (see Fig. 5.2) in the first and second channels, for this case may be written in the form:

$$E_1(t, \theta) = E_{m1}F(\theta_0 - \theta_1)\exp(i\omega t) + E_{m2}F(\theta_0 - \theta_2)\exp[i(\omega t + \alpha)]$$
$$E_2(t, \theta) = E_{m1}F(\theta_0 + \theta_1)\exp(i\omega t) + E_{m2}F(\theta_0 + \theta_2)\exp[i(\omega t + \alpha)]$$

where $F(\theta_0 - \theta)$ and $F(\theta_0 + \theta)$ are the antenna patterns for the corresponding channels; $\theta_0$ is the off-axis squint of the beam; $\theta_{1,2}$ are the off-axis angles of the first and second signal sources (see Fig. 5.2); and $\alpha$ is the phase shift induced by the distance between the targets.

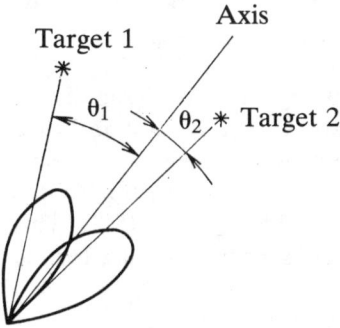

**FIGURE 5.2.** Geometry for angle sensing with two-point targets.

Employing sum-and-difference processing, at the output of the identical receiver channels (to within a multiplicative constant) we obtain [21]

$$\underline{u}_s(t, \theta) = (E_{m1}/\sqrt{2})[F(\theta_0 - \theta_1) + F(\theta_0 + \theta_1)]\exp(i\omega_c t)$$
$$+ (E_{m2}/\sqrt{2})[F(\theta_0 - \theta_2) + F(\theta_0 + \theta_2)]\exp[i(\omega_c t + \alpha)]$$
$$\underline{u}_d(t, \theta) = (E_{m1}/\sqrt{2})[F(\theta_0 - \theta_1) - F(\theta_0 + \theta_1)]\exp(i\omega_c t)$$
$$+ (E_{m2}/\sqrt{2})[F(\theta_0 - \theta_2) - F(\theta_0 + \theta_2)]\exp[i(\omega_c t + \alpha)]$$

If the signal is normalized by a fast AGC and averaged, the error signal at the output of a multiplier phase detector is given by

$$S(\theta) = \{\mathrm{Re}[\underline{u}_s(t, \theta)\underline{u}_d^*(t, \theta)]\}/[\underline{u}_s(t, \theta)\underline{u}_s^*(t, \theta)] \tag{5.1}$$

With the appropriate substitutions and transformation, this becomes

$$S(\theta) = \frac{E_{m1}^2 W_1 + 2E_{m1}E_{m2}W_2 - 2E_{m1}E_{m2}W_3 + E_{m2}^2 W_4}{E_{m1}^2 W_5 + 2E_{m1}E_{m2}W_6 + E_{m2}^2 W_7} \tag{5.2}$$

where

$$W_1 = F^2(\theta_0 - \theta_1) - F^2(\theta_0 + \theta_1)$$
$$W_2 = F(\theta_0 - \theta_1)F(\theta_0 - \theta_2)\cos \alpha$$
$$W_3 = F(\theta_0 + \theta_1)F(\theta_0 + \theta_2)\cos \alpha$$
$$W_4 = F^2(\theta_0 - \theta_2) - F^2(\theta_0 + \theta_2)$$
$$W_5 = [F(\theta_0 - \theta_1) + F(\theta_0 + \theta_1)]^2$$
$$W_6 = [F(\theta_0 - \theta_1) + F(\theta_0 + \theta_1)]$$
$$\times [F(\theta_0 - \theta_2) + F(\theta_0 + \theta_2)]\cos \alpha$$
$$W_7 = [F(\theta_0 - \theta_2) + F(\theta_0 + \theta_2)]^2$$

The phase shift $\alpha$ changes relatively rapidly even when the relative target motions are small, and the AGC system has a significant time constant compared with these phase changes. As a result, the low frequency filters at the phase detector output and in the AGC circuit filter out all signal components with the factor $\cos \alpha$ in (5.2). This allows us to simplify the expression for the angle-sensing response for the case of two targets within the beam:

$$S(\theta) = \frac{[F^2(\theta_0 - \theta_1) - F^2(\theta_0 + \theta_1)] + a^2[F^2(\theta_0 - \theta_2) - F^2(\theta_0 + \theta_2)]}{[F(\theta_0 - \theta_1) + F(\theta_0 + \theta_1)]^2 + a^2[F(\theta_0 - \theta_2) + F(\theta_0 + \theta_2)]^2}$$

$$\tag{5.3}$$

where $a = E_{m2}/E_{m1}$.

We can find the angle-sensing condition for two-point targets by setting the numerator equal to zero. It is clear that in this case the radar will track the energy center of the sources, with the axis closer to the source with the greater power.

The most interesting case is when one of the targets, for example target 1, is located at a small angle $\theta = \Delta\theta$ from the axis, and a second target at a relatively large angle $\theta_2$ from the axis affects the sensitivity of the measurement on the first. The two squinted beams in this case are given by

$$F(\theta_0 \pm \theta_1) \approx F(\theta_0)(1 \mp \mu\Delta\theta) \tag{5.4}$$

where $\mu$ is the angle-sensing sensitivity corresponding to the linear portion of the angle-sensing response when tracking a single target:

$$\mu = \frac{1}{F(\theta_0)} \left| \frac{dF(\theta_0 \pm \theta_1)}{d\theta_1} \right|_{\theta_1=0}$$

Substituting (5.4) in (5.3), we obtain through elementary transformations:

$$S(\theta) = \frac{\mu\Delta\theta + a^2\{[F^2(\theta_0 - \theta_2) - F^2(\theta_0 + \theta_2)]/[4F^2(\theta_0)]\}}{1 + a^2\{[F(\theta_0 - \theta_2) + F(\theta_0 + \theta_2)]/[2F(\theta_0)]\}^2}$$

Inasmuch as targets are usually resolved for angular distances exceeding the beamwidth, and the angular errors under normal conditions do not exceed a tenth of the beamwidth, we may assume $\theta_2 \gg \Delta\theta$, and consider the function $F(\theta_0 \pm \theta_2)$ to be independent of $\Delta\theta$ in a first-order approximation. Then the angle-sensing sensitivity at the tracking point can be placed in the following form:

$$\mu' = \frac{dS(\theta)}{d\Delta\theta}\bigg|_{\Delta\theta=0} = \mu \bigg/ \left\{ 1 + \left[ a\frac{F(\theta_0 - \theta_2) + F(\theta_0 + \theta_2)}{2F(\theta_0)} \right]^2 \right\}$$

from which it follows that

$$\frac{\mu - \mu'}{\mu'} = a^2 \left[ \frac{F(\theta_0 - \theta_2) + F(\theta_0 + \theta_2)}{2F(\theta_0)} \right]^2 \tag{5.5}$$

Equation (5.5) allows us to estimate the change in the angle-sensing sensitivity, due to the influence of an interfering target located at an angle $\theta_2$ from the tracked target.

Setting $S(\theta)$ to zero, we can find the angle error:

$$\mu\Delta\theta + a^2 \frac{F^2(\theta_0 - \theta_2) - F^2(\theta_0 + \theta_2)}{4F^2(\theta_0)} = 0$$

from which we obtain

$$\Delta\theta = -\frac{a^2}{\mu} \frac{F^2(\theta_0 - \theta_2) + F^2(\theta_0 + \theta_2)}{4F^2(\theta_0)} \tag{5.6}$$

Equations (5.5) and (5.6) show that the influence of the second target on the angle-sensing sensitivity introduces errors in locating the first target. The conditions for resolving the targets can be determined with these equations.

Curves for the normalized values of $\delta\theta_a$ and $\delta\mu_a$, as calculated from (5.5) and (5.6), are shown in Fig. 5.3; for an antenna pattern approximated by [43]:

$$F(\theta) = \frac{1}{2} \frac{\sin(k_\lambda \cdot \tfrac{1}{2}d_p\theta)}{k_\lambda \cdot \tfrac{1}{2}d_p\theta}$$

where $d_p$ is the diameter of the parabolic antenna.

The squint of the beams $\theta_0$ is then found from

$$k_\lambda \cdot \tfrac{1}{2}d_p\theta_0 = \frac{\pi}{2}$$

The quantities are normalized in Fig. 5.3 as follows:

$$\delta\theta_a = \Delta\theta/[a^2/(k_\lambda \cdot \tfrac{1}{2}d_p)]$$
$$\delta u_a = (1/a^2)(\mu - \mu')/\mu'$$

With the approximation used, the 3-dB beamwidth is $\theta_3 = \lambda/d_p$, and $\theta_0 = \theta_3/2$.

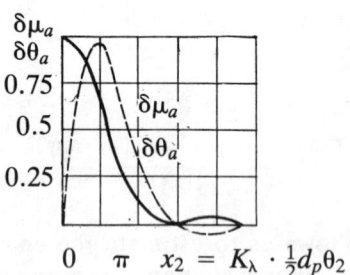

**FIGURE 5.3.** Change induced by an interfering target in the angle-sensing sensitivity $\delta u_a$ and angular error $\delta\theta_a$ in an amplitude-sum-and-difference monopulse radar.

It can be seen in Fig. 5.3 that as $\theta_2$ is increased, the normalized angle-sensing sensitivity decreases and falls to zero at $x_2 = k_\lambda d_p \theta_2 / 2 = 3\pi/2$, for which $\theta_2 = 1.5\theta_3$. For $x_2 = \pi/2$ ($\theta_2 = \theta_0$), the angular error is greatest, and with $a = 1$ reaches the value $\Delta\theta = 0.6\theta_0$, which is approximately equal to half the angle between the sources. Further increase in $\theta_2$ leads to a decrease in the angular error, which falls to zero for $\theta_2 = 1.5\theta_3$. It follows that the targets are completely resolved when the angle between them is approximately $\theta_3$. This is close to the resolution obtained from the width of the correlation function.

Naturally, if a less rigorous resolution criterion is used, the relative target displacement required to resolve the two targets is reduced accordingly.

### 5.2.2 Phase-Sum-and-Difference Monopulse Systems for Angle Sensing in One Coordinate

An analysis similar to that above shows that with a phase-sum-and-difference system, a second interfering source also affects the angle-sensing sensitivity and causes angular errors, the values of which may be calculated by the formulas [21, 43]:

$$\delta\theta_\phi = \frac{\Delta\theta}{a^2/(k_\lambda l)} = \frac{F^2(\theta_2)}{F^2(0)} \sin(k_\lambda l \sin\theta_2) \tag{5.7}$$

$$\delta\mu_\phi = \frac{\mu - \mu'}{\mu' a^2} = \frac{F^2(\theta_2)}{F^2(0)} \cos^2(k_\lambda l \sin\theta_2) \tag{5.8}$$

The normalized angle-sensing sensitivity and angular errors in a phase-sum-and-difference system, as calculated from (5.7) and (5.8), approximate the antenna pattern by

$$F(\theta) = \frac{1}{2} \frac{\sin(k_\lambda l\theta/2)}{k_\lambda l\theta/2}$$

For simplicity, the diameter of the phase antenna is taken to be the same as that of the amplitude antenna, and is $d_p = 2l$.

Comparison of Figs. 5.3 and 5.4 shows that an amplitude-comparison monopulse system is more sensitive to the presence of a second target within the main beam of the sum pattern, while a phase comparison system is more sensitive to a second target located outside the beam.

This analysis of monopulse system resolution capabilities was performed for the receiving mode. If the radar transmits and receives, then the echo

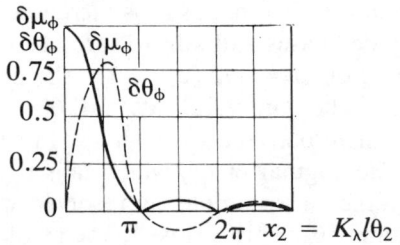

**FIGURE 5.4.** Change induced by an interfering target in the angle-sensing sensitivity and angle error in a phase-sum-and-difference monopulse radar.

amplitudes $E_{m1}$ and $E_{m2}$ will be proportional to the corresponding transmission patterns, which should be taken into account when determining the angle-sensing response $S(\theta)$ and the quantities $\Delta\theta$ and $\mu'$.

It was also assumed above that one target was close to the axis. If this condition is not satisfied, the analysis is made much more complex, and it is easier to assess the system target resolution with graphs of the responses resulting for various target displacements, given by the pattern approximations used.

As can be seen in (5.3), the angle-sensing response is determined by the sum of the difference and sum patterns, each weighted by the ratios of the signal powers reflected from the corresponding sources. By plotting the sum and difference patterns on separate scales (one on graph paper, and the other on tracing paper), and summing them for the various angular target displacements, in accordance with the expression for the angle-sensing response, the resulting angle-sensing response for each interesting case may be constructed, and from its behavior the target resolution performance may be derived.

The graphical method gives directly visible results and is relatively simple. We will examine the results of the graphical analysis of target resolution for four radar types: amplitude monopulse, paired lobing, conical-scanning, and sequential lobing systems [21]. The patterns are shown in Fig. 5.5, approximated by the function $\cos^2(56.25\theta)$ for the main beam, where $\theta$ is the angle in degrees. The 3-dB beamwidth with this approximation is about $1.2°$. We will assume in this analysis that the target echo amplitudes are equal ($a = 1$).

Figure 5.6 shows curves for the normalized amplitudes of the angle-sensing response of an amplitude-comparison monopulse system receiving the signals from two targets, for three values of the angular target separation. It is apparent that for $\psi_s = 1°$ the targets are not resolved, and the

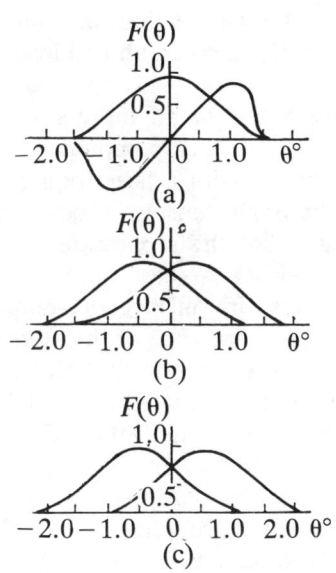

FIGURE 5.5. Antenna patterns
of compared systems:
  (a) monopulse and paired
  lobing
  (b) conical-scanning system
  (c) sequential lobing system

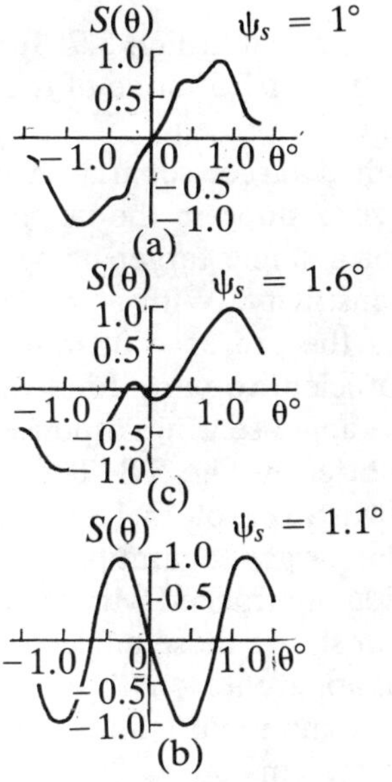

FIGURE 5.6. The normalized
angle-sensing response of a sum-
and-difference system with two
targets in the beam, for three
angular target separations:
  (a) 1°    (b) 1.1°    (c) 1.6°

form of the angle-sensing response is typical for the case of a single target.
The existence of the second target is apparent only in the presence of
slight perturbations in the response near the axis.

With a target separation of 1.1°, the angle-sensing slope goes negative
through the middle null point at $\theta = 0$, which indicates the presence of
two targets in the beam. However, the slope at each target is low, which
makes it impossible to track either target reliably. Only with $\psi_s = 1.6°$ do
two stable points with steep slope appear, allowing the radar to fully resolve
the two targets and reliably track either of them, giving the exact angular
coordinates. Thus the angle 1.6°, equal to $1.33\theta_3$, can be specified as the
target resolution for this case.

The target resolution performance was determined analogously for
paired lobing and sequential lobing systems, and also for conical-scanning
systems. The resultant angle-sensing responses are similar to one another,

and are shown in Fig. 5.7. It is apparent from the curves that for a target separation of 1.5° none of the systems resolves the targets, and all instead indicate the presence of a single target.

With a target separation of 1.6°, the responses for all three systems have zero slope in the neighborhood of $\theta = 0$. This corresponds to the case of a single target being tracked by a system with a large region of low sensitivity. With a separation of 1.7° the angle-sensing response indicates the presence of more than one target, but the separation is still not sufficient to permit individual tracking of either target.

The angle-sensing responses for all the systems with full target resolution are plotted in Fig. 5.8. The relative resolving performance of the systems is shown in Table 5.1. The quantity $\psi_{s\ min}$ in the table is the minimum angular target separation for which the targets are resolved and may be individually tracked with errors not exceeding the accepted value. The last column shows these angles normalized by the 3-dB beamwidth.

It is apparent from the data that the difference in resolution performance between monopulse and paired lobing systems is not great (less than 10%). These systems outperform conical-scanning and sequential lobing systems by about 30%.

These results are valid when the signal amplitudes are stable and not subject to the effects of various forms of interference. Radars usually operate in the presence of several forms of random interference, including internal receiver noise and natural background noise (thermal and cosmic noises, and noise from clutter and multipath effects). The signal fluctuations originating at the target have the greatest effect.

With these fluctuations, a statistical approach is required to determine the target resolution performance. The probability distributions of estimates of the angle $\theta$ for the two signals for various separations is shown in Fig. 5.9 [45]. This figure suggests that the resolving capability may be estimated by setting an acceptable measurement error. For example, we can consider the targets to be resolved if the sum of the variances in the measurements of $\theta_1$ and $\theta_2$ is less than the absolute value of the difference of the average values of these quantities, *i.e.*,

$$\sigma_{\theta_1} + \sigma_{\theta_2} \leq |\theta_1 - \theta_2|$$

Other criteria may also be used.

The signal parameter estimates are usually calculated in a computer, with the following steps:

- separation of the signals by their energy parameters;
- extraction (selection) of the signals with a single-valued relation to the measured parameter;
- measurement (estimation) of the parameters by which the targets are resolved.

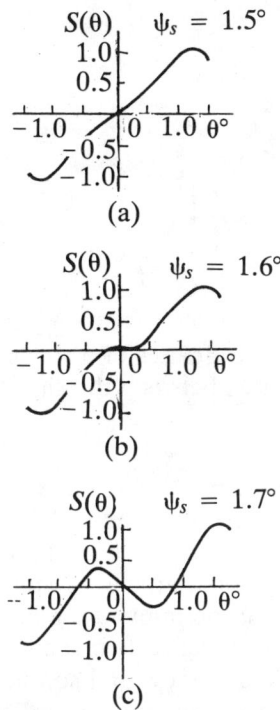

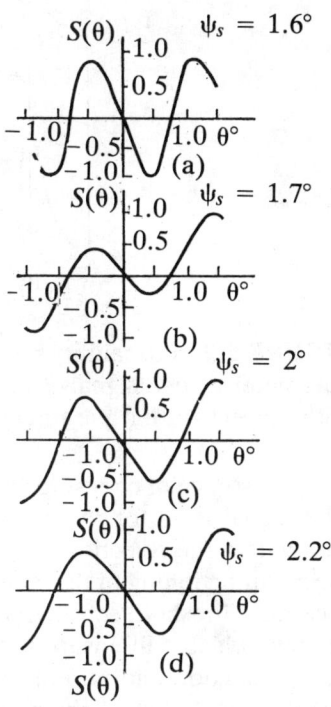

**FIGURE 5.7.** The normalized angle-sensing response representative of the performance of paired lobing, conical-scanning, and sequential lobing systems with two targets in the beam, for three angular target separations:

   (a) 1.5°   (b) 1.6°   (c) 1.7°

**FIGURE 5.8.** The normalized angle-sensing response at the minimum target separation required to obtain accurate tracking of either target:

   (a) monopulse system
   (b) paired lobing system
   (c) conical-scanning system
   (d) sequential lobing system

**TABLE 5.1**

| Radar type | $\psi_{s\ min}$ | $\psi_{s\ min}/\theta_3$ |
|---|---|---|
| Monopulse | 1.6° | 1.3 |
| Paired lobing | 1.7° | 1.4 |
| Conical-scanning | 2.0° | 1.7 |
| Sequential lobing | 2.2° | 1.8 |

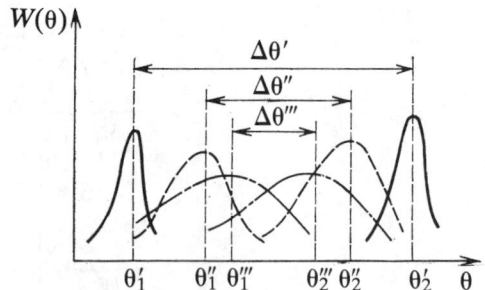

**FIGURE 5.9.** Probability density function for estimates of a single parameter of two signals, for different distances between the signals with respect to that parameter.

The essence of choosing an objective resolution criterion is the selection of a signal level for the measured parameter. The targets will clearly be resolved if the selection system presents signals of sufficient strength for normal functioning of the measurement circuit. Measurement systems have a certain threshold sensitivity corresponding to the minimum input signal energy, which will ensure a measurement satisfying certain requirements (true detection and false alarm probabilities, accuracy, and so one). The angular resolution of the angle measurement system may accordingly be taken to be the minimum angular target separation for which the input signal exceeds the threshold. The internal system noise, present at the input to the measurement circuit, is often used as the threshold.

When the input signal lies below the threshold, the measurement error increases sharply. In tracking systems, the tracking may then be interrupted, inasmuch as the dependence of the probability of broken track on the signal-to-interference ratio itself has a thresholding character. It follows that an estimate of the resolution can be found by determining the conditions under which the signal power (energy) at the input of each single-channel measurement device will exceed the threshold.

The process of resolution in angle, as described in its physical aspects above, is very similar to the detection process. Considering the statistical character of the detection problem, in determining the qualitative angular resolution performance it is necessary to use the probability distributions for the signals and noises present at the input of the angle measurement system, and to calculate the probability of determining the number of targets for a given false alarm probability. Inasmuch as the *a priori* distributions of the resolvable signal parameters are usually unknown and must be considered uniform, and because internal system noise with a Gaussian distribution is usually considered for the interference, the maximum likelihood method is usually used in the calculations [45].

Thus, estimating the resolution performance of a monopulse radar on a group of targets, with randomly fluctuating amplitude relationships and various noises, is a complex statistical problem. As a result, the angular resolution can only be expressed in the form of some statistical quantity.

Figure 5.10 shows the probability distributions for the tracking system direction to each of two targets for various target separations, expressed in units of beamwidths, using actual amplitude fluctuations [21]. These curves indicate that the two targets may be resolved if they are separated by at least 0.85 of the beamwidth. This approximately corresponds to the condition when, in the method described above, the presence of two targets is first discerned, but they cannot be tracked individually due to the low slope of the angle-sensing response at the tracking points.

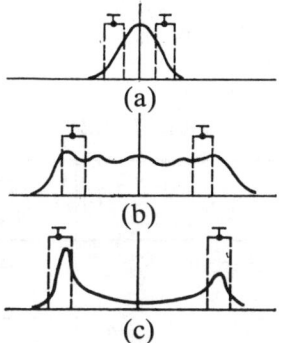

FIGURE 5.10. Probability density function for antenna direction when tracking two targets at various separations:
   (a) $\psi_s = 0.3\theta_3$    (b) $\psi_s = 0.75\theta_3$    (c) $\psi_s = 0.85\theta_3$

The antenna sidelobes will degrade the resolution performance of any system, and may be responsible for the appearance of spurious targets.

## 5.3 THE ANGLE-ERROR SLOPE OF MONOPULSE SYSTEMS

The angle-sensing sensitivity as measured by the slope of the angle-sensing response at the axis is as important a characteristic of a radar system as its angular resolution. The normalized responses to a point target for several systems are plotted in Fig. 5.11, derived from [21]. It was assumed in these calculations that the various channels were associated with identical antenna patterns and signal-to-noise ratios.

The angle-error slopes for the various systems may be compared with the data in Table 5.2. The values are normalized with respect to the value

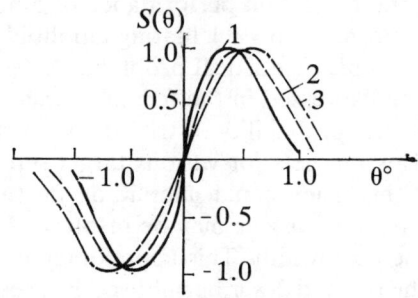

**FIGURE 5.11.** Normalized angle-sensing response in the case of tracking a point target, for various radars:
1  monopulse system
2  sequential and paired lobing
3  conical-scanning system

**TABLE 5.2**

| Radar type | slope 1/deg | normalized slope |
|---|---|---|
| Monopulse-sum-and-difference | 4.6 | 1 |
| Sequential lobing | 3.3 | 0.72 |
| Paired lobing | 3.1 | 0.67 |
| Conical-scanning | 2.7 | 0.59 |

of the slope for an amplitude-comparison sum-and-difference monopulse system. The figures show that with a relatively low SNR, a monopulse system has the greatest angle-error slope, exceeding the slope of a conical-scanning radar by a factor of two. Inasmuch as the angle-sensing accuracy, taking account of the internal receiver noises, is inversely proportional to the angle-error slope, a monopulse system provides the greatest accuracy under the given conditions.

When the signal is sufficiently strong to ensure satisfactory operation of a tracking system, there is no noticeable difference in the angle-error slope between monopulse and other systems.

Table 5.3 provides a comparison of the angle-error slope for various monopulse systems [35]. It can be seen in the table that monopulse radars with sum-and-difference discriminators, regardless of the angle-sensing method used, have an angle-error slope one half that for the other systems. This does not mean that the actual accuracy of sum-and-difference systems

**TABLE 5.3**

| Type of angle discriminator | Angle-sensing method | | | Comment |
|---|---|---|---|---|
| | amplitude | phase | combination | |
| Amplitude | $\mu_a$ | $\mu_\phi$ | $\mu_a$ | $\mu_a$ and $\mu_\phi$ are the angle- |
| Phase | $\mu_a$ | $\mu_\phi$ | $\mu_\phi$ | sensing sensitivities for |
| Sum-and-difference | $0.5\mu_a$ | $0.5\mu_a$ | $0.5\mu_a$ | amplitude- and phase- |
| | | | $0.5\mu_\phi$ | sensing, respectively. |

is lower, however, because the accuracy is also determined by several other important factors. [*The factor of 0.5 here and in the Sviridov reference appears to result from an arbitrary normalization constant, which also affects noise and external interference. Hence, it does not lead to increased tracking error.* Trans.]

It should be kept in mind that specification of the resolution and angle-error slope is not a sufficient criterion for choosing the best type of system for a given mission. In choosing the system configuration, it is also necessary to consider the range sensitivity, design complexity, immunity to natural and man-made interference, the tactical operating conditions, and several parameters.

## 5.4 METHODS FOR IMPROVING THE ANGULAR RESOLUTION OF MONOPULSE RADARS

The need for improved angular resolution arises for operation against multiple targets, and also when the enemy employs various countermeasures which significantly complicate the radar situation, such as AM jamming and false targets. Improved resolution is also required for operation against low-flying targets, when the effects of specular reflection from the ground become a factor. We will examine several methods for improving the angular resolution of monopulse systems.

### 5.4.1 Increasing Antenna Dimensions and Operating at Short Wavelengths

Inasmuch as the angular target resolution is inversely proportional to $\theta_3$, and $\theta_3 = \lambda/d_p$, the angular resolution can be improved by increasing the antenna dimensions and decreasing the wavelength. The intensive development of millimeter wave systems in the US in recent years is thus partly explained by a desire for improved angular resolution.

Increasing the antenna aperture in order to improve the resolution is not always appropriate, because it increases the size and weight of the antenna system and does not always give the desired improvement. It has been shown [106] that when propagation through the ionosphere is involved, for example, a complex wave is formed at the receiver, composed of a roughly plane wave formed by reflection or refraction, and a wave resulting from scattering within the ionosphere over a range of angles of several degrees. The resulting wave is not planar and undergoes amplitude fluctuations with time constants, which depend on the structure and drift rate of the ionospheric irregularities. As a result, the apparent direction to the source differs from the actual direction by some constantly varying angle. Postwar radar investigations exposed two forms of ionospheric fluctuations: fast fluctuations measured in seconds, and slow ones measured in minutes. The slow variations alter the angle of arrival in the meter waveband by an angle ranging from tenths of a degree to several degrees. These fluctuations present a severe limitation on the resolution performance, since they cause the received signals to overlap a good deal in the resolvable parameters.

If the ratio of reflected energy to scattered energy is sufficiently high, the angular resolution is determined, for the most part, by the beamwidth. When this ratio is small, the scattering component dominates, and the resolution is determined by the angular spread, which may exceed the beamwidth. Since the apparent angular distribution of the sources may overlap one another, even if the source separation is several degrees, the system resolution is limited by ionospheric scattering, and not the antenna dimensions.

The spectra of the scattering fluctuations may be as much as 100 times as wide as the spectra of the echo fluctuations. This allows the scattering components to be filtered out to a large extent, by methods employed in pulse radar (for moving target indication). This decreases the scattered energy level and increases the reflected energy to scattered energy ratio, allowing the resolution to be improved by enlarging the antenna aperture [106].

Figure 5.12 shows the calculated variation of the angular resolution with antenna dimension for $\lambda = 20m$. In the absence of ionospheric effects the resolution improves linearly with antenna dimension (curve 1), so that increasing the antenna aperture increases the resolution by the same factor. Diffraction in the ionosphere places a limit on the resolution improvement, as illustrated by curve 2. Curve 3 shows the effects of filtering the scattered signal, and is the result of averaging over a wide range of parameters characterizing the state of the ionosphere and the target properties. Averaging the results with the reflected-to-scattered signal ratio over a 20-dB range, and the effective target cross section over a 40-dB range, the

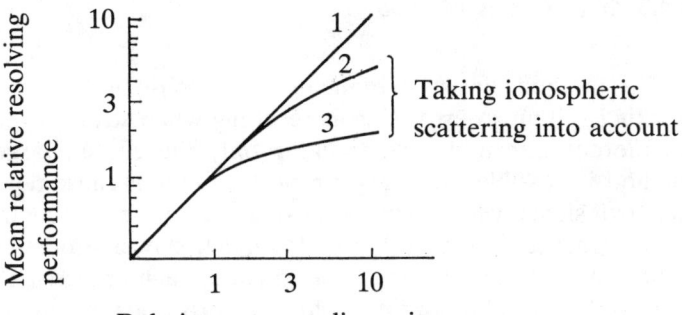

**FIGURE 5.12.** Resolution data for two targets *versus* antenna dimensions:
1   with no ionospheric scattering
2   with optimal signal processing
3   without optimal signal processing

resolution is found to be improved by a factor of 2.5 as the antenna dimension is increased by a factor of three relative to a standard, and increasing the aperture ten times gives a resolution gain of 5 [106].

### 5.4.2 Circuit Methods

Recently much attention has been given to improving system resolution through more advanced processing circuitry, an approach for which mono-pulse radars are better suited than other types of angle-sensing systems. Monopulse systems not only offer better angular resolution than single-channel systems, but can also resolve multiple targets within the beam under certain conditions. This capability is achieved in monopulse systems through the use of multiple receiver channels, which provide additional target angle information.

The failure to resolve multiple targets results in two types of system response. When the two target echoes have equal amplitudes, the radar will track a point midway between the targets. If there are low frequency fluctuations in the individual target signals, one or the other signal will be stronger, and the indicated target angle will jump between the two actual targets. In either case, the radar cannot give the proper target locations, and missiles with radar homing devices can be counteracted by flying aircraft in groups [21].

Inasmuch as there are usually fluctuations in the target signals, one method for resolving multiple targets is the incorporation of angle gating into the system.

### 5.4.3 Angle Gating Method

This method consists of reducing the response of the angle-measurement tracking system to error signals resulting when strong signals arrive from a different direction, and tracking only the chosen target. The gating method is possible since target echo fluctuations cause rapid variations in the error signal, whereas the error signal variations arising from the motion of the tracked target are part of a much slower process, for which the indicated angle error is relatively small in each processing cycle. Thus, it is possible to establish a threshold angle error; error signals exceeding this threshold will be caused by signals from other targets, and will be ignored in tracking the chosen target.

The block diagram for a monopulse system which resolves multiple targets with angle gating is shown in Fig. 5.13. A lens antenna with four-horn feeds is used in the system, and two transmitters transmit either on different frequencies, or on the same frequency with a time delay. It should be noted that because the reflecting properties of the targets will vary with frequency, if the frequency difference is too great, the error signal will be affected by the differing signatures and not just the angle error. Therefore, it is better to adopt the second approach, using two transmitters on the same frequency but offset in time; one receiver may then be used instead of two. The drawback to this method is that target range measurement is made more complicated, as additional ambiguity is introduced by the time offset.

The frequency or time shift in the transmitted signals results in the formation of two independent error signals in the receiver, each of which contains information on the location of the multiple targets.

For the sake of simplicity, we will consider a system for angle measurement in one coordinate. As can be seen in the drawing, the antenna pattern is made up of two beams, squinted from the axis by equal angles, and generated by two separate transmitters (A and B). Two targets on either side of the axis can thus be illuminated primarily by signals from the different transmitters, providing additional target position data.

Two separate channels are used for receiving and processing the reflected signals. The receivers are similar, and, therefore, only one of them (A) is shown in detail. The reflected signals pass from the antenna, through the receive/transmit switches, to a balanced tee, where they are processed just as in an amplitude-sum-and-difference monopulse system.

Sum and difference signals are thus formed at the inputs of the corresponding receivers. These signals are then down-converted, amplified at IF, normalized with respect to the sum signal with AGC circuits, and

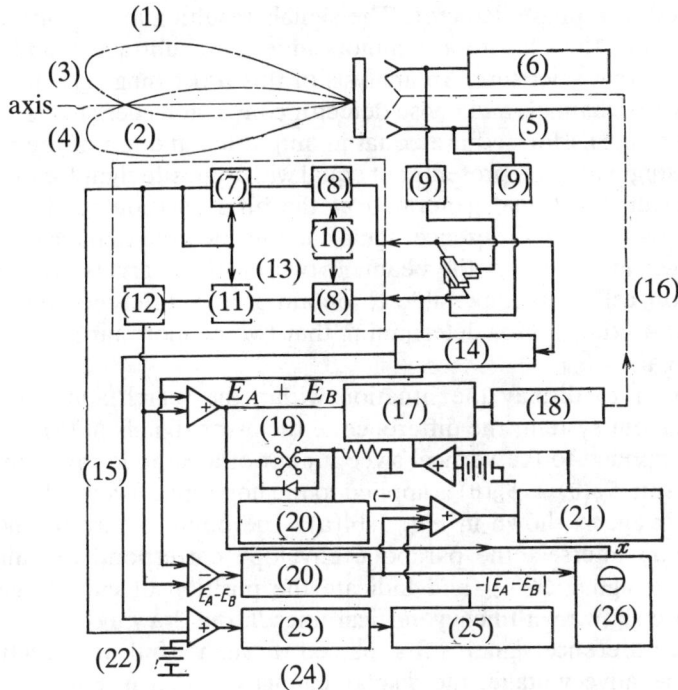

**FIGURE 5.13.** Block diagram of a monopulse system with enhanced resolution.

| | | | |
|---|---|---|---|
| (1) | transmitter A radiation pattern | (14) | receiver B |
| (2) | transmitter B radiation pattern | (15) | receivers' AGC voltage |
| | | (16) | mechanical antenna device |
| (3) | target 1 | (17) | gated memory |
| (4) | target 2 | (18) | servomotor |
| (5) | transmitter A | (19) | target selector |
| (6) | transmitter B | (20) | full-wave rectifier |
| (7) | IF amplifier with AGC | (21) | constant component restorer |
| (8) | mixer | (22) | threshold bias |
| (9) | receiving transmitting switching | (23) | phase inverter |
| | | (24) | threshold signal channel |
| (10) | local oscillator | (25) | half-wave rectifier |
| (11) | IF amplifier | (26) | oscilloscope |
| (12) | phase detector | | |
| (13) | receiver A | | |

compared in a phase detector. The signals resulting at the outputs of the receivers are then fed to a common adder and subtracter, and are then used for target selection. An analysis of this processing [21] shows that if the receiver channels and phase detector conversion coefficients are equal, the error signals formed are equal in amplitude and are of the same sign when a single target is present, but equal with opposite signs for two targets. As a result, the signal generated at the difference output is zero for a single target; a zero difference signal can thus be used to indicate that only one target is present in the beam. If two or more targets are within one resolution cell, it is impossible to obtain a zero difference signal, which provides a criterion for determining that two or more targets are present in the beam.

In order to display the situation within the search zone of an angle measurement system, the difference of the error signals $\Delta S(\theta) = S_A(\theta) - S_B(\theta)$ is applied to the vertical sweep input of a cathode-ray tube display, and the sum $S_A(\theta) + S_B(\theta)$ is applied to the horizontal sweep. The resultant display image is shown in Fig. 5.14(a). The points where the horizontal sweep line intersect the parabolic envelope correspond to nulls in the difference signal $\Delta S(\theta)$, and indicate the position of each target (these points are shown with heavy vertical lines in the drawing).

If the difference signal is first passed through a full-wave rectifier producing negative voltage, the display changes to that in Fig. 5.14(b), and is more convenient for determining the position of and tracking any of the targets. If there are three targets, the display will appear as in Fig. 5.14(c). The outer edges of the oscillogram indicate the outer targets, allowing the operator to track these targets automatically.

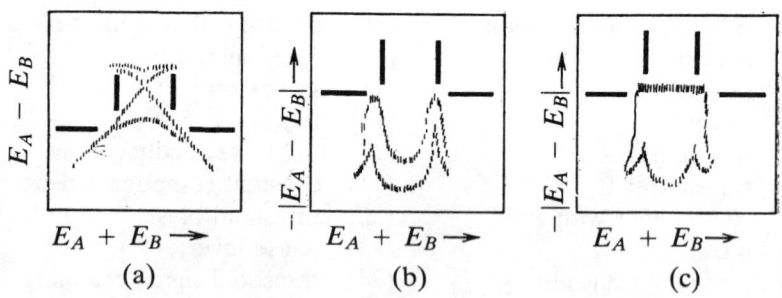

(a)          (b)          (c)

**FIGURE 5.14.** Display screen images:
  (a)   tracking a pair of targets (initial version of system)
  (b)   tracking a pair of targets (improved version)
  (c)   tracking a group of three targets

As a selected target is automatically tracked, the sum signal is fed to the servomotor, which controls the antenna position through a gated memory device which blocks signals not meeting the resolution criterion, and passes signals from the selected target. In this system, the gating device is triggered by positive pulses and shut off by negative pulses, formed by undesired or interfering error signals. The device simultaneously stores the signal for some specified period, which eliminates the problem of the track being interrupted by error signals from undesired targets.

As was stated, the criterion for resolution is that the difference of the error signals $\Delta S(\theta)$ be zero. At these moments the sum signal may be either positive or negative, depending on which target is producing the stronger signal. Therefore, a target may be selected for autotracking by choosing the operating polarity of the sum signal in a target selector. Such a selector consists of a switch and a detector connected in series with the input load of the full-wave rectifier for the sum error signal. The polarity of the detector and the signal passing through it is selected by the switch.

It should be noted that this method of solving the problem of resolving multiple targets cannot provide high accuracy, for instance, the accuracy achieved in tracking single targets, due to the fact that it involves blocking the tracking system for specified time intervals, during which the antenna continues to move by inertia. Nevertheless, the advantages of this approach are clear, especially for jamming protection, as will be shown in Chapter 8.

If high accuracy is required, a different method for improving the target resolution performance is advisable [45, 21], and is described below. The essence of this method is the functional processing of signals received simultaneously in several independent channels.

### 5.4.4 Functional Signal Processing

The field at a point in space resulting from $M$ sources is given by the expression:

$$\underline{E}_\Sigma = \sum_{m=1}^{M} \underline{A}_m \exp\{-ik[(x \cos \epsilon_m)\sin \theta_m + y \sin \epsilon_m]\} \tag{5.9}$$

where $\underline{A}_m = E_{0m} \exp[i(\omega_m t + \phi_{m0})]$ is the complex amplitude of the signal from the $m$th source; $\omega_m$ and $\phi_{m0}$ are the frequency and initial phase of the $m$th signal; and $\epsilon_m$ and $\theta_m$ are the elevation and azimuth of the $m$th source.

Equation (5.9) shows that the voltage resulting at the given point can be determined by the sum of $M$ complex equations, each of which is determined by the location of the corresponding source.

It is clear that if the radar can form and solve the corresponding number of independent equations, relating the positions of the signal sources, the parameters describing the direction to each source could be obtained, which would thus solve the problem of resolving multiple targets. This is the basis of the functional method for improving the resolution performance of the radar. This method requires the specified number of linearly independent receiver channels and a computer to solve the system of equations.

The required number of linearly independent receiver channels depends on the number of resolved sources and the characteristics of the radiated signals. Studies show [45] that with reception of coherent signals, the number of resolved targets with two-coordinate angle measurement is found from the expression $n_t \leq m_a/2$, where $m_a$ is the number of independent reception points (antennas).

With noncoherent sources we have

$$n_t \leq m_a^2/3 \quad \text{for amplitude-phase systems}$$
$$n_t \leq (C^2 + m_a)/3 \quad \text{for an amplitude angle sensor}$$
$$n_t \leq (2C^2 + 1)/3 \quad \text{for a phase angle sensor}$$

where $C^2$ is the number of combinations of $m_a$ taken two at a time.

A typical monopulse radar for angle sensing in two coordinates has four receiving channels. In accordance with the calculated data, such a system can, in principle, resolve two targets reflecting coherent signals, even if the targets are not resolved in range or velocity. The resolving performance also improves for incoherent sources, inasmuch as the number of independent equations is increased. In this case, it is possible to resolve 3, 4, or 5 targets when using amplitude, phase, or amplitude-phase sensors, respectively. It can be shown that the performance of phase systems exceeds that of amplitude systems by a factor which increases with the number of separate reception points. An amplitude-phase system offers the best resolution performance, and can resolve a given number of targets with fewer receiving antennas [45].

The effectiveness of the functional method for improving the radar resolution has been supported mathematically for the example of a phase angle sensor with four linearly independent reception points [21, 45]. In comparison with the other methods considered above, this method provides much more accurate target angle measurements, but requires additional system complexity and more stringent requirements for balanced and stable channel characteristics.

The capabilities of the functional method improve with multifrequency operation, because more linearly independent equations may then be obtained than with a single frequency, and, as a result, the number of simultaneously tracked points increases for a given number of antennas. With $m_a$ antennas, the number of incoherent sources which may be tracked in two coordinates is [45]

$$n_t \leqslant Q_y m_a^2/(2 + Q_y) \quad \text{for an amplitude-phase system}$$
$$n_t \leqslant (C_m^2 + m_a)Q_y/(2 + Q_y) \quad \text{for an amplitude system}$$
$$n_t \leqslant (2C_m^2 + 1)Q_y/(2 + Q_y) \quad \text{for a phase system}$$

where $Q_y$ is the number of frequencies used.

With coherent signals, the number of tracked sources in two coordinates, which is independent of the type of sensor, is

$$n_t \leqslant Q_y m_a/(1 + Q_y)$$

A comparison of these results with those given earlier will show that using many frequencies increases the number of sources which can be tracked in two coordinates, but by factors not greater than two and three for coherent and incoherent signals, respectively.

# Chapter 6

# *The Accuracy of Monopulse Angle Measurement*

## 6.1 SOURCES OF ANGLE ERRORS

Errors in measuring angular coordinates may be categorized as either external, associated with the target and propagation medium, or instrumental, caused by imperfect radar apparatus, inadequate measurement methods, and internal noises in the receiver and tracking system. These errors may also be either systematic or random. The basic sources of error are presented in Table 6.1.

**TABLE 6.1**

| Source of error | Causes of error |
|---|---|
| Propagation medium | Reflection by land and water surfaces; tropospheric and ionospheric refraction, diffraction, and depolarization of the waves; external noises. |
| Target | Reflected signal amplitude fluctuations; wandering of the center of reflection (angle noise); depolarization of waves reflected by complex targets. |
| Radar | Nonidentical amplitude-phase responses in the receiver channels; antenna deformation; cross-correlation of the receiving antennas; nonlinearity and backlash in the antenna drives; internal receiver noises; manufacturing defects in the radar equipment, and deviations caused by age; mechanical and climatic effects, and so on. |

The angle errors caused by differences in the amplitude-phase responses of the receiving channels are analyzed in Chapter 7.

## 6.2 EFFECTS OF THE PROPAGATION MEDIUM ON ANGLE ACCURACY

The effect of the propagation medium on the accuracy of angle measurements has been examined in many works [21, 36, 38]. We will study only two of these sources of angle errors, those associated with the Earth and with external noises.

### 6.2.1. Influence of the Earth on Angle Accuracy

As illustrated in Fig. 6.1, the energy radiated by the antenna reaches the target either by a direct path from the radar, or by reflection at the Earth's surface. The reflected wave may be considered the direct radiation from an imaginary source which is the mirror image of the real source; therefore, the range to the real source is somewhat less than the range to the image source. The voltage at the receiver will be determined by the amplitude and phase of the direct and reflected signals, and will be a maximum when they are in phase, and a minimum when they are out of phase.

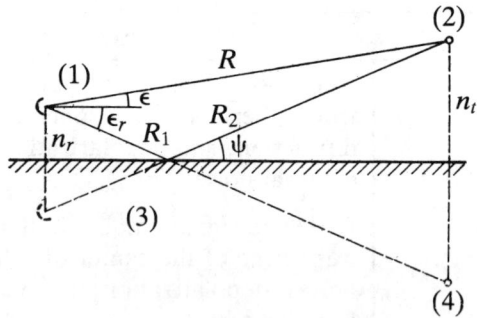

**FIGURE 6.1.** Multipath propagation.
　　　　　(1) radar antenna　(3) mirror image of radar antenna
　　　　　(2) target　　　　　(4) mirror image of target

Due to the reflecting surface, a single lobe pattern in the vertical plane becomes a multilobe pattern. As a result of these additional lobes, the number of equisignal zones in the angle-sensing response increases, and there are spurious equisignal directions at various angles to the main axis.

In certain cases these secondary equisignal directions will be sufficiently stable to result in automatic target tracking with substantial errors. The primary axis will also be displaced [21].

The angle errors will be further increased by the manner in which the multilobe pattern varies with the propagation and reflection changes accompanying target motion. This causes changes in the position of the equisignal direction and somewhat jerky antenna motion while tracking the target. The limits within which the tracking antenna will move depend on the reflection coefficient, the beamwidth, and the target elevation.

Many works address the influence of reflections from the underlying surface (ground or water) on the angle accuracy [49, 52, 62, 63, 70, 78, 79, 102, 105]. It has been well established that these multipath effects cause the most interference when tracking low-flying targets.

The structure of the signals reflected from the surface is rather complex. Both specular and diffuse components will arise, the overall nature of the reflections being determined by the roughness of the surface and the grazing angle. For the case of low-flying targets, the specular component dominates, and the reflection, therefore, the reflection may be analyzed with the methods of geometric optics. The influence of the specular component is illustrated in Fig. 6.2, which shows the dependence of the position of the line of sight to the target in the elevation plane on the relative target range [62]. The actual range may be calculated by specifying $\theta_3$, $h_r$, and $h_t$.

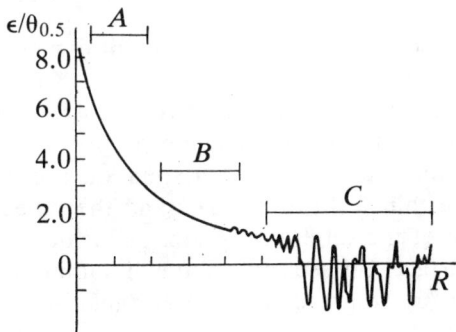

**FIGURE 6.2.** Dependence of the apparent target elevation with range.

The interfering signals (signals from the image target) begin to be received by the sidelobes and have an effect in region $B$ (in region $A$, the system is free of multipath interference). In region $C$, the target and its image can no longer be resolved in angle. The antenna system undergoes oscillations reaching $\frac{3}{4}\theta_3$ in elevation relative to the horizon, and the target track may then be broken.

The performance degradation resulting from multipath effects is most severe when automatically tracking low-flying targets in elevation, and is caused by the surface reflections and accompanying interference phenomena. There are three basic cases to be considered [52]:

(1) the signal reflected from the surface reaches the receiver through a sidelobe of the pattern;
(2) the reflected signal is received in the main lobe, but the direction to the target is significantly different from that to its image, which corresponds to the condition $0.3\theta_3 < \epsilon \le 1.5\theta_3$;
(3) the reflected signal is received by the main lobe, and the direction to the target is approximately the same as that to the image ($\epsilon < 0.3\theta_3$).

Case (1) corresponds to region $B$ in Fig. 6.2, when the specularly reflected signal has virtually no effect on the direct signal. In monopulse systems, this signal causes sinusoidal interference in the difference channel, with an amplitude smaller than that of the direct signal in the sum channel by a factor of $\rho\sqrt{G_\Delta(-2\epsilon)/G_\Sigma}$, where $\rho$ is the surface reflection coefficient, $G_\Delta$ and $G_\Sigma$ describe the difference and sum power patterns, and $\epsilon$ is the target elevation angle. The rms error which results is given by

$$\sigma = \frac{\rho\theta_3/\mu}{2G_{sl}} \qquad (6.1)$$

where $G_{sl}$ is the ratio of the sum main lobe and difference sidelobe levels, and $\mu$ is the angle-sensing slope.

When the level of the sidelobes drops off very rapidly, diffuse reflection from the radio horizon will affect the angle errors.

The transition to the second multipath situation occurs when the reflection sources are within $1.5\theta_3$ of the axis and the reflected signal falls within the main lobe of the difference pattern. Excluding the case of an extremely smooth surface, diffuse reflection will dominate the process up to an elevation $\epsilon \approx 0.7\theta_3$, where angle signals originating near the specular point enter the main lobe.

Diffuse scattering and weak specular reflection ($\rho < 0.5$) lead to errors described by (6.1). Plots of the angle error *versus* the difference in target and reflection point elevation angles is shown in Fig. 6.3. When strong specularly reflected signals begin reaching the main lobe, the linear approximation to the angle-sensing response becomes inaccurate, and it is necessary to use vector representations for the signals and interference in the sum-and-difference channels. It is then possible to derive

$$S(\theta) = \frac{[F_d(\epsilon) + \rho F_d(\epsilon_r)\cos \alpha][F_s(\epsilon) + \rho F_s(\epsilon_r)\cos \alpha]}{[F_s(\epsilon) + \rho F_s(\epsilon_r)\cos \alpha]^2 + [\rho F_s(\epsilon_r)\sin \alpha]^2}$$

$$+ \frac{\rho^2 F_d(\epsilon_r)F_s(\epsilon_r)\sin^2 \alpha}{[F_s(\epsilon) + \rho F_s(\epsilon_r)\cos \alpha]^2 + [\rho F_s(\epsilon_r)\sin \alpha]^2}$$

(6.2)

where $F_d(\epsilon)$, $F_d(\epsilon_r)$, $F_s(\epsilon)$, and $F_s(\epsilon_r)$ are the sum and difference pattern voltages at the target and reflection point elevations; $\rho = \rho_0\rho_s$ is the specular reflection coefficient for the surface; $\rho_s$ is the rms value of the rough surface specular scattering factor; and $\alpha$ is the phase shift of the reflected signal.

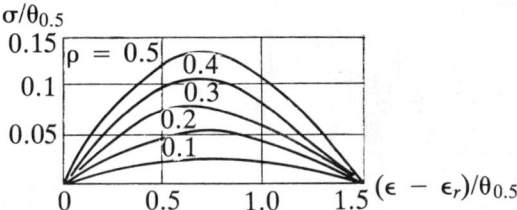

**FIGURE 6.3.** Dependence of the angle error due to multipath on the elevation difference between the target and point of reflection, for weakly reflected signals.

The angle error will vary within limits determined by $\alpha$ and $\rho$. The dependence of the limits of the tracking angle on the target elevation with specular reflection is shown in Fig. 6.4, and the regions of unstable tracking for $\epsilon < 0.75\theta_3$ and $\rho \to 1$ are shown. When the reflected and direct signals are in phase, the centroid of the two-point target (the target and its image), located on the horizon, is tracked. As $\alpha$ approaches 180°, the tracking point jumps to $0.7\theta_3$ when $\rho < 1.0$. If, for some reason, the reflected signal is stronger than the direct signal, the tracking angle will be $-0.7\theta_3$.

The variation of the elevation error *versus* target elevation for an amplitude-comparison monopulse system is shown in Fig. 6.5 for various reflection coefficients. The following parameters were used for the calculations: operating frequency 10 GHz; tracking system passband 2 Hz; half-power beamwidth 1°; height of the antenna above ground 5 m; target range 10 km; target altitude 50 m [105]. The calculations corroborate the oscillatory nature of the elevation error with low-flying targets. Thus, target motion causes fluctuations in the angle errors arising from multipath effects.

As the elevation angle goes from $0.7\theta_3$ to $0.1\theta_3$, the tracking process becomes unstable. When the reflection coefficient reaches 0.7 and the target elevation is below $0.7\theta_3$, the radar will try to track the point midway between the target and its image; this point lies on or near the horizon. When the beam elevation is less than $0.3\theta_3$, strong specular reflection can completely cancel the direct signal in the sum channel and it is impossible to ensure reliable system operation, even with a strong smoothing filter.

Studies have shown that the tracking errors resulting from multipath effects are the same in both monopulse and single-channel scanning radars [105]. In monopulse systems, however, these errors may be reduced by certain methods we will now describe.

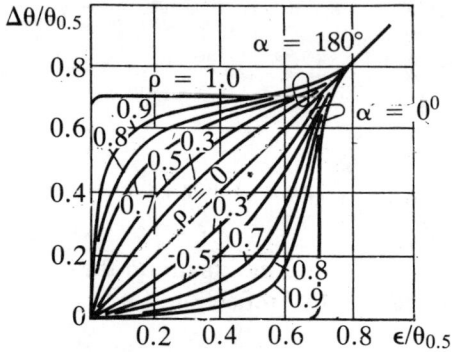

**FIGURE 6.4.** Dependence of the tracking angle limits on target elevation for specular reflection.

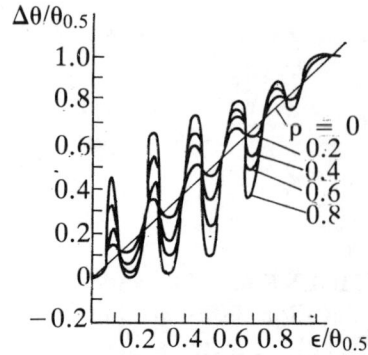

**FIGURE 6.5.** Dependence of the multipath elevation error in amplitude-comparison monopulse tracking on elevation and specular reflection coefficient.

### 6.2.2. Methods for Reducing Multipath Effects

There are many methods for reducing the effects of multipath phenomena on the accuracy of low-flying target tracking. These include improving the resolution in elevation, employing frequency agility or several simultaneous frequencies, off-axis monopulse tracking, the use of symmetrical sum-and-difference or asymmetrical antenna patterns, complex angles, multitarget measurements, radar fences, using circular polarization and altitude data obtained from other sources, and employing a pair of antennas in the elevation plane [52, 62, 70, 72, 78].

Improving the elevation resolution and reducing the sidelobe levels are effective methods for countering the effects of multipath propagation, but they entail increasing the antenna dimensions and operating at shorter wavelengths, which is not always possible. Any attempt to suppress the sidelobes without enlarging the antenna results in a wider main lobe, which then increases the influence of multipath effects.

Improving the range and velocity resolution reduces multipath degradation in radars with antennas located far enough off the surface to make the difference between the direct and reflected path lengths significant (i.e., to make the time delay of the reflected signal relative to the direct signal exceed the transmitted pulse duration). The Doppler difference between the direct and reflected signals may also be significant, allowing the interfering signals to be filtered out with narrowband filters inserted either in front of the detector or directly in the receiver tracking loop. Such filtering is not usually possible in radars located close to the ground (because the range and Doppler differences are then insignificant).

When carrier-frequency agility is employed, the angle errors fluctuate at the frequency of the modulation, and, thus, may be reduced by filtering. The errors arising from multipath may then be reduced by a factor of $2\Delta f \Delta \tau$, where $\Delta \tau$ is the time delay of the reflected signal. The time delay factor makes the frequency agility improvement insignificant at long ranges, when $\Delta \tau$ is small.

In off-axis monopulse tracking, the beam is fixed at the elevation $\epsilon = 0.8\theta_3$ and target positions below this angle are measured by the error signal. The measured error signal is converted to an angle and added to the fixed antenna angle, and the target is tracked in an open loop. With this approach, the signal from the image target is weaker than the direct signal since the target is always closer to the axis than its image, which improves the angle measurement and tracking stability. However, the errors may still be large, especially at long ranges. This approach is unnecessary at higher target elevations, when normal closed-loop tracking can be used. Therefore, the system may be designed to switch over to open-loop off-axis operation when the axis is at the threshold elevation and the target lies below it, as indicated by a negative error signal [62].

In the symmetrical pattern method, symmetrical sum and difference patterns are centered on the horizon (more exactly, on the point midway between the target and its image). If the first null of the difference pattern is positioned on the null of the sum pattern, there will be no error from specular reflection. Large errors will arise, however, due to diffuse reflection near the horizon.

When employing asymmetrical patterns, two asymmetrical beams are formed and offset from the horizon by different angles. The necessary pattern can be formed by placing an additional feed above or below a four-horn (or two-horn) monopulse feed. The ratio of the asymmetrical pattern values is symmetric about the horizon, which eliminates systematic errors arising from specular reflection. In addition, long range tracking stability is improved.

The use of complex angles is associated with the phase distortions introduced in the received signals by the reflected signals. As an example of this approach, we will consider a typical monopulse system with sum-and-difference angle measurement [62, 70]. The sum and difference patterns for such a system are shown in Fig. 6.6(a). The signals from the target and its image received in the corresponding channels are also shown. A vector diagram of the resulting signals and their components is shown in Fig. 6.6.(b).

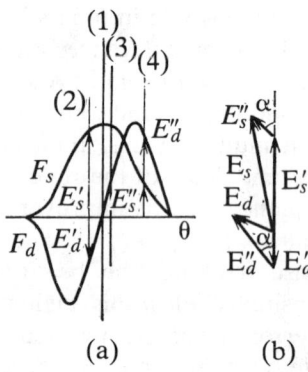

(a)                              (b)

**FIGURE 6.6.** (a) Sum and difference channel responses

    (1) axis                 (3) horizon

    (2) direction to target    (4) direction to target image

  (b) vector diagram of the signals received in these channels

When the radar is operating under ideal conditions, with no multipath reflections, the sum and difference signals ($E_s'$ and $E_d'$) are either in phase or antiphase. The signal reflected from the underlying surface differs from the direct signal in both amplitude and phase. When these reflected signals are picked up by the tracking beam, they produce components $E_s''$ and $E_d''$ at the antenna sum and difference outputs, which are added to the direct signals to form the signals $E_s$ and $E_d$; these signals differ in both phase and amplitude from the direct target signals. As a result, quadrature components appear in the angle error, which not only attest to the existence of multipath interference, but can be used to reduce the effects of this interference.

One of these methods is based on zeroing the quadrature components [62, 72]. The antenna is rotated until the vectors $E_s$ and $E_d$ are either in phase or antiphase; the quadrature components go to zero, and the difference-to-sum ratio is the same as it would be in the absence of multipath effects. The antenna axis, when corrected by the error voltage, will then indicate the target direction, and the target image will exert no influence on the system.

No change in the antenna structure is necessary to extract the quadrature components. It is sufficient to add another phase detector in the receiver, one input of which is shifted 90° [72]. Multipath effects can also be eliminated with a quadrature detector by employing special calibration. If the antenna elevation is fixed in the range of $0.4\theta_3$ to $0.7\theta_3$, and the real and imaginary components of the error signal are plotted in rectangular coordinates for various target elevations, it will be seen that a spiral is formed, each point of which corresponds to a specific target elevation [52, 72]. With this calibrated error-signal spiral, it is possible to measure the target angle with sufficient accuracy even in the presence of multipath effects.

If the radar is tracking targets over water, only a single such spiral need be calculated. When operating over rough ground, however, several spirals will be needed, each associated with a given azimuthal sector. This consideration, along with the necessity for eliminating the ambiguities caused by overlapping turns of the spirals, complicates the implementation of this approach, but experiments have demonstrated that these difficulties can be surmounted [78].

Multiple-target measurement is based on estimating the target coordinates using the maximum-likelihood methodology. Analyses of monopulse systems have shown that the angular coordinates of two targets separated by an angle $0.5\theta_3$ can be separately measured with an error $\sigma = 0.1\theta_3$, as long as their phase shift changes and the signal-to-noise ratio (SNR) is near 20 dB. With a higher SNR, the coordinates of targets separated by smaller angles can be measured to the same accuracy.

Shielding the radar in its operating region, with grounded metal grids (fences), for example, can reduce the ability of the sidelobes to pick up reflected signals, and of the main lobe to receive specular reflections from outside the expected range of off-axis target angles. This creates diffraction effects at the edges of the fences, however, which add to the diffraction losses of targets near the horizon.

The effects of multipath can also be reduced with an intelligent choice of transmitted polarization. In tracking radars, for example, vertical polarization is preferable, because it is associated with a lower reflection coefficient at grazing angles close to the Brewster angle. This significantly reduces specular reflection for $\epsilon > 1°$ over water and for $\epsilon > 2°$ over dry land. Thus, it follows that when tracking at low angles, it is best to use vertical polarization to reduce the effects of multipath signals received in the main lobe and sidelobe.

It is sometimes suggested that circular polarization could also be used to good effect, because the orientation of the polarization reverses upon reflection. This is true only for elevations exceeding the Brewster angle (greater than 5°), where multipath effects do not usually arise; at small target elevation angles, the polarization is not reversed upon reflection.

The use of target altitude data obtained from other sources can also reduce the angle errors introduced by multipath phenomena.

Automatic elimination of the effects of multipath reflections can be achieved with monopulse systems employing two antennas in the vertical plane positioned at different heights (see Fig. 6.7) [70]. These two monopulse antennas have identical patterns with the same orientation in elevation. The distance between them is automatically adjusted during tracking to ensure that the phase shift between the direct and surface-reflected signals received by the separate antennas is $\pi$, and that the phase shift of the direct signal between the antennas is different from $\pi$. The signals received by the antennas are used to generate signals controlling the phase shift and the height difference of the antennas. When the elevation control signal and the antenna height difference signal are both zero, the system will be tracking the target with comparatively small errors.

The theoretical analysis in [70], for an amplitude-sum-and-difference system with two antennas, verifies the effectiveness of this method, and shows how such a system can be constructed with mechanical and electronic control of the vertical separation of the antenna beams. Electronic beam control is, of course, realized with phased array antennas. On the whole, this method results in a fairly complex system design.

An interesting method for reducing the effects of reflections from the surrounding medium uses the harmonics of the signal reflected from the target. Studies have shown that natural objects, such as vegetation and the surfaces of land and water, do not generate harmonics of signals they reflect, whereas most man-made objects of interest in radar applications reflect transmitted signals at various harmonics of the carrier frequency. If the receiver is tuned to one of the harmonics, it is possible to suppress the interference from the surroundings.

There are several drawbacks to this method. There is a significant reduction in detection range resulting from the low efficiency of most artificial objects in reflecting the harmonics, and the requirements for linearity in the radar transmitter and receiver become more stringent. It is also necessary to suppress the parasitic harmonics in the transmitted signal, as they would be reflected from natural objects, thus eliminating the necessary distinction between man-made targets and their surroundings.

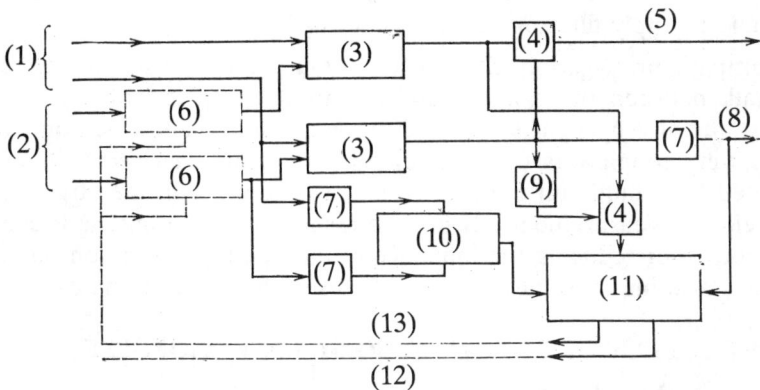

**FIGURE 6.7.** Simple block diagram of a monopulse system with two antennas at different heights.

(1) from upper antenna     (8) AGC
(2) from lower antenna     (9) $\pi/2$
(3) adder     (10) subtraction circuit
(4) phase detector     (11) antenna displacement control
(5) elevation control signal       system
(6) phase shifter     (12) antenna displacement control
(7) amplitude detector       signal
      (13) phase-shift control signal

The following conclusions may be drawn from an examination of these various methods.

1. There are no methods for reducing the degradation in accuracy caused by multipath effects which would be effective for all low-flying target tracking scenarios.

2. Most of the known methods do not give reliable target altitude data, but are intended more to maintain continuous tracking near the horizon.

3. Certain methods make it possible to maintain a tracking accuracy on the order of $0.1\theta_3$ in the region of the main beam, when ordinary tracking would be unstable at best. The most effective method is the utilization of asymmetrical patterns, which counters the effects of specular reflections, and minimizes the effects of diffuse scattering near the horizon. This entails a corresponding design complexity, as does the off-axis monopulse method.

4. When specular reflections are received in the sidelobes, problems arise when high accuracy (on the order of $0.01\theta_3$) is required. In this case, it is useful to lower the sidelobe levels of the difference pattern and to apply filtering and frequency agility.

Multipath propagation also causes azimuth errors. This is caused by crosstalk between the azimuth and elevation channels with large error angles, and also by signals reflected in the azimuth plane when operating over a very rough area (hills, mountains, structures, *etcetera*). It should be noted that multipath effects also give rise to error in the target range and velocity measurements. It may be assumed that the methods considered above for reducing the influence of multipath propagation on angle accuracy also improve the accuracy of these other measurements.

## 6.3 THE EFFECT OF EXTERNAL NOISE ON MONOPULSE ANGLE MEASUREMENT

External radio noises include cosmic and thermal noises. Cosmic noises, originating in the galaxy, cover an extremely wide band, but are not constant and drop off fairly rapidly at high frequencies. Above 200–300 MHz, galactic noise may be neglected in comparison with other interference sources. Thermal noise arises from the thermal emissions of objects around the radar antenna and of the atmosphere. The spectral density of these noises is roughly uniform to frequencies on the order of $10^4$ GHz, extending well beyond the limits of frequencies employed in radars.

Thus, there will always be noise present at the antenna, which will distort the useful signals in proportion to its strength. This noise contributes to the random nature of the signal fields in the aperture, and, therefore, to the statistical character of the various measurement processes performed by the radar. Statistical methods, therefore, must be used to derive general estimates of the effects of these external noises on monopulse system accuracy.

A Monte-Carlo computer simulation of random homogeneous fields with normally distributed parameters in monopulse systems has shown the following:

- among systems using amplitude-comparison angle detection, amplitude-amplitude systems with logarithmic amplifiers offer the highest accuracy; linear measurement systems are nearly as accurate;
- among systems using phase-comparison angle detection, those incorporating logarithmic amplitude-angle measurement again are most accurate (phase-amplitude systems). They are followed by linear measurement systems, then phase-phase and phase-combination systems;

• the greatest accuracy for each type of angle sensor is obtained with logarithmic angle measurement, and the worst results with sum-and-difference measurement circuits.

The relative accuracy of the best systems for each type of angle sensor are shown in Fig. 6.8, where $\sigma_{sPA}^2$ and $\sigma_{sAA}^2$ are the variances of the tracking errors for phase-amplitude and amplitude-amplitude systems; $\alpha_n$ is the noise-to-signal ratio; $\beta_0' d_a$ is the degree of field correlation in the equipment; $q_k^2$ is the coherence parameter; $d_a$ and $d_p$ are the aperture widths for the amplitude and phase methods, respectively; and $l$ is the antenna separation in the phase method, so that $d_a = d_p + l$ (the antenna dimensions are the same for both methods). The angle accuracy of an amplitude-amplitude system with logarithmic amplifiers is shown in Fig. 6.9.

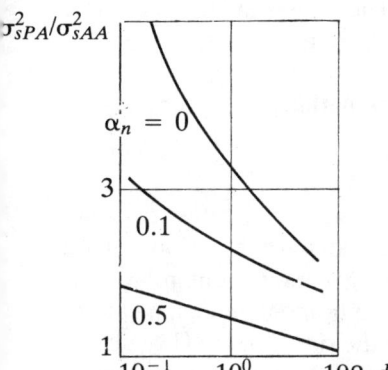

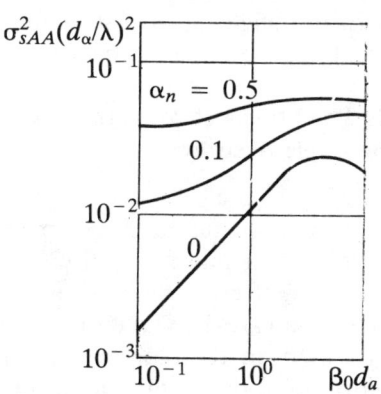

**FIGURE 6.8.** Comparison of amplitude-amplitude and phase-amplitude angle-sensing accuracies for $q_k^2 = 2$.

**FIGURE 6.9.** Angle-sensing accuracy of an amplitude-amplitude system with logarithmic measurement.

The noise background will be inhomogeneous. Systematic angle errors will then arise in addition to random errors. An analysis of the influence of inhomogeneous external noises on the angle accuracy of phase monopulse systems showed that if internal receiver noises are neglected, the rms phase measurement error can be reduced by decreasing the integration time of the detected signals, i.e., by reducing the bandwidth of the low-pass filter [62]. The systematic phase measurement error can be reduced by narrowing the receiver bandwidth prior to detection.

## 6.4 THE EFFECTS ON ANGLE ACCURACY OF AMPLITUDE AND ANGLE FLUCTUATIONS

Fluctuations of the target signals are caused by various interference phenomena associated with the illumination of a complex target, the dimensions of which are significantly greater than the operating wavelength. Modern radars operate for the most part in the decimeter to millimeter wavebands, and target dimensions can be hundreds or thousands of wavelengths. As a result, the interference pattern of the scattered target radiation can be rather complex, and can undergo significant changes for small target movements, causing fluctuations in the target signals.

Detailed studies have shown that these fluctuations may be of two types: fluctuations in the amplitude and phase of the target signals, and fluctuations in the angle of arrival of the signals. These two fluctuations have different characters, and affect the radar in different ways.

### 6.4.1. The Effects of Amplitude and Phase Fluctuations on the Angle Accuracy

Amplitude fluctuations are apparent in changes of the reflected signal amplitude from pulse to pulse. They are caused by fluctuations in the effective target cross section resulting from target movement relative to the radar, and also by variations in atmospheric inhomogeneities, including changes in the meteorological conditions along the signal path. The spectral composition of the amplitude fluctuations varies widely, depending on the type of target.

Because monopulse systems can, in theory, perform angle measurement with a single pulse, and amplitude fluctuation is a pulse-to-pulse phenomenon, it might be expected that amplitude fluctuations should not affect the angle measurement accuracy of monopulse radars. This is not entirely correct, however; amplitude fluctuations actually interfere with monopulse systems, although substantially less than with conical scanning or sequential lobing systems.

This is primarily a result of the fact that modern monopulse systems do not make use of their potential bandwidth by obtaining rapid angle measurements with single pulses, and instead make use of a series of pulses to measure the target angle. To eliminate the dependence on the input signal amplitude, the signals are usually normalized with AGC systems which act on the sum signal. Since AGC systems have finite passbands, and therefore have a limited response time, there is a lag in the signal normalization. As a result, the error signal is, to some extent, dependent on the received signal amplitudes and thus modulated by the signal amplitude fluctuations.

Components of this error-signal amplitude modulation, with frequencies lying in the radar tracking system passband, unavoidably increase the dynamic target angle errors. To counter these fluctuations, the AGC system must be fast, and capable of suppressing at least those fluctuations lying within the tracking system passband. As will be shown later, however, increasing the AGC bandwidth leads to the increase of other angle errors, particularly glint. The AGC passband must, therefore, be selected on the basis of a compromise between these two considerations.

The effects of AGC systems on angle errors of various sources are examined in more detail in Section 6.7.

There are also phase fluctuations in the received signals. Under actual conditions, however, with a complex target moving relative to the radar, these fluctuations are of higher frequencies than the amplitude fluctuations, and their components in the error signal are filtered out by the AGC system, and have very little effect on the angle accuracy. The low-frequency components of the phase fluctuations have the same effect on the angle measurement accuracy as amplitude fluctuations.

### 6.4.2 The Effects of Glint on Angle Accuracy

Glint means the fluctuations in the angle of arrival of the signals reflected from the target; its occurrence was established during the experimentation with the first monopulse radars [21].

It was found during the ensuing theoretical and experimental investigations that glint (angle noise) arises from fluctuations in the slope of the phase front. Because the operation of any system that measures the angle of small targets is based on determining the position of the normal to the phase front, variation in the position of the phase front during these measurements is a factor limiting the angle accuracy; this applies to monopulse systems as well. These considerations are related to the problem of tracking extended targets, although in that case there is not complete correspondence between the null axis of the antenna and the target direction (see Section 3). Due to the fact that the tilt of the phase front is identical to the angle error, surveillance radars are also subject to angle noise.

Angle noise is connected with changes in the apparent position of the target relative to its physical center. This is a result of the multipoint structure of the reflecting surface of an extended target, which causes wandering (flickering) of the apparent center of reflection within limits, which can exceed the target dimensions.

Analyses have shown [21, 38] that when measuring the angle to two coherent signal-point sources displaced from one another by the distance $L$ (see Fig. 6.10), angle errors result which can be calculated from the formula:

$$\Delta\psi/\psi_s = \frac{1 - a^2}{2\{1 + a^2 + 2a \cos[2\pi L \sin(\psi/\lambda) + \alpha]\}} \tag{6.3}$$

where $a$ is the ratio of the signal amplitudes; $L$ is the distance between the sources; $\Delta\psi$ is the angle error as measured from the point midway between them; $\psi_s$ is the angular separation of the sources; and $\alpha$ is the phase difference between the signals.

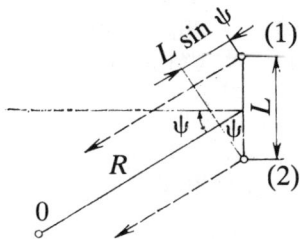

**FIGURE 6.10.** Geometry for analysis of two-point signal source.
(1) source 1    (2) source 2

The form of the phase front formed by a two-point source with identical signals is shown in Fig. 6.11 [21]. When the line joining the sources is perpendicular to the line to the receiver, $\psi = 0$ and (6.3) may be simplified as

$$\Delta\psi/\psi_s = (1 - a^2)/[2(1 + a^2 + 2a \cos \alpha)] \tag{6.4}$$

It follows from (6.4) that the distortion of the wave front caused by the two-point nature of the source depends on the amplitude ratio and the phase shift of the signals, and is greatest when the signals are directly out of phase ($\alpha = 180°$) and have the same amplitude ($a = 1$).

In general, when observing the signals reflected from a two-point target, the amplitude ratio and phase shift of the signals will fluctuate, in turn causing fluctuations in the slope of the phase front reflected from the target. It is convenient to express the changes in the slope of the phase front with probabilistic methods, by determining the probability of the error exceeding a given level. We will assume for the sake of simplicity that there are no amplitude fluctuations and that the phase shift is uniformly distributed from $-\pi$ to $\pi$.

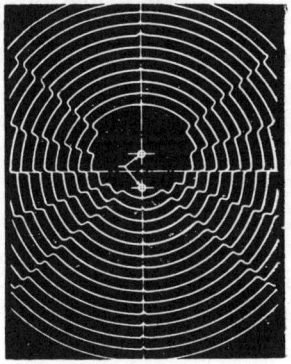

**FIGURE 6.11.** Calculated phase front radiated by a two-point source.

It is known from the theory of random processes that the probability of errors exceeding a threshold $M$ is given by the expression:

$$P(M) = \int_{M}^{M_{\max}} W(M)\, dM \tag{6.5}$$

where $W(M)$ is the probability density of the random quantity $M$; $M = |\Delta\psi/\psi_s|$; and $M_{\max}$ is the maximum possible error.

In the case under consideration:

$$W(M) = W(\alpha)\, d\alpha/dM$$

Calculating the integral (6.5), we obtain the following formula:

$$P\left(\frac{\Delta\psi}{\psi_s}\right) = \tfrac{1}{2} - (1/\pi)\arcsin\left[(1 + a^2)/2a - (1 - a^2)/4Ma\right] \tag{6.6}$$

Plots of the probability of exceeding a given value of the phase front deflection for various signal amplitude ratios are shown in Fig. 6.12. The deflection of the normal to the phase front is measure relative to the normal from a corresponding spherical front generated by a point source, and is normalized by the angular separation of the two-point sources.

Because the antenna of any radar reacts to the phase distribution of the waves at its aperture, distortion of the phase front generates unavoidable angle measurement errors. The nature of these errors is determined by the structure of the target, and the parameters of the antenna and tracking system. The process of tracking a small target may be identified with the tracking of the normal to the phase front of the signals at the antenna, and the tracking error will be given by (6.4). Accordingly, if the spectrum of the phase fluctuations is not wider than the passband of the tracking system, which will be the case for systems tracking slow or stationary targets, the curves of Fig. 6.12 can be used to obtain the probability of the angle error exceeding a certain value for a given amplitude ratio. For example, the probability that the difference between the indicated target angle and the actual angle to the midpoint of the two-point target exceeds the angular separation of the sources, will be approximately 17%, 14%, and 7%, for $a = 0.6$, 0.8, and 0.95, respectively. For $a = 0.2$, the probability of the angle error exceeding the source separation is equal to zero. As the amplitude ratio decreases, the stronger source dominates, and the phase front then appears to originate at that source, which is half the source separation from the midpoint. In the limiting case $a = 0$, the phase front is stable and centered at the source with nonzero amplitude, resulting in a 100% probability of the angle error being one half the source separation, as indicated by the curve $a = 0$ in Fig. 6.12.

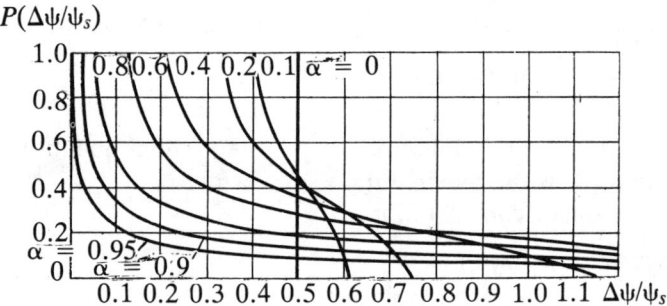

**FIGURE 6.12.** Probability of a phase front error exceeding at a given level for various signal intensity levels.

Knowing the distribution $W(M)$, it is possible to calculate the rms errors. Table 6.2 gives the rms angle errors normalized by the source separation for various two-point source amplitude ratios [38].

If the spectrum of the fluctuations is wider than the tracking system passband, the method of calculating the errors changes. In order to find the expression describing the tracking conditions, it is necessary to first average the received signal according its statistical characteristics and the system bandwidth.

It should be noted that identifying the tracking angle with the direction of the normal to the phase front is correct only when the angle-sensing response is linear. When tracking point targets and small targets, the

**TABLE 6.2**

| $\alpha$ | 0.95 | 0.9 | 0.8 | 0.6 | 0.4 | 0.2 | 0.1 |
|---|---|---|---|---|---|---|---|
| $\sqrt{\overline{\theta^2}}/\psi_s$ | 2.2 | 1.6 | 1 | 0.8 | 0.6 | 0.52 | 0.505 |

angular deviation of the axis from the direction to the geometric center of the reflected signals is sufficiently small that the angle-sensing response can be considered linear.

When tracking targets of large angular extent, it is necessary to consider the actual antenna patterns. Therefore, we will consider the general case of measuring complex extended target angles.

### 6.4.3. Angle Errors for Complex Targets

In the majority of practical cases it is not possible to consider the targets to possess simple geometric forms which lend themselves to mathematical descriptions. Therefore, in analyzing complex target angle errors, certain target models are used. The most common model is an $n$-point target, for which the signal reflected from the complex target is constructed as the combination of the signals reflected from the points on the model, corresponding to reflecting points of the actual target. The amplitudes and phases of the individual signals are assumed to be mutually independent.

By studying the angle-sensing process for a complex target in one plane under static conditions, it is possible to obtain the angle-sensing response for an amplitude-sum-and-difference radar in the form [21, 33]

$$S(\theta) = \frac{\displaystyle\sum_{m=1}^{M} \sum_{n=1}^{N} E_m E_n [F(\theta_0 - \theta_m)F(\theta_0 - \theta_n) - F(\theta_0 + \theta_m)F(\theta_0 + \theta_n)] \cos(\phi_m - \phi_n)}{\displaystyle\sum_{m=1}^{M} \sum_{n=1}^{N} E_m E_n [F(\theta_0 - \theta_m)F(\theta_0 - \theta_n) + F(\theta_0 + \theta_m)F(\theta_0 + \theta_n)] \cos(\phi_m - \phi_n)}$$

(6.7)

where $E_m$ and $E_n$ are the amplitudes of the signals from the $m$th and $n$th target reflectors, and $\theta_m$ and $\theta_n$ are the angular positions of these sources relative to the radar.

Making use of a suitable approximation to the antenna pattern and setting the numerator of (6.7) to zero, we can find the equilibrium position of the tracking system when receiving signals from the given point sources on a complex target, and hence the position of the point which the system tracks. Comparing the position of the equilibrium position with the geometric target center, it is possible to determine the tracking error for each concrete case.

For small off-axis target angles, a linear approximation to the antenna pattern may be used:

$$F(\theta_0 \pm \theta_k) = F(\theta_0)(1 \mp \mu\theta_k)$$

We can then simplify (6.7), describing the equivalent angle-sensing response of the system:

$$S(\theta) = \mu \frac{\displaystyle\sum_{m=1}^{M} \sum_{n=1}^{N} E_m E_n (\theta_m + \theta_n) \cos(\phi_m - \phi_n)}{\displaystyle\sum_{m=1}^{M} \sum_{n=1}^{N} E_m E_n \cos(\phi_m - \phi_n)} \qquad (6.8)$$

It can be shown that (6.8) is the same as the expression for the phase front resulting from an $n$-point target [21]. Analysis also shows that the equation describing the angle sensing of a combination of point sources for small angles is identical to the equation for the resultant Poynting vector.

For the general case, when the angular deviations of the sources are large and the antenna pattern cannot be considered to be linear, the target angle will not correspond with the Poynting vector direction and the normal to the phase front. This can be explained by comparing (6.7) and (6.8), determining the angle-sensing conditions for both cases.

As can be seen from (6.7), when taking account of the nonlinearity of the antenna pattern, the null axis direction is determined primarily by three factors: the field strength of each of the points, the spatial position of the points, and the antenna pattern. The position of the null axis, when measuring the angles of complex targets, will vary with the parameters of the antenna pattern. This leads to a corresponding change in the angle error. The Poynting vector, on the other hand, as described by (6.8), does not depend on the antenna pattern, but is determined by the strength and position of the various sources. Therefore, there is an unavoidable discrepancy between the null axis and the Poynting vector (normal to the phase front) when measuring the angle of a target whose extent exceeds the linear portion of the angle-sensing response. This difference increases with the target dimensions and the narrowness of the antenna pattern.

Under actual conditions, the amplitude and phase relations of the signals from the target sources change in a random fashion during the angle-sensing process, resulting in angle noise and fluctuating angle errors. To determine the angle errors in this case, it is necessary to consider the spectral composition of the angle noise, the AGC parameters, and the passband of the tracking system.

The fundamental conclusions concerning angle sensing of a complex target, with an amplitude-sum-and-difference monopulse system, are applicable to other monopulse systems, and conical scanning systems as well. Methods for measuring and interpreting angle noises are discussed in [21].

### 6.4.4 Reducing the Influence of Glint on Angle Accuracy

Two methods were proposed in 1956 for reducing the effects of angle errors resulting from fluctuations in the slope of the phase front reflected by the target [104]. Both methods are based on the premise that the maximum angle errors occur when the reflected signal amplitude is lowest. The first method involves maintaining the signal level at a specified level with automatic gain control. The second is based on registering only those angle errors formed by signals exceeding the given threshold; weaker signals are rejected by the tracking system.

The use of frequency-modulated signals was investigated in [54], and it was shown that the accuracy of a monopulse radar may be significantly improved if the carrier frequency is changed from pulse to pulse. This conclusion was supported by later experimental studies. It was noted that due to the negative correlation between the slope deviation of the phase front and the amplitude of the reflected signal, the accuracy of target angle tracking could be increased by suppressing error signals resulting from weak input signals. Employing frequency agility from pulse to pulse succeeds in reducing the angle errors only when the carrier frequencies are sufficiently far apart so that the angular signal fluctuations at each frequency are uncorrelated.

The coefficients of correlation between the tracking error and the carrier frequency are shown in Table 6.3, as obtained from a computer simulation based on an advanced model of the fluctuations of the apparent radar target angle [104]. From the results of this simulation, it can be seen that it is advisable to use more than two carrier frequencies for target angle measurement.

The simulation in [104] also supports the strong negative correlation between the occurrence of large tracking errors and the amplitude of the received signals when operating at different carrier frequencies. It was shown that the incidence of large positive or negative angle errors cor-

**TABLE 6.3**

| $f_1$, GHz | $f_2$ GHz | | | | | |
|---|---|---|---|---|---|---|
| | 10.00 | 9.98 | 9.96 | 9.94 | 9.92 | 9.90 |
| 10.00 | 1.0 | 0.138 | 0.404 | 0.186 | 0.326 | 0.163 |
| 9.98 | | 1.000 | −0.023 | 0.611 | −0.076 | 0.163 |
| 9.96 | | | 1.000 | 0.025 | 0.457 | 0.156 |
| 9.94 | | | | 1.000 | 0.079 | 0.008 |
| 9.92 | | | | | 1.000 | 0.102 |
| 9.90 | | | | | | 1.000 |

responds exactly with the reception of weak signals from the tracked source. It was established through computer calculations that when operating at two carrier frequencies, the minimum rms target angle errors are obtained with the combination of 10 and 9.9 GHz.

The results of a computer analysis for spatially separated receivers are presented in Table 6.4, where $\sigma_1$ and $\sigma_2$ are the rms target angle errors at the two reception points, and $\sigma_{1,2}$ is the rms angle error which is produced by joint processing of the data generated at the two receivers. It was demonstrated that, if the receivers are separated by a distance exceeding the dimension of the distorted phase front of the signals, then each receiver was subject to independent target angle fluctuations. As a result, averaging the respective angle errors reduces the target angle errors resulting from glint. These conclusions are supported by the results of the simulation, which are presented in Table 6.4.

Analysis shows that the sensitivities of conically scanned, and phase and amplitude monopulse systems to angular fluctuations are approximately equal. The angle errors resulting when tracking complex targets with an

**TABLE 6.4**

| distance between receivers (meters) | correlation coefficient | $\sigma_1$ | $\sigma_2$ | $\sigma_{1,2}$ |
|---|---|---|---|---|
| 1 | 0.383 | 12.0 | 11.8 | 6.1 |
| 3 | 0.001 | 12.9 | 11.7 | 5.5 |
| 6 | 0.045 | 15.5 | 11.7 | 6.4 |

amplitude phase monopulse system, are significantly larger due to crosstalk between the angle-sensing channels. Therefore, it is not advisable to use amplitude-phase systems for tracking complex targets.

## 6.5 THE INFLUENCE OF THE INTERNAL NOISE OF MONOPULSE RECEIVERS ON ANGLE ACCURACY

There are various types of internal system noise. The basic noises are: thermal noise, associated with the random motion of electrons in conductors at temperatures above absolute zero; shot noise, caused by the random nature of the arrival of electrons at the anode (grid) of a tube, or the random barrier crossings in a semiconductor; and flickering noise resulting from the nonuniform emission of electrons in electronic and semiconductor devices.

There is always noise in a receiver, and it is possible to establish a natural limit to the attainable sensitivity and measurement accuracy. As it is amplified along with the useful signal, the noise destroys the structure of the angle error signal, giving rise to fluctuations and creating a region in which the system is insensitive to the measured parameter.

As has been shown [21, 38], when measuring a single angular coordinate with an amplitude-sum-and-difference monopulse system which normalizes the error signal with an ideal AGC system, the variance of the angle errors is given by the expression:

$$\sigma_\theta^2 = n_0 \Delta f / k^2 E^2 \mu^2 \tag{6.9}$$

where $\Delta f$ is the effective passband of the IF amplifier; $n_0$ is the spectral power noise density at the IF amplifier input; $\mu$ is the slope of the angle-sensing response.

As was shown earlier, for amplitude-comparison systems, the optimum squint angle is such that the beams intersect at their half-power level, with $\theta_0 = \theta_3/2$ and $\mu = 2/\theta_3$. Using these parameters, along with the value of the noise-power density and the tracking system passband in (6.9), the rms tracking error resulting from receiver noise can be shown to be

$$\sigma_\theta = 0.5\theta_3 / \sqrt{P_s/P_n} \tag{6.10}$$

where $P_s$ is the average signal power at the receiver output, and $P_n$ is the noise power in the tracking system passband $\Delta F_{ts}$.

It should be noted that the internal noises of modern receivers are associated with causes which cannot be completely eliminated, so that the signal processing is unavoidably degraded, and there is a natural limit to the improvement in angle accuracy. Therefore, the angle errors resulting from internal receiver noise constitute a limit on the angle accuracy attainable in angle-sensing radar systems. For large signal-to-noise ratios, (6.10) determines the potential accuracy for such systems. To establish the connection between the limiting accuracy and the passband of the receiver and tracking system, we determine the signal-to-noise ratio, taking account of the radar waveform. With pulse transmission, the ratio is

$$P_s/P_n = P_p \tau_s f_p \Delta f / n_0 \Delta F_{ts} \Delta f \qquad (6.11)$$

where $P_p$ is the signal pulse power at the receiver input; $\tau_s$ is the pulse duration; $f_p$ is the pulse repetition frequency.

When the receiver is matched ($\tau_s \Delta f \approx 1$) and $P_{nr} = n_0 \Delta f$ is the noise power in the receiver passband, (6.11) takes the form:

$$P_s/P_n = (P_p/P_{nr}) f_p / \Delta F_{ts}$$

The ratio $f_p/\Delta F_{ts}$ is the number of integrated pulses in the received signal.

With continuous wave transmission:

$$P_s/P_n = P_s/n_0 \Delta F_{ts}$$

Consequently,

$$\sigma_\theta = \theta_3/k_n$$

where $k_n = 2\sqrt{(P_p/P_{nr}) f_p / \Delta F_{ts}}$ for pulse radars, and $k_n = 2\sqrt{P_s/n_0 \Delta F_{ts}}$ for CW radars.

Clearly, the higher the factor $k_n$, the greater the possible radar angle accuracy.

It should be noted that receiver noise produces different effects in conically scanned radars than in monopulse radars. This is explained by the quantitative difference in the receiver noises penetrating the passband of the tracking system, and the difference between the antenna structures. In monopulse systems, where the received signals are not artificially modulated, the only frequency components of the receiver noise affecting the tracking system are those located within a band twice the width of the passband and centered at each frequency component of the received signal (see Fig. 6.13). In conically scanned systems, the received signals are modulated at the scanning frequency, and there are now two bands, each twice the passband width, located on either side of each frequency component at a distance equal to the scanning frequency (see Fig. 6.14). As a result, the effective noise bandwidth of a conically scanned system is twice that of a monopulse system, so that the noise power at the output of a conically scanned system will be twice that of a monopulse system, even if they both have equal signal-to-noise ratios at the IF amplifier and identical tracking system passbands. Therefore, monopulse systems have a greater theoretical accuracy capability than conically scanned systems.

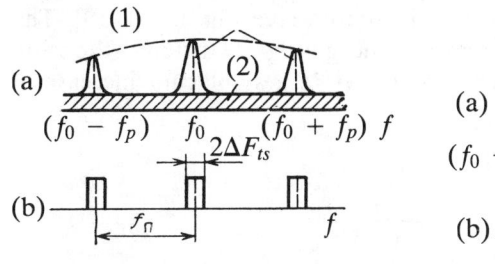

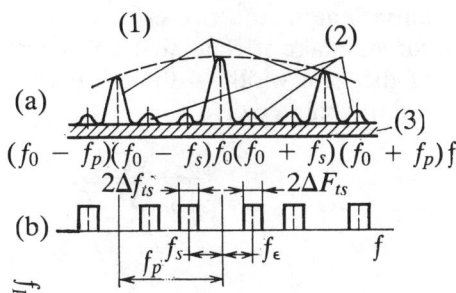

**FIGURE 6.13.** Passage of signal and noise in a monopulse radar:
    (a) signal and noise spectra
    (b) tracking system passband

    (1) signal spectrum lines
    (2) noise

**FIGURE 6.14.** Passage of signal and noise in conical-scan radar:
    (a) signal and noise spectra
    (b) tracking system passband

    (1) basic signal spectral lines
    (2) sidebands caused by scanning
    (3) noise

Monopulse systems also offer the advantage of illuminating the target with the maximum beam power, greatly reducing the signal losses associated with the fact that conically scanned beams are not aligned with the antenna's electric axis. As a result, monopulse receivers have a higher signal-to-noise ratio than conically scanning systems.

Considering the greater slope of the angle-sensing response, along with the factors discussed above, it is possible to obtain a value of 5.2 dB for the energy gain of a monopulse system over a conically scanned system relative to receiver noise [21].

At large ranges, the signal-to-noise ratio may be too low for normal tracking, and the target may be lost. The performance of some receivers is estimated according to the "loss rate," which is the frequency (number of times per second) with which the tracking error exceeds the acceptable threshold [21].

In deriving (6.10) it was assumed that the gating process is ideal, so that the noises enter only during the time in which the useful signal is received. If the duration of the gating pulse, in fact, exceeds the signal pulse length, the rms error $\sigma_\theta$ will be increased (see Fig. 6.15) [38]. The curves make it clear that an increase in the gating coefficient—the ratio of the gate width to the pulse width—results in a significant increase in the tracking error.

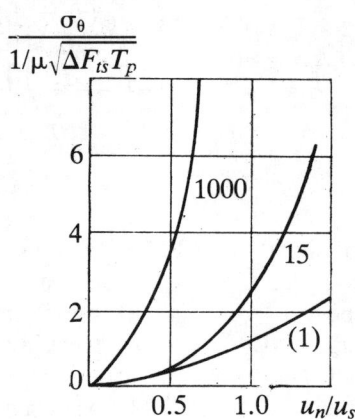

**FIGURE 6.15.** Dependence of autotracking error on gating coefficient, $p$ = (gate width)/(pulse width).

(1) $p = 1$   $u_n/u_s$ : ratio of rms noise voltage to signal

This method of estimating the tracking errors, valid for $P_{sr}/P_n \gg 1$, may also be used for other types of monopulse systems. The angle-sensing errors, arising from internal receiver noises in phase and amplitude monopulse systems, are on the same order.

It was shown in [21] that for a phased array amplitude monopulse radar with square-law detection in the receiver, the variance of the phase error in the angle-sensing system, arising from internal receiver noises, is determined by the expression:

$$\sigma_{\Delta\phi}^2 \approx (\sigma_n^2/2N^3E_m^2)F(\psi) = F(\psi)/4N^3q^2$$

where

$$F(\psi) = \psi^3[4\psi \sin^2(\psi/2) + \sin 2\psi - 2 \sin \psi]/[\psi \sin \psi - 4 \sin^2(\psi/2)]^2$$

$\psi = N\delta\phi$ ($N$ is the number of array elements); $\delta\phi$ is the phase increment which steers the beam through $\theta_0$; $\sigma_n^2$ is the variance of the noises (it is assumed that the noises in the channels are independent and Gaussian);

and $q^2 = P_s/P_n = E_m^2/2\sigma_n^2$ is the signal-to-noise power ratio in each receiver channel.

The function $F(\psi)$ is plotted in Fig. 6.16. For beam steering angles less than half the beamwidth, which corresponds to $\psi < \pi$, we have $F(\psi) \approx 24$, and, with large $N$, we may write

$$\sigma_{\Delta\phi}^2 = 6/q^2 N^3$$

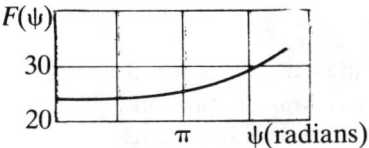

**FIGURE 6.16.** Graph of function $F(\psi)$.

## 6.6 THE TOTAL ANGLE ERROR RESULTING FROM INTERNAL AND EXTERNAL ANGLE NOISES

In order to grasp better the character of the angle error components resulting from various noise sources, the rms values of these components *versus* target range are plotted qualitatively in Fig. 6.17 [21]. The position of the curves depends on many parameters and the characteristics of the actual radar and target. The curves in the figure correspond to a typical tracking radar, and show that only the error components arising from glint and receiver noise vary with the range. The influence of the receiver noise grows in proportion to the square of the range as long as the receiver is unsaturated.

The amplitude noise acts to modulate the signal amplitude about its average value, and, therefore, the angle errors it causes do not depend on the target range when the input signal level is maintained at a constant value over a wide dynamic range by an AGC system in the receiver. Glint is a function of the angular dimension of the target, and its contribution, therefore, is inversely proportional to the target range.

In addition to these noises, there are also errors caused by zones of low sensitivity, backlash and friction in the antenna drives, zero drift in the dc amplifiers, imbalance in the drive amplifiers, and so forth.

The angle errors caused by these noises in the tracking system are independent of the target characteristics and range, and are wholly determined by the system design particulars, the accuracies of the employed devices, and the stability of the corresponding circuits.

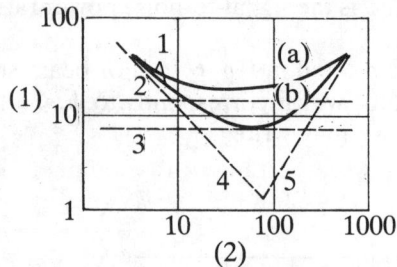

**FIGURE 6.17.** Dependence of rms autotracking errors on target range.
  1:  total error for (a) conical-scan, and (b) monopulse radar
  2:  error due to amplitude fluctuations
  3:  tracking system error
  4:  errors due to shifts in center of reflection
  5:  errors due to receiver noise

(1) relative rms error   (2) relative range

With small $N$ the variance of the phase error is given by the formula:

$$\sigma^2_{\Delta\phi} = 6/q^2(N^3 - N)$$

Knowing the phase error $\Delta\phi$, the angle error may be estimated with the basic equation:

$$\Delta\phi \equiv k_\lambda l_a \cos\theta_r \tag{6.12}$$

where $\theta_r$ is the angle of arrival of the signals; and $l_a$ is the distance between the array elements.
  Solving (6.12) for $\theta_r$, we obtain

$$\theta_r = \cos^{-1}(\Delta\phi/k_\lambda l_a)$$

Differentiating with respect to $\Delta\phi$ and changing from incremental to rms errors, it is possible to estimate the angle measurement error depending, on the phase measurement errors caused by internal receiver noises:

$$\sigma_{\theta_r} = \sigma_{\Delta\theta}/\sqrt{(k_\lambda l_a)^2 - (\Delta\phi)^2} \tag{6.13}$$

The total angle measurement error in a radar system may be described by the rms value of all of the mutually independent errors. The heavy curves in Fig. 6.17 show the qualitative dependence of the total angle error on target range for a conical-scan radar (a) in which the amplitude noise exceeds the tracking system noise, and for a monopulse radar (b). If the tracking system noise is greater than the amplitude noise, then the advantage of a monopulse system over sequential systems is insignificant.

The curves of Fig. 6.17 show that the autotracking error is determined by target glint at short ranges. At intermediate ranges, the angle errors are caused primarily by amplitude fluctuations and distortion originating in the tracking system. At large ranges, the dominant component of the angle error is internal receiver noise.

## 6.7 THE INFLUENCE OF AUTOMATIC GAIN CONTROL ON ANGLE ACCURACY

A target being tracked by a radar may be thought of as an element of a closed tracking loop. Therefore, any changes in the strength of the signal reflected from the target may be considered to be changes in the gain of the closed loop, and, hence, have a direct effect on the accuracy of the target coordinate measurements. The receiver's AGC system is, therefore, extremely important, as it must maintain the gain of the loop at a constant level with the required accuracy.

With a slow AGC system, the gain may be held at a constant average value. Such a system does not react to rapid signal variations, however, which still affect the closed loop gain and cause fluctuations in the error signal. A fast AGC reacts to most amplitude fluctuations and can keep the tracking loop gain constant over a wide range of reflected signal strength.

We will consider an amplitude-sum-and-difference monopulse system to examine the influence of the AGC system on the angle accuracy. We will limit the discussion to tracking in one coordinate, with errors not lying outside the linear portion of the angle-sensing response.

It was shown earlier that the signal reflected from a complex target may be modeled as the sum of the signals from $M$ elementary reflectors comprising the target:

$$\underline{E}(t) = \sum_{m=1}^{M} E_m \exp[i(\omega t + \phi_m)]$$

Accordingly, the signals received in the two channels of an amplitude-comparison monopulse system may be expressed as follows:

$$\underline{E}_1(t, \theta) = \sum_{m=1}^{M} E_m F(\theta_0 - \theta_m) \exp[i(\omega t + \phi_m)] \tag{6.14}$$

$$\underline{E}_2(t, \theta) = \sum_{m=1}^{M} E_m F(\theta_0 + \theta_m) \exp[i(\omega t + \phi_m)] \tag{6.15}$$

with the approximation:

$$F(\theta_0 \mp \theta_m) = F(\theta_0)(1 \pm \mu\theta_m)$$

The angular position $\theta_m$ of the $m$th reflector, relative to the antenna axis, may be expressed as the vector sum (see Fig. 6.18):

$$\theta_m = \theta + \Delta\theta_m$$

where $\theta$ is the angular position of the target center relative to the axis, and $\Delta\theta_m$ is the angular position of the $m$th reflector relative to the target center.

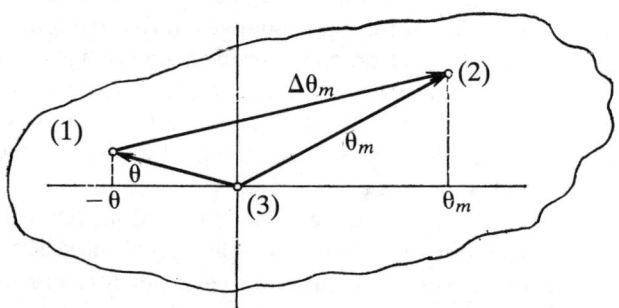

**FIGURE 6.18.** Target model.
   (1) target center   (2) $m$th reflector   (3) axis

The angle $\theta_m$ may be expressed as the algebraic sum of the projections of the corresponding angles onto the tracking plane:

$$\theta_m = -\theta + \Delta\theta_m \tag{6.16}$$

Using (6.16) in (6.14) and (6.15), and setting $F(\theta_0) = 1$ for simplicity, we obtain

$$\underline{E}_1(t, \theta) = (1 - \mu\theta)\underline{E}(t) + \underline{E}_g(t) \tag{6.17}$$

$$\underline{E}_2(t, \theta) = (1 + \mu\theta)\underline{E}(t) - \underline{E}_g(t) \tag{6.18}$$

where

$$\underline{E}(t) = \sum_{m=1}^{M} E_m \exp[i(\omega t + \phi_m)] = E \exp[i(\omega t + \phi)]$$

$$\underline{E}_g(t) = \sum_{m=1}^{M} \mu\Delta\theta_m E_m \exp[i(\omega t + \phi_m)] = E_g \exp[i(\omega t + \phi_g)]$$

$E$ and $E_g$ are the resulting signal amplitudes; $\phi$ and $\phi_m$ are the resulting signal phases.

It is easy to see that the first terms of (6.17) and (6.18) contain useful information about the position of the target relative to the axis, while the second terms are conditioned by the finite extent of the target and are interfering components (glint). With slow changes in the relative positions of the radar and target, and with a sufficiently large number of elements $M$, the process $E(t)$ and $E_g(t)$ may be considered to be stationary processes [38], and

$$\overline{\underline{E}(t)\underline{E}_g(t)} = 0$$

Neglecting the effects of internal receiver noises, with (6.17) and (6.18) we find at the input of the sum and difference channels:

$$\underline{E}_s(t, \theta) = \sqrt{2\underline{E}(t)} \qquad (6.19)$$

$$\underline{E}_d(t, \theta) = \sqrt{2[\mu\theta\underline{E}(t) - \underline{E}_g(t)]} \qquad (6.20)$$

We will consider the further signal processing with two types of AGC: wideband and narrowband.

With instantaneous gain control, the normalized signal may be written as

$$\underline{E}_{AGC} = \sqrt{2}Ek_sk_{AGC}$$

where $k_s$ is the gain of the sum channel, and $k_{AGC}$ is the AGC gain.

After the signals have been normalized, the output of the sum-and-difference channels may be expressed, to within a constant, by

$$u_s(t, \theta) = \underline{E}_s(t, \theta)/\underline{E}_{AGC} = (1/k_{AGC})\exp[i(\omega_c t + \phi)]$$

$$u_d(t, \theta) = \underline{E}_d(t, \theta)/\underline{E}_{AGC}$$

$$= (k_d/k_s k_{AGC})\{\mu\theta \exp[i(\omega_c t + \phi)]$$

$$- (E_g/E)\exp[i(\omega_c t + \phi_g)]\}$$

where $k_d$ is the gain of the difference channel. This leads to a phase-detector output given by

$$U(\theta) = \text{Re } \underline{u}_s(t, \theta)\underline{u}_d^*(t, \theta) \qquad (6.21)$$

$$= (k_d/k_s k^2_{AGC})[\mu\theta - (E_g/E)\cos(\phi - \phi_g)]$$

With a narrowband AGC, when the passband of the AGC system is many times narrower than the spectrum of the reflected signal fluctuations, the normalized signal takes the form:

$$E_{\text{AGC}} = \sqrt{2}k_s k_{\text{AGC}}\overline{E}$$

where $\overline{E}$ is the average value of the envelope obtained by passing the signal through a narrowband filter.

Then, with (6.19) and (6.20), we obtain

$$\underline{u}_s(t, \theta) = \underline{E}_s(t, \theta)/E_{\text{AGC}} = (E/k_{\text{AGC}}\overline{E})\exp[i(\omega_c t + \phi)]$$

$$\underline{u}_d(t, \theta) = \underline{E}_d(t, \theta)/E_{\text{AGC}}$$

$$= (k_d/k_s k_{\text{AGC}})\{(E/\overline{E})\mu\theta\exp[i(\omega_c t + \phi)]$$

$$- (E_g/\overline{E})\exp[i(\omega_c t + \phi_g)]\}$$

For the error signal with a narrowband AGC, we then obtain

$$u(\theta) = \text{Re}[\underline{u}_s(t, \theta)\underline{u}_d^*(t, \theta)] \tag{6.22}$$

$$= [(k_d/k_s k^2_{\text{AGC}})/\overline{E}^2][E^2\mu\theta - EE_g\cos(\phi - \phi_g)]$$

Comparison of (6.21) and (6.22) shows that with $\theta = 0$, when the center of a complex target is on the axis, the error signal does not equal zero, but, with a wideband AGC system, is given by

$$u'(\theta) = (k_d/k_s k^2_{\text{AGC}})(E_g/E)\cos(\phi - \phi_g)$$

and, with a narrowband AGC system, by

$$u''(\theta) = (k_d/k_s k^2_{\text{AGC}})(EE_g/\overline{E}^2)\cos(\phi - \phi_g)$$

When tracking a point target:

$$\Delta\theta = 0, \quad E_g = 0, \quad \text{and} \quad u'(\theta) = u''(\theta) = 0$$

Thus, these expressions confirm the fact that errors, arising from glint, result when tracking a complex target. These errors depend on the type of AGC system.

The influence of glint on the angle accuracy is somewhat greater with a rapid AGC than with a narrowband AGC. In practical systems, this difference is usually small, and the glint errors, with a rapid AGC, do not exceed those which result with slow AGC, by more than a factor of 2 or 3 [21].

With a narrowband AGC, as follows from (6.22), the sum signal fluctuations affect the angle accuracy. Limiting the sum signal will reduce these fluctuations; in this case, the error signal is

$$u(\theta) = (k_d u_0/k_{\text{AGC}}k_s\overline{E})[E\mu\theta - E_g\cos(\phi - \phi_g)] \tag{6.23}$$

where $u_0$ is the limiting level at the output of the sum channel.

It is evident that the signal amplitude fluctuations affect the angle accuracy of a system with narrowband AGC.

Representing the reflected signal amplitude as the sum of an average and fluctuating components, (6.23) may be written

$$u(\theta) = (k_d u_0 / k_{AGC} k_s \overline{E})[\overline{E}\mu\theta + \tilde{E}u\theta - E_g \cos(\phi - \phi_g)] \qquad (6.24)$$

where $\tilde{E}$ is the fluctuating component of the signal relative to the average value $\overline{E}$.

For the case of a fluctuating point target signal:

$$U(\theta) = (k_d u_0 / k_{AGC} k_s \overline{E})[\overline{E}\mu\theta + \tilde{E}u\theta]$$

Therefore, signal fluctuations introduce systematic and random errors in an angle measurement system which degrade the target angle accuracy. The system, in this case, is no longer a true monopulse system, because it does not satisfy the signal normalization requirement. The angle errors resulting from amplitude fluctuations, as opposed to glint errors, depend on and increase with the systematic target tracking error (lag).

With the appropriate AGC system parameters and fluctuation distributions, the angle errors caused by amplitude fluctuations and glint may be estimated quantitatively. This is done by the formulas which result by setting (6.21) and (6.24) equal to zero:

for wideband AGC:

$$\mu\theta = (E_g/E)\cos(\phi - \phi_g) \qquad (6.25)$$

for narrowband AGC:

$$\mu\theta = [E_g/(\overline{E} + \tilde{E})]\cos(\phi - \phi_g) \qquad (6.26)$$

Examination of (6.25) and (6.26) shows that, in a system with fast AGC, amplitude fluctuations cause no error, while the errors caused by changes in the slope of the phase front are somewhat greater than with slow AGC. With slow AGC, however, the error resulting from amplitude fluctuations increases.

The curves in Fig. 6.19 represent the experimental dependencies of the total error on the systematic error of a monopulse system tracking the target center, where the systematic error $\Delta L$ is normalized with respect to the target dimension $L_t$ [21]. These experiments showed that when the systematic error is about one half the target length, the total error arising from amplitude fluctuations and glint is about equal, regardless of the AGC bandwidth. In this case, when the target's angular extent is small and the influence of internal noises in the automatic tracking system increases, the effects of glint on the angle accuracy are insignificant. For large systematic tracking errors, the fluctuation errors are greater in a system with fast AGC than in one with slow AGC. In some cases of short range operation, it is advisable to use narrowband AGC to offset the increasing glint effects.

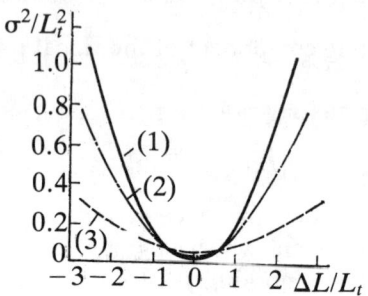

**FIGURE 6.19.** Dependence of rms error caused by amplitude and angle fluctuations on the antenna pointing error, for various AGC bandwidths.

  (1) without AGC   (2) AGC—1 GHz   (3) AGC—12 GHz

During the experiments, it was also noted that low-frequency amplitude fluctuation components penetrated in the tracking system passband, and caused additional errors. This reaffirms the fact that very low-frequency amplitude fluctuations, lying in the passband of the tracking system, affect all angle-sensing radars, including monopulse systems, to an extent determined by the AGC characteristics and the tracking system passband.

  Due to the fact that the systematic tracking errors are increased by the longer tracking time constants of systems with slow AGC, the passband of the tracking system must be extended in order to minimize the time needed to smooth the angle error. Widening the passband, however, increases the influence of glint on the system, and, under certain conditions, the target may be lost (see Fig. 6.20). The influence of internal automatic tracking noises increases, and, in turn, the effect of amplitude fluctuations grows with slow AGC. The greater the internal radar noise and the wider the passband necessitated for tactical reasons, the greater the advantage of fast AGC over slow AGC.

  The effects of glint have thus been shown both theoretically and experimentally to decrease as the time constant of the AGC is increased. However, with slow AGC, the errors, caused by low-frequency amplitude fluctuations modulating the error signal, are increased. This error is proportional to the tracking error and grows rapidly as the tracking error increases. Internal automatic tracking noises further degrade the tracking performance with slow AGC.

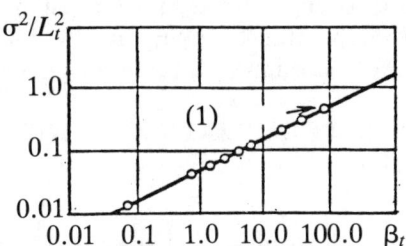

**FIGURE 6.20.** Experimental dependence of rms tracking errors on tracking system passband with a slow AGC and zero lag error.

$\beta_t$:  ratio of tracking system passband to half-power bandwidth of amplitude fluctuations

(1) target loss

The operation of a tracking system is improved by using fast AGC, which effectively eliminates the influence of amplitude fluctuations in the target signals. Because this increases the effects of glint on the angle accuracy, a compromise must be struck in selecting AGC parameters. The tracking system passband should be kept as narrow as possible, because the sensitivity of the system to internal and external sources of error increases with the width of the passband, and the accuracy suffers accordingly. The block diagram in Fig. 6.21 aids in obtaining good accuracy characteristics [56].

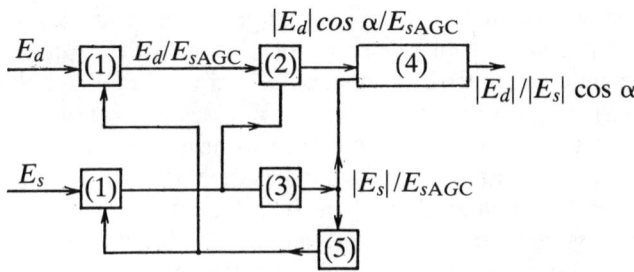

**FIGURE 6.21.** Block diagram of a sum-and-difference monopulse system with fast (wideband) AGC.

(1) IF amplifier  (4) division circuit
(2) phase detector  (5) AGC
(3) amplitude detector

In a typical system, as has been shown, the IF amplifier outputs are normalized with respect to the average component of the sum signal. As a result, the instantaneous sum-to-difference signal ratios may not be accurate, which gives rise to angle errors. In the system of Fig. 6.21, this shortcoming is corrected by transforming (by division) the normalized difference signal into the true sum-to-difference signal ratio.

## 6.8 THE EFFECTS OF DEPOLARIZATION OF THE TARGET ECHO ON ANGLE ACCURACY

The polarization of an electromagnetic wave is determined by the orientation of the electric field vector. Most radar antennas are designed for linear polarization, in which the electric field vector is either vertical or horizontal. This is partly explained by the fact that such antennas are the easiest to construct.

Much more rarely, radar systems will utilize circular polarization, in which the electric field vector rotates in the plane perpendicular to the direction of propagation, in either a right- or left-handed sense, at the signal frequency. The use of circular polarization is advantageous when the signals must propagate through ionospheric layers and in the presence of interfering meteorological effects (rain, snow, thunder clouds, *etcetera*). As a rule, however, the signal will not retain its original polarization characteristics, but will undergo one or another distortion based on the operating conditions. We will now examine the causes of signal depolarization, and its effects on the system angle accuracy.

### 6.8.1 The Causes of Reflected Signal Depolarization

One source of signal depolarization is the propagation medium. If the signals propagated in a vacuum or plasma with no magnetic fields, then they would retain their original polarization. Radar signals must propagate through the atmosphere, however, where various nonuniformities are encountered, including plasma clouds (ionized gas) in the Earth's magnetic field. As a result, the signal polarization is distorted. For example, the slope of the polarization plane of a linearly polarized wave will change when passing through plasma-filled areas. This rotation of the polarization plane is the same for both the transmitted and reflected signal, so that the overall rotation is twice as large for two-way propagation as for a one-way path. Because this rotation is not constant and cannot be determined in advance, it is possible for the polarization of the reflected signal at the antenna to be orthogonal to that of the transmitted signal, and, as a result, signal reception and target tracking are either impossible, or attended by large errors. To avoid this problem, most long range radars designed for tracking targets in space, necessitating ionospheric propagation, use circular polarization. The additional rotation caused by ionospheric effects does not substantially alter the polarization of the signal, leaving the system's tracking capability unimpaired.

The propagation medium is not the sole cause of signal depolarization. Another, even more important source of radar signal depolarization, is the radar target itself. In the overwhelming majority of cases, a radar target has a complex reflecting surface, along which complex currents are induced under illumination. As a result of the interfering surface currents, the reflected signal will itself have a complex structure, which extends to the signal polarization. Instead of possessing a stationary polarization component, the reflected signal will be "partially polarized," with a stable polarization component, but also fluctuating components with no representation on the Poincaré polarization sphere. The extent of depolarization resulting from this process is largely determined by the structure of the target and the parameters, and dynamics of its motion. This has led to a widespread investigation of target depolarization properties in an attempt to use them in target classification and recognition [16].

In operation against low-flying targets, when the elevation angle is sufficiently small, the land or water surface may also contribute to depolarization of the radar signal. As is well known, the coefficients of reflection and refraction of a given surface are strongly dependent on the polarization of the incident radiation. Decomposing the signal polarization into two components, one parallel to the surface and the other perpendicular, it can be shown that the reflected signal will have a distorted polarization (usually elliptical). Interference of the signal reflected directly from the target with that reflected from the land or water surface will further distort the polarization of the combined signal incident on the radar antenna.

Thus, in general, the received target signal will be depolarized, i.e., have a polarization different from that with which the radar is operating. In addition to components coinciding with the operating polarization, there will be significantly different stable components, and also irregular (random) polarization components.

## 6.8.2 The Effects of Depolarization on Angle Accuracy

The influence of signal depolarization makes itself apparent in two ways: reduction in signal strength at the receiver input, and cross-polarization of the signals relative to the receiving antennas. The reduction in signal power results from polarization mismatch of the received signal. The sensitivity of the receiver is reduced accordingly, which, in turn, reduces the detection range and tracking accuracy through a lower signal-to-noise ratio. When the polarization of the received signal is orthogonal to the operating polarization, target detection and tracking, in general, become impossible.

The mechanism by which antenna cross-polarization affects angle accuracy is more complex. It is connected with the imbalances in the amplitude-phase characteristics of the receiver channels when tracking a target with signals of mismatched polarization, which is examined in more detail in Chapter 7.

The reduction in angle accuracy resulting from reduced signal strength may be lessened by employing radars with adjustable polarization or orthogonally polarized receiving channels. With the first approach, the operating polarization may be adjusted until it matches that of the signal, eliminating the power loss resulting from depolarization. The second method achieves the same effect as receiving with matched and cross-polarization simultaneously. This substantially complicates the structure of the radar, because additional orthogonal channels are required. Realization of this design in a monopulse radar is even more complicated [21].

# Chapter 7
# Angle Errors Caused by Unequal Response of Receiver Channels

## 7.1 BASIC SOURCES OF INSTRUMENTATION ERRORS

As has been shown, the monopulse method is based on the simultaneous reception of signals by two or more receiver channels, with subsequent comparison by amplitude or phase. The angle measurement accuracy is accordingly determined largely by the extent to which the characteristics of each pair of channels are identical.

Throughout the preceding analysis of the reception and processing of signals in monopulse systems, we have assumed that the receiving channel characteristics were identical. This allowed us to examine the design principles of monopulse systems in conditions under which the full potential of the monopulse method could be realized. In practice, the construction and circuitry of monopulse systems are subject to various shortcomings which disturb the equality of the receiver channels. In some cases, these variations develop over the operating lifetime of the system as a result of component aging, and climactic or mechanical effects. In these cases the angle accuracy will be determined by the character and extent of the instrumentation errors.

In addition to the basic sources of instrumentation errors in monopulse systems, there are also deficiencies in antenna pattern generation, variations in the pattern with received signal polarization, receiver channels with nonidentical amplitude-phase responses, and improper operation of the frequency and automatic gain controls.

## 7.2 THE EFFECTS OF IMPERFECT ANTENNA PATTERNS ON MONOPULSE ANGLE ACCURACY

The character of the antenna pattern is determined by the amplitude-phase distribution of the fields excited over the aperture. Inasmuch as the antennas of monopulse systems must form symmetrical pairs of patterns in each plane of measurement, the aperture fields must meet certain conditions. Thus, in amplitude-comparison systems it is important to obtain a symmetric amplitude distribution in the aperture, and in phase-comparison systems, an odd-symmetric phase distribution. Failure to meet these conditions leads to antenna patterns which are not matched to the angle-sensing method employed, asymmetry in amplitude and phase, and a number of other problems. These, in turn, lead to additional angle measurement errors.

One of the fundamental causes of asymmetric field excitation is inaccurate antenna manufacturing. Deficiencies in antenna construction cause distortion in the aperture fields which can not be corrected, and, as a result, lead to the formation of patterns which do not match those desired.

Errors in the aperture excitation may be either systematic or random. Systematic errors may be caused by shadowing of the aperture by feeds and mounting components, diffraction at the edges of the reflector, and crosstalk between the monopulse channels. Random errors may result from random deformations in the reflector surface, construction errors, temperature or wind force gradients around the antenna causing phase oscillations, and so on.

Systematic errors are, as a rule, identical in antennas of the same type built with established technology, and therefore, can be calculated beforehand with a specified accuracy. It is not possible to determine the random errors in advance, because they vary within wide limits from antenna to antenna. Estimates of the random errors are thus usually obtained with statistical methods.

The effects of systematic errors on the pattern may be determined by introducing them into the aperture field distribution. It may be shown that a linear phase error over the aperture leads to a corresponding deflection of the pattern. If the phase error exhibits a quadratic spatial distribution, the beam is defocused (broadened).

Random phase and amplitude errors in the aperture excitation cause increased sidelobes, angular pattern deflections, and reduced directivity. The effects of random phase and amplitude errors in the aperture distribution on the intended antenna characteristics are examined from a statistical viewpoint in several works. In order to illustrate the results, we will examine two examples—a two-dimensional antenna array and a parabolic antenna [21].

### 7.2.1 The Effect on Angle Accuracy of Errors in the Distribution over an Array Aperture

In our analysis of the angular errors introduced by the elements of a phased array, we will assume that the antenna is a rectangular array in the $XY$-plane [21], and is used with an amplitude-sum-and-difference monopulse system.

A shift in the axis caused by errors in the array excitation leads to an angular error in the $XZ$-plane:

$$\Delta\theta_{x0} = \theta'_{x0} - \theta_{x0} \tag{7.1}$$

where $\theta'_{x0}$ and $\theta_{x0}$ are the boresight angles with and without the errors, respectively.

Monopulse systems utilizing array antennas usually track by means of the value of the error signal, without zeroing $S(\theta)$. This results in an additional source of errors, since the angle-sensing sensitivity changes. If the angle-sensing response is taken to be linear, given by $S(\theta) = \mu\theta_x$, then the tracking error is given by the expression:

$$\Delta\theta_x = \theta'_{xt} - \theta_{xt} = \Delta\theta_{x0} + \theta_x\Delta\mu/\mu \tag{7.2}$$

where $\theta'_{xt}$ is the actual target angle; $\theta_{xt}$ is the apparent target angle, including the error; $\Delta\mu/\mu$ is the relative change in sensitivity during the track; and $\theta_x$ is the target position relative to the axis in the measurement plane $XZ$, equal to $\theta'_{xt} - \theta'_{x0}$.

It may be seen from (7.2) that the tracking error consists of two components. The first component results from the shift in the axis, and the second from the change in the angle-sensing sensitivity.

The maximum tracking error will occur when the target is farthest from the axis. If this maximum target angle is taken to be half the beamwidth, then (7.2) may be written as

$$(\Delta\theta_x)_{max} = \Delta\theta_{x0} + (\Delta\mu/\mu)[\lambda/(2X_a \sin\theta_{x0})] \tag{7.3}$$

where $X_a$ is the array dimension in the $x$ direction.

The total tracking error depends on the type of amplitude-phase distribution in the excitation fields, and the errors in these fields. We will now examine each component of the tracking error. As an example, we will consider an even excitation forming the sum pattern $F_s(\theta)$, with an amplitude-phase distribution along the array of the form:

$$\hat{\gamma}_{nm} = \gamma_{nm}(1 + \Delta\gamma_{nm})\exp[-i(\beta x_n \cos\theta_{x0} + \beta y_m \cos\theta_{y0} - \vartheta_{nm})]$$

where $x_n$ and $y_m$ are the coordinates of the array elements; $\gamma_{nm}$ is the excitation field amplitude without errors; $\Delta\gamma_{nm}$ is the amplitude error; and $\vartheta_{nm}$ is the phase error.

Analogously, we find for the aperture excitation forming the difference pattern $F_d(\theta)$ in the plane $XZ$:

$$\hat{\eta}_{nm} = \eta_{nm}(1 + \Delta\eta_{nm})\exp[-i(\beta x_n \cos\theta_{x0} + \beta y_m \cos\theta_{y0} - \xi_{nm})]$$

The phase error for any array element consists of the error in its position and the phase error in the signal driving it:

$$\vartheta_{nm} = \vartheta'_{nm} + \beta(\Delta x_{nm} \cos\theta_{x0} + \Delta y_{nm} \cos\theta_{y0} + \Delta z_{nm} \cos\theta_{z0})$$

where $\Delta x_{nm}$, $\Delta y_{nm}$, and $\Delta z_{nm}$ are the errors in the location of the driven element, and $\vartheta'_{nm}$ is the excitation phase error at the element.

The phase error in the difference pattern excitation is found analogously:

$$\xi_{nm} = \xi'_{nm} + \beta(\Delta x_{nm} \cos\theta_{x0} + \Delta y_{nm} \cos\theta_{y0} + \Delta z_{nm} \cos\theta_{z0})$$

For antenna arrays with a large number of elements, the excitation may be represented as continuous functions $\gamma(x, y)$ and $\eta(x, y)$. In this case, the two components of the tracking error will take the form [21]:

$$\Delta\theta_{x0} = \frac{1}{\beta \sin\theta_{x0}} \frac{\iint \xi(x, y)\eta(x, y)dA}{\iint x\eta(x, y)dA} \tag{7.4}$$

$$\frac{\Delta\mu}{\mu} \frac{\lambda}{2X_a \sin\theta_{x0}} = \frac{1}{2}\left(\frac{\iint \vartheta^2\gamma\, dA}{\iint \gamma\, dA} - \frac{\iint \xi^2 x\eta\, dA}{\iint x\eta\, dA}\right)\frac{\pi}{\beta X_a \sin\theta_{x0}} \tag{7.5}$$

$$= -\left(\frac{\iint \Delta\gamma\gamma\, dA}{\iint \gamma\, dA} - \frac{\iint \Delta\eta x\eta\, dA}{\iint x\eta\, dA}\right)\frac{\pi}{\beta X_a \sin\theta_{x0}}$$

The integration is performed over the equivalent array aperture with dimensions $X_a$ and $Y_a$, and not all the amplitude and phase errors enter into (7.4). Equations (7.4) and (7.5) may be used to calculate the tracking errors arising from both systematic and random amplitude and phase errors in the aperture excitation.

We will examine the systematic tracking error for the case of uniform amplitude excitation $\eta$, where

$$\eta(x, y) = \begin{cases} -1 & \text{for } x < 0 \\ 1 & \text{for } x > 0 \end{cases}$$

and the systematic phase error has the form:

$$\xi = -\xi_0(2x/X_a)^3$$

where $\xi_0$ is the maximum value of the phase error at the edge of the array.

The tracking error components corresponding to (7.4) and (7.5) will then be given by

$$\Delta\theta_{x0} = [-\lambda/(X_a \sin \theta_{x0})]\xi_0/(2\pi)$$
$$(\Delta\mu/\mu)\lambda/(2X_a \sin \theta_{x0}) = [-\lambda/(X_a \sin \theta_{x0})](\xi_0/4)^2$$

Consequently, the total error, expressed in beamwidths, is determined by the expression:

$$\Delta\theta_x/[\lambda/(X_a \sin \theta_{x0})] = [\xi_0/(2\pi)][1 + (\pi/8)\xi_0]$$

It may be seen in the plot of the tracking error *versus* cubic phase error (see Fig. 7.1) that displacement of the axis is the predominant tracking error due to phase errors.

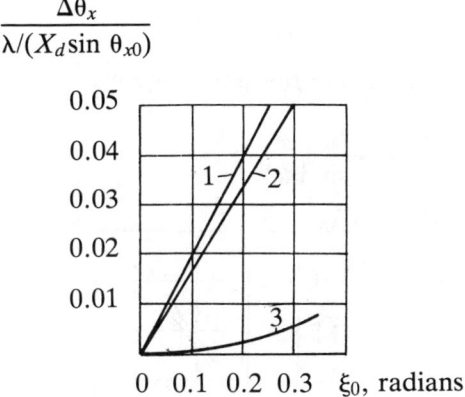

$$\frac{\Delta\theta_x}{\lambda/(X_d \sin \theta_{x0})}$$

0.05
0.04
0.03
0.02
0.01

0   0.1  0.2  0.3   $\xi_0$, radians

**FIGURE 7.1.** Normalized tracking error as a function of the cubic antenna excitation error:

1: total error
2: error due to shift of the null axis
3: error due to change in the angle-sensing sensitivity

We will examine the random tracking errors for the case where the amplitude and phase errors in the excitation have normal distributions. With zero-mean phase errors, uniformly distributed $\gamma$, and stepped-uniform $\eta$, the rms total tracking error and its components due to phase errors will be [21]

$$\sigma_{\theta x 0} = \left(\frac{\lambda}{X_a \sin \theta_{x0}}\right) \frac{0.64}{\sqrt{NM}} \sigma_\xi$$

$$\frac{\sigma_\mu}{\mu} \left(\frac{\lambda}{2X_a \sin \theta_{x0}}\right) = \left(\frac{\lambda}{X_a \sin \theta_{x0}}\right) \frac{0.54}{\sqrt{NM}} \sigma_\xi^2$$

$$\sigma_{\theta x} = \sigma_{\theta x 0}^2 + \frac{\sigma_\mu^2}{\mu^2} \left(\frac{\lambda}{2X_a \sin \theta_{x0}}\right)^2$$

where $N$ and $M$ are the number of elements along the $x$- and $y$-axes, respectively, and $\sigma_\xi^2$ is the variance of the phase error in radians.

Plots for these errors as functions of $\sigma_\xi$ are presented in Fig. 7.2. In this case the predominant tracking error is again the displacement of the axis.

The rms total tracking error caused by random amplitude errors is determined by the expression:

$$\sigma_{\theta x} = [\lambda/(X_a \sin \theta_{x0})](1.56/\sqrt{NM})\sigma_a \tag{7.6}$$

where $\sigma_a = \sigma_{\Delta\gamma} = \sigma_{\Delta\eta}$ is the rms amplitude error.

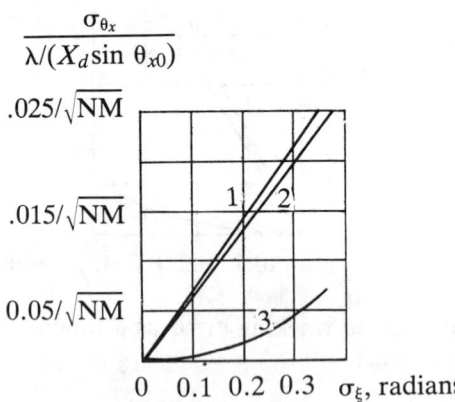

**FIGURE 7.2.** Normalized rms tracking error as a function of the rms antenna excitation phase error:

1: total error
2: error due to shift of the null axis
3: error due to change in the angle-sensing sensitivity

Plots of this error as a function of $\sigma_a$ are shown in Fig. 7.3.

These expressions for the tracking error were obtained for the case of tracking in a single plane, but are also valid for tracking in other planes.

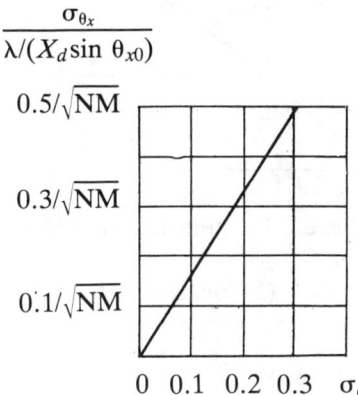

FIGURE 7.3. Normalized rms tracking error as a function of the rms antenna excitation amplitude error.

The effects of amplitude and phase errors may also be estimated from the perspective of changes in the antenna pattern. For this it is convenient to use statistical antenna calculations. Thus, with a normal distribution of currents in the array elements and a normally distributed phase error, calculations [21] give the following generalized expression for the average array power pattern:

$$\overline{F(\theta, \vartheta)} = F(\theta, \vartheta) + u(\theta, \vartheta)\sigma_0^2 \sum_{m=1}^{M} \sum_{n=1}^{N} I_{mn}^2 \Big/ \left( \sum_{m=1}^{M} \sum_{n=1}^{N} I_{mn} \right)^2 \quad (7.7)$$

where $F(\theta, \vartheta)$ is the antenna array pattern with no errors; $u(\theta, \vartheta) = (\cos\theta)(\cos^2\theta \cos^2\vartheta + \sin^2\vartheta)$ is the pattern slope; $\sigma_0^2 = \sigma_a^2 + \sigma_\xi^2$ is the mean square total error; $\sigma_a^2$ is the rms amplitude error; $\sigma_\xi^2$ is the rms phase error; $I_{mn}$ is the current in the $mn$th element; and $\theta$ and $\vartheta$ are the angles giving the orientation of the beam in the chosen coordinate system (see Fig. 7.4).

It may be seen from (7.7) that the effect of amplitude and phase errors is to cause spurious radiation (determined by the second term in the expression), resulting in reduced directivity. This last effect may be determined with the following approximate formula:

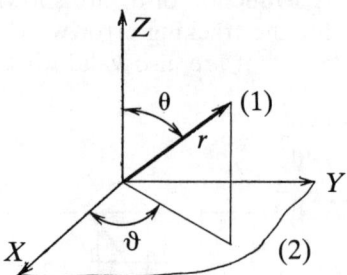

**FIGURE 7.4.** Coordinate system for determining the angles $\theta$ and $\vartheta$.

(1) beam direction    (2) plane of array elements

$$G/G_0 \approx 1/[1 + (3\pi/4)(l_e/\lambda)^2\sigma_0^2] \qquad (7.8)$$

where $G$ and $G_0$ are the directive gains with and without errors, respectively, and $l_e$ is the distance between the array elements.

### 7.2.2 The Effects of Aperture Field Errors on Angle Accuracy in a Parabolic Antenna

The statistical approach for small normally distributed phase errors in a parabolic reflector gives

$$\overline{F(\theta, \vartheta)} = F(\theta, \vartheta) + \mu(\theta, \vartheta) \frac{4\tau_0^2\pi^2\sigma_\xi^2}{\lambda^2 G_0} \exp\left(-\frac{\pi^2 u^2 \tau_0^2}{\lambda^2}\right) \qquad (7.9)$$

where $u = \sin\theta$; $\tau_0$ is the correlation length in wavelengths, corresponding to the length at which errors in the excitation currents cannot be considered to be independent.

As in (7.7), the first term in (7.9) characterizes the pattern with no errors, and the second term describes the distortion caused by phase errors in the aperture field excitation. As in an array antenna, the spurious radiation is proportional to the mean square error, and is also proportional to the square of the correlation length, expressed in wavelengths.

For small errors and a small correlation length, the reduction in directive gain may be calculated with the approximate expression:

$$G/G_0 \approx 1 - (3/4)\sigma_\xi^2\tau_0^2\pi^2/\lambda^2, \qquad \tau_0/\lambda \ll 1 \qquad (7.10)$$

and, for a large correlation length, by the expression:

$$G/G_0 \approx 1 - \sigma_\xi^2, \qquad \tau_0/\lambda \gg 1 \tag{7.11}$$

Thus, amplitude and phase errors in the aperture excitation lead to distortion of the pattern, manifested in part by a reduction in the antenna's directive gain. As a result, the amplitude responses of the receiver channels may not be identical, and the angular errors will increase accordingly.

To estimate the effects of unequal patterns on the target angle accuracy, we will suppose that the gain of one of the receiving antennas differs from the other by a factor $\Delta G$. As can be seen in Fig. 7.5, such a change in the gain of one of the monopulse antennas causes a shift of $\Delta\theta_0$ in the null axis. The value of this systematic angular error may be determined in the following manner. Let $G_0$ be the gain on the axis with identical patterns; $G_0'$ is the gain on the axis if the gain of one of the patterns differs by $\Delta G$, and $\mu$ and $\mu'$ are the pattern slopes in the tracking region for the same conditions. Then we may write the following:

$$G_0' = G_0(1 + \mu\Delta\theta_0) \tag{7.12}$$

$$G_0 + \Delta G = G_0'(1 + \mu'\Delta\theta_0) \tag{7.13}$$

Using (7.12) in (7.13) and disregarding second-order perturbations, after elementary transformations we obtain

$$\Delta\theta_0 = \Delta G_0/[G_0(\mu + \mu')] \approx (\Delta G_0/G_0)/(2\mu)$$

or

$$\mu\Delta\theta_0 \approx \Delta G/(2G_0)$$

Inserting the values of the slope of the angle-sensing response and the antenna gain difference, the angular error may be determined. Calculations show that with an angle-sensing slope of 0.25/degree and a 10% difference in gain, the systematic tracking error will by 0.2°. If the requirement is for the angular error not to exceed one minute, then the gain difference must be less than 0.8%.

(1)

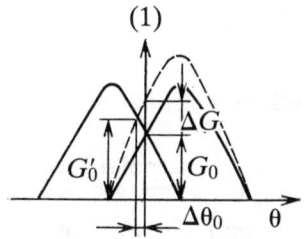

**FIGURE 7.5.** Shift in null axis direction due to change in gain of one of the antennas.

Using (7.11), we may calculate the acceptable rms excitation phase error for a parabolic antenna. Thus, with a large correlation length:

$$\sigma_\xi^2 = 1 - G/G_0 = \Delta G/G_0, \qquad \sigma_\xi = (\Delta G/G_0)^{1/2}$$

With $\Delta G/G_0 = 0.1, \sigma_\xi = 0.32$ radians $= 18.3°$. If the acceptable error is $\Delta\theta_0 = 1$ minute, then $\Delta G/G_0$ must be less than 0.8%. The acceptable rms phase error in this case becomes

$$\sigma_\xi = (0.008)^{1/2} = 0.09 \text{ radians} = 5°$$

This is a rather stringent requirement on the antenna excitation. Thus, there are higher demands on the accuracy of construction and rigidity of monopulse antennas. Equations (7.8), (7.10), and (7.11) may be used to calculate the acceptable phase errors if there are random variations in the gain imbalance.

In addition to the antenna gain, the beamwidth and squint angle may also vary. Analysis of a typical four-channel amplitude monopulse system with logarithmic receivers shows that for such variations the null axis shifts, the error sensitivity changes, and the shape of the angle-sensing response changes (the linear portion becoming concave or convex depending on which pattern has the greater variation) [48].

The pattern parameters may also vary with frequency in wideband applications. In this case, it is necessary to devise a scheme to compensate for the tracking errors by varying the frequency according to some experimentally determined pattern.

## 7.3 THE EFFECTS ON MONOPULSE ANGLE ACCURACY OF CROSS POLARIZATION OF THE RECEIVING ANTENNAS

It was shown in Chapter 6 that the signal reflected by a target will, in general, become depolarized, *i.e.*, have a polarization different from that at which the radar is operating. Therefore, it is of interest to consider the polarization properties of modern antennas and their effects on angle accuracy.

### 7.3.1 Cross-Polarized Radiation of Reflector Antennas

Many works have addressed the polarization structure of reflector antenna patterns [17, 27, 53, 55, 68, 75, 82]. It has been established that in all antennas containing components with curvilinear surfaces, such as reflectors, lenses and feeds, there is a cross-polarized component in the radiation. The cross-polarized component has a polarization orthogonal to the operating polarization. The level of cross-polarized radiation and its structure are determined by the antenna geometry, the type and location of the feeds, diffraction phenomena at the reflector and feed edges, and imperfect reflecting surfaces.

The fields in a paraboloid aperture for two forms of electric dipole excitation are illustrated in Fig. 7.6. The vertical arrows represent the field strength of the operating polarization, while the cross-polarized component is indicated with horizontal arrows. Ideally, an axially symmetric antenna will have no cross-polarized radiation in the main planes, because the cross-polarized components in neighboring quadrants will have opposite phase and equal amplitudes, and will cancel out in the paraboloid's planes of symmetry.

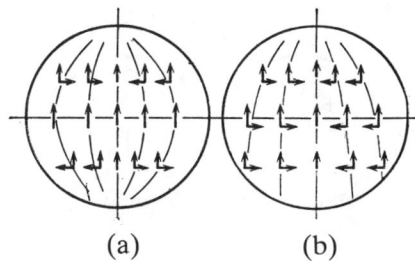

(a)                    (b)

**FIGURE 7.6.** Field distribution in the aperture of a paraboloid:

(a)  with frontal axially symmetric excitation by electric dipole
(b)  with excitation by offset dipole

When the feed is moved below the paraboloid, the current lines are seen to converge along the direction to the upper edge of the reflector. Of the cross-polarized components resulting, only those on opposite sides of the vertical plane of symmetry will have opposite phase. Therefore, the cross-polarized radiation is absent only in this plane, which is parallel to the orientation of the dipoles, as opposed to the case for an axially symmetric antenna, in which all of the cross-polarized components in neighboring quadrants cancel. The phase distribution of the cross-polarized components for these parabolic antennas are approximated in Fig. 7.7 with + and − signs. It should be noted that comparatively small shifts of the feed from the geometric focus, as in an amplitude monopulse antenna, for example, do not lead to significant changes in the maximum cross-polarized radiation.

Cross-polarized radiation results in a complex pattern determined by the contributions of components of the operating and cross polarizations.

Thus, if a single-lobed axisymmetric antenna is switched from the basic polarization to cross polarization, a four-lobed pattern results, with lobes in all four quadrants lying in planes skewed 45° from the main antenna planes, and with a null along the main axis. A parabolic antenna with an offset feed will form a two-lobed pattern for cross polarization, with phase values corresponding to those in Fig. 7.7.

The level of cross-polarized radiation is determined primarily by the curvature of the reflector; the cross polarization level increases with the curvature. An increase in the ratio of focal length to reflector diameter ($f/d$ ratio) reduces the intensity of the cross polarization. Table 7.1 presents the calculated maximum cross-polarized radiation levels of parabolic antennas with frontal short dipole excitation, for various values of $f/d$ [73, 82].

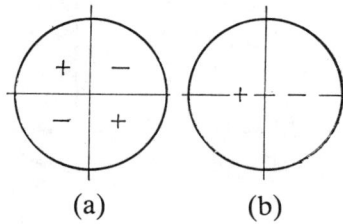

(a)                    (b)

**FIGURE 7.7.** Phase distribution of cross polarization field components in paraboloid aperture:

(a)  with frontal axially symmetric excitation by electric dipole
(b)  with excitation by offset dipole

**TABLE 7.1**

|  | maximum cross polarization level, dB | |
|---|---|---|
| $f_d/d_p$ | data from [73] | data from [82] |
| 0.25 | −15.8 | −14.6 |
| 0.30 | −18.1 | −17.5 |
| 0.40 | −22.2 | −22.1 |
| 0.46 | −24.3 | −24.2 |
| 0.60 | −28.0 | −28.4 |

If a magnetic dipole is used as the feed, then the paraboloid aperture distribution will have the same character, but the cross-polarized components will have phases opposite to those for the electric dipole case.

Consequently, if the feed is a combination of electric and magnetic dipoles at right angles to one another, and they illuminate the paraboloid with the same intensity (equivalent to a Huygens source), then the cross polarization may be completely eliminated [68]. Such a feed may be realized with a small rectangular horn, but, in practice, the cross polarization is not completely removed.

Cassegrain antennas have much lower cross polarization levels. Measurements on such an antenna with $f/d = 0.33$ and a rectangular horn feed gave a maximum cross polarization level of $-42$ dB [72].

Even when fed with a Huygens source, an antenna in the form of a paraboloid of rotation will have high cross polarization levels, approximating those resulting with electric dipole illumination. Displacement of the feed from the focus, which is necessary in amplitude monopulse systems, causes an asymmetric amplitude distribution in the currents on the reflector, which raises the cross polarization level in the direction of the feed offset [21]. In this case, the cross-polarized pattern is deformed: the minima become deeper, the maxima less peaked, and the intensity and width of the cross-polarized lobes increase. This deformation increases as the feed is offset further from the focus. The compensation afforded by using a Huygens source at the focus is lost if the feed is moved.

Cross polarization also results when circular polarization is used with a reflector antenna; this cross polarization is circular, but oriented opposite to the basic polarization. In [21], the case of a square horn excited by two orthogonal $H_{11}$ waves in quadrature was considered. The calculations confirmed the existence of cross polarization radiation, with a maximum level 16 dB below the basic polarization maximum.

It must be noted that cross-polarized radiation is absent along the axis of a parabolic antenna only when the reflector surface is ideally symmetric. Errors in constructing the antenna surface ruin the amplitude-phase symmetry of the fields at the reflector surface, and cross-polarized radiation along the axis is then possible. The effects of antenna surface faults have been studied in several works [73, 74]. It has been shown that when the errors are small, not exceeding $0.04\lambda$, the cross-polarized intensity on the axis is directly proportional to the rms error in the reflector surface.

An investigation was carried out at 10 GHz with a paraboloid, fed with a pyramidal horn with an aperture of 4 cm $\times$ 4 cm, and $f/d = 0.25$. It was shown that for variations in the reflector profile within the limits $(0.021 \rightarrow 0.029)\lambda$, the decoupling in polarization along the axis was worsened by 4–8 dB, and the loss in gain reached 0.3–0.35 dB.

It should be noted that the maximum acceptable deviation from an ideal parabolic profile in accordance with mechanical tolerances during antenna construction is $\pm 0.03\lambda$ [36].

It should further be noted that lens antennas also exhibit cross-polarized radiation. These antennas, however, are free from the drawbacks associated with the feed and support shadowing in reflector antennas. This allows much lower cross polarization levels to be obtained in some cases. It has been shown that if the lens is of the planar convex type with constant coefficient of refraction, and a symmetric dipole at the focus is used for the feed, the field in the lens aperture will have no cross-polarized component [82]. Cross-polarized radiation will result if the feed is moved from the focus.

### 7.3.2 The Effects on Monopulse Radar Angle Accuracy of Cross Polarization of the Receiving Antennas

Due to the depolarization of the reflected target signals, the receiving antenna pattern will assume its usual form only when the signal polarization corresponds to the basic (operating) polarization. In the remaining cases, the pattern is distorted to an extent which increases with the degree to which the signal polarization differs from the operating polarization. When the signal polarization corresponds to cross polarization of the antenna, the expected pattern distortion is a maximum. In this case, the antenna pattern is determined solely by the structure of the cross-polarized radiation of the given antenna.

Inasmuch as the cross-polarized pattern has neither the same form nor position as the operating polarization pattern, such a deformation unavoidably degrades the accuracy of angle measurement systems, including monopulse systems. We will consider the example of an amplitude-sum-and-difference monopulse system. The typical patterns of the sum and difference channels at operating and cross polarization for such a system are shown in Fig. 7.8. The signs + and − approximate the phase relations.

The angle-sensing response of the system, as is known, is given by the expression:

$$S(\theta) = \{\mathrm{Re}[E_s^*(\theta, \vartheta)E_d(\theta, \vartheta)]\}/[E_s(\theta, \vartheta)E_s^*(\theta, \vartheta)]$$

where $E_s(\theta, \vartheta)$ and $E_d(\theta, \vartheta)$ are the complex expressions for the sum-and-difference channel signals.

Investigations have shown that if the cross polarization radiation of the sum channel is small in comparison with that at the operating polarization, and if the depolarization of the reflected target signal is comparatively small, then the angular error, for targets near the axis, is determined by the expression [75]:

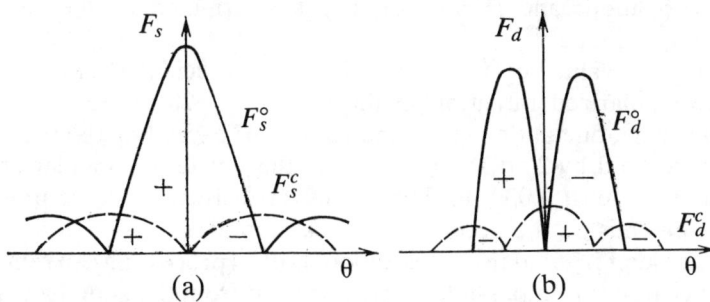

**FIGURE 7.8.** Typical monopulse antenna patterns at operating (index *d*) and cross (index *s*) polarizations:

(a)   sum channel   (b)   difference channel

$$\Delta\theta = (\theta_3/\mu)[E_s^0 E_d^0/(E_s^0)^2 + E_s^0 \rho_t E_d^c/(E_s^0)^2]$$

where $E_s^0$ and $E_d^0$ are the amplitudes of the operating polarization signal components in the sum and difference channels; $E_d^c$ is the amplitude of the cross-polarized signal component in the difference channel; $\rho_t$ is the complex coefficient of depolarization of the reflected signal; and $\overline{\mu}$ is the pattern slope.

From this, for the expression for the cross polarization angular error we obtain

$$\Delta\theta^c = (\theta_3/\mu)E_s^0 \rho_t E_d^c/(E_s^0)^2$$

$$\underline{\rho_t} = \sqrt{\sigma^c/\sigma^\circ}\,\exp(i\Phi)$$

where $\sigma^c$ and $\sigma^\circ$ are the cross polarization and basic polarization components of the target cross section, respectively (real values), and $\Phi$ is the phase difference between the components $\sigma^c$ and $\sigma^\circ$.

If we assume that $\sigma^c$ and $\sigma^\circ$ are independent fluctuating quantities with chi-square distributions, and the phase $\Phi$ is uniformly distributed in the interval $\pm\pi$, then the rms target angular error caused by depolarization of the reflected target signal and cross polarization of the radar antenna will be [75]

$$\sigma_\theta^c = (\theta_3/\sqrt{2}\mu)/(r_0 g_c f_c)$$

where $r_0 = \overline{\sigma^c/\sigma^\circ}$ is the ratio of the mean target cross section components at cross polarization and basic polarization; $g_c$ is the antenna power gain at cross polarization; and $f_c$ is a coefficient determined by the chosen fluctuating target model.

With a typical value for the depolarization coefficient near $-3$ dB, antenna cross-polarized radiation in the range of $-20$ dB to $-30$ dB, $f_c = 4.0$ (Rayleigh fluctuating target model with $\sigma°$ exceeding the receiver sensitivity threshold by 20 dB) and $\mu = 1.6$, the rms target angular error will lie in the range of $0.03\theta_3$ to $0.06\theta_3$. This error is rather large in light of existing demands on the accuracy of tracking radars.

It should be kept in mind that no additional signal processing can remove this error, because it is impossible to distinguish it from the actual angular errors caused by a discrepancy between the axis and the target direction. It is also impossible to compensate for this error, since the amplitude and phase of the complex depolarization coefficient of the target echoes are unknown *a priori*. Thus, when the depolarization of the reflected signal exceeds the level used above, the angular error grows accordingly, and can reach values at which the monopulse system loses the ability to track the target with the necessary accuracy.

Plots of monopulse angle-sensing responses, as calculated for various signal polarizations and cross polarization levels, are presented in Fig. 7.9 [21]. For single-point signal sources, the variable parameter $a_c b$ corresponds to the ratio of the amplitudes of the signals received at cross polarization and the operating polarization. Hence $a_c b = \infty$ with $a_c < 1$ corresponds to the case of complete polarization mismatch, when the received signals are cross polarized; $a_c b = 0$ corresponds to reception of matched (operating) polarization. As can be seen in the drawing, an increase in the depolarization of the received signals ($a_c b$ increases) leads to an increase in the distortion of the angle-sensing response. This distortion manifests itself in an increase in the shift of the null axis, and changes in the pattern slope and shape. The null-axis shift for $a_c b \rightarrow \infty$ approaches the beamwidth. Starting with $a_c b = 0.5$, there arises a significant asymmetry in the angle-sensing response. The lobe of the response on the side, toward which the null axis is displaced, is greatly lowered in amplitude, narrowing the region of stable tracking.

For $a_c b = \infty$ the angle-sensing response at the axis is discontinuous. The systematic and dynamic target angular errors are increased with shifts in the null axis and reduction in the angle-sensing slope.

The phenomenon of depolarization of the reflected signals and the associated additional target angular errors, due to cross polarization of the receiving antenna, has been observed experimentally. The results of measurements of the patterns of an amplitude-sum-and-difference monopulse system at operating and cross polarization were cited in [21] (see Fig. 7.10). As can be seen in the drawing, with reception of signals cross-polarized with respect to the receiving antenna, the partial monopulse

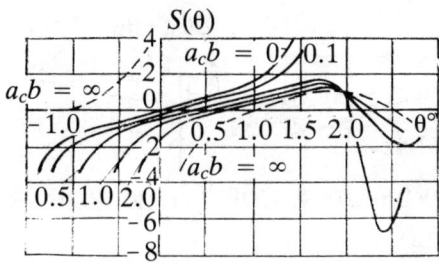

**FIGURE 7.9.** Angle-sensing response of an amplitude-sum-and-difference monopulse system for tracking in one plane as a function of received signal polarization mismatch.

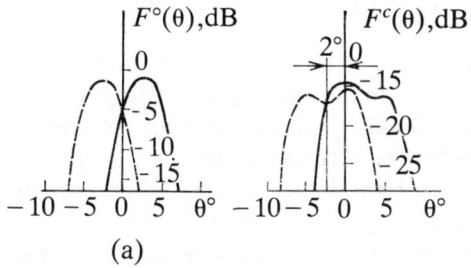

(a)

**FIGURE 7.10.** Experimental patterns for an amplitude-sum-and-difference monopulse system:

(a)   at operating polarization   (b)   at cross polarization

patterns are deformed, and, as a result the null axis is shifted by $2°$, which corresponds to a shift of $0.45\theta_3$. The existence of amplitude and phase imbalances in the receiving channels and initial tracking errors can increase the influence of antenna cross polarization on the angle accuracy.

Shown in Fig. 7.11 are the experimentally obtained variations of the null axis with orientation of the polarization and operating frequency, obtained with a modified AN/FPS-16 monopulse radar with circular polarization when receiving linearly polarized signals.

The figure shows that with the change in the orientation of the linear polarization of the operating frequency, the null-axis shift varies in both amplitude and sign, and reaches 1.5 minutes of arc. If the tracking accuracy requirement of the AN/FPS-16 radar is estimated to be on the order of tenths of a minute of arc [21], the resulting shifts of the null axis cannot be ignored.

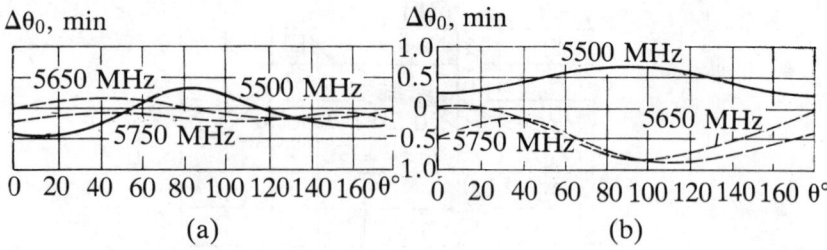

**FIGURE 7.11.** Experimentally obtained values for shift of the null axis as a function of the rotation of the polarization plane of the received signals:

(a)   in azimuth   (b)   in elevation

It should be noted that a change in the orientation of a linearly polarized wave affects the variation in the depth of the difference pattern null, as well as the sidelobe levels. In actual conditions, the reflected signals will have a partially polarized structure, and, as a result the final effect will be determined by the spectral density of the polarization fluctuations and the passband of the radar's tracking system.

The cross-polarized pattern of the antenna which results from imperfect manufacturing makes the receiving antennas sensitive to the signal polarization, and, in combination with the depolarization of the signal upon reflection from the target, is thus one of the fundamental factors limiting the accuracy of monopulse tracking systems.

### 7.3.3 Methods for Reducing the Influence of Signal Depolarization and Antenna Cross Polarization on Angle Accuracy

The depolarization of radar signals is caused by the physical properties of the target, its motion, and the propagation conditions. It is not possible to control these factors in an attempt to reduce the effects of signal depolarization on target angle accuracy. The main emphasis in striving to reduce these effects is accordingly placed on lowering the influence of the antennas' cross-polarized patterns. With this approach, there are the following possibilities [21, 53, 70, 71, 89, 91, 98, 110]:

- reducing the cross polarization caused by the antenna reflector;
- reducing the cross polarization caused by the antenna feeds;
- reducing the cross polarization caused by diffraction phenomena at the reflector edges;

- reducing the cross polarization caused by the mounting hardware in the aperture and shadowing of the aperture caused by feeds and subreflectors.

Reduction of the cross polarization caused by the reflector may be achieved by maintaining stringent controls during construction of the reflector, reducing its curvature and using special components which compensate for the cross polarization components in the electromagnetic fields. The greater the curvature of the reflector, the greater the cross polarization. It is, therefore, preferable to use long-focus antennas with large f/d values, because they have lower cross-polarized radiation levels.

Installing special correcting plates at the maxima of the cross-polarized electric field, to lower the cross-polarized effects, was proposed in [21, 71]. One version of such a compensator is shown in Fig. 7.12. It is a complex antenna feed, consisting of the main active dipole $(D_0)$ and four passive elements $(D_1-D_4)$, located at the maxima of the antenna cross-polarized pattern and perpendicular to the active dipole. Currents, presented by the active dipole at cross polarization, flow along the reflector surface and induce currents in the passive elements (dipoles). The amplitude and phase of these currents are set by choosing the dimensions of the passive elements (width $w$ and length $l$) such that the cross-polarized field components are cancelled. Experiments carried out at 9.99 GHz, with a 31-cm diameter reflector with $f/d = 0.25$, showed that the cross polarization could be lowered 10 dB by this method [71]. The reduction in cross-polarized effects is lessened at nonoptimum frequencies.

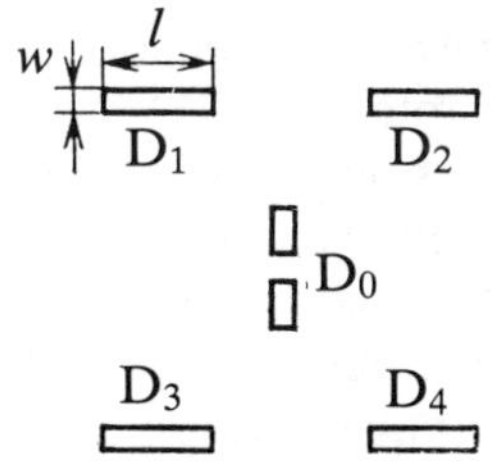

**FIGURE 7.12.** Geometry for a complex antenna feed which produces low cross polarization levels.

Active (auxiliary) elements could be used in place of passive elements at the locations of the cross-polarized maxima (see Fig. 7.12). With this design, however, it is necessary to ensure close correspondence of the amplitudes and phases of the auxiliary currents to those in the main dipole in order to obtain optimum performance. As a result, antenna construction becomes more complicated and expensive, because equalization of the amplitude-phase relations must be calibrated with precise attenuators and phase shifters. The antenna bandwidth is further reduced with this approach.

Antenna feeds can also be designed to cancel the cross polarization. Investigations have shown [53] that, for any paraboloid with given value of $f/d$, it is possible to find both a single-mode and double-mode horn with circular aperture, which provides low cross-polarization levels at the given frequency, even over a band of frequencies. It was established that a single-mode feed gives cross polarization levels below $-30$ dB for $f/d = 0.45$–$0.6$, and levels below $-35$ dB can be obtained with dual-mode feeds for $f/d = 0.7$–$0.8$.

The theoretical feasibility of building a three-reflector antenna with no cross polarization has been established [89]. Such an antenna must consist of a main parabolic reflector and two auxiliary reflectors, formed by second-order surfaces of rotation. There is no data on the use of such antennas in monopulse radar systems.

In constructing the feeds, special attention should be given to obtaining low cross polarization levels and good symmetry, because it has been established that the maximum off-axis cross polarization in the 45°-plane is directly proportional to the asymmetry in the feed pattern. The symmetry in the principal planes may be improved with dielectric inserts placed in the feed aperture. Experiments reported in [93] showed that at 10 GHz, a 10-dB reduction in cross polarization can be obtained with this method. The experiments were carried out with a 30° conical horn of 6-cm diameter. Dielectric paraffin spheres, with 5.8-cm diameter were attached in the horn aperture with conforming adapters.

In order to reduce the cross polarization induced by feeds, wide use is made, in the US, of corrugated waveguides and conical horns with quadruple choke grooves [70, 91]. A monopulse feed consisting of four circular waveguides connected to a circular corrugated waveguide is shown in Fig. 7.13; this feed provides difference patterns with low cross polarization.

Figure 7.14 shows a circular waveguide feed with a single corrugated choke ring (a) and its pattern (b) at a frequency of 8.5 GHz for $l_1 = 0.011$ m [91]. Increasing the number of choke rings leads to a further reduction in the maximum cross-polarized radiation level. This is confirmed by the data in Table 7.2, showing the maximum cross polarization level for the given feed (in decibels) as a function of frequency and number of choke rings. The distance between the feed aperture and choke rings is chosen to obtain symmetrical patterns in the orthogonal planes.

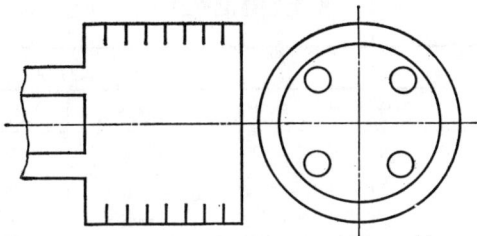

**FIGURE 7.13.** Corrugated waveguide monopulse feed.

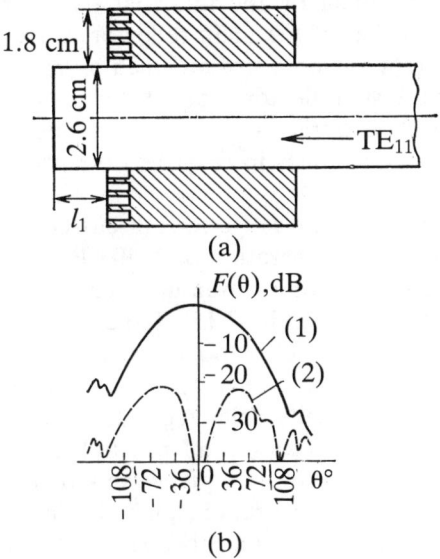

**FIGURE 7.14.** Waveguide feed with one corrugated ring:

(a)  sectional view of feed
(b)  cross polarization pattern of feed
　　(1)  at matched polarization  (2)  at cross polarization

Theoretical and experimental investigations carried out with ribbed feeds at 10 GHz [98] have shown that the choke attenuates the cross-polarized radiation by 6 dB. This type of feed is often used in deep parabolic antennas, offering a reduction in the secondary cross-polarized radiation and an extended antenna bandwidth.

**TABLE 7.2**

| Number of choke rings | Frequency, GHz | | | | |
|---|---|---|---|---|---|
| | 8.0 | 8.25 | 8.5 | 8.75 | 9.0 |
| one | −20 | −22 | −22 | −22 | −22 |
| two | −20 | −23 | −23 | −23 | −20 |
| three | −25 | −25 | −25 | −25 | −25 |

Analysis shows that, for off-axis reception, the best cross-polarized radiation characteristics are obtained with two-reflector Cassegrain antennas [110]. This is explained by the fact that the subreflector, having a curvature opposite to that of the main reflector, corrects the cross polarization of that reflector to some extent. Such asymmetry as is obtained, for example, when the subreflector is offset, leads to some degradation in the cross-polarized characteristics. But even in this case the Cassegrain antenna has definite advantages in regard to the problem of cross polarization.

Theoretical experimental investigations discussed in [46] show the feasibility of obtaining cross polarization of − 40 dB and lower at SHF with a Cassegrain antenna. Flat polarization grids are sometimes used as subreflectors in dish antennas (see Fig. 7.15). Experiments with such antennas at 12–14 GHz have achieved cross-polarized reductions of 3–4 dB. The advantage of this construction is that it allows the main paraboloid to be illuminated with an orthogonally polarized field with a feed located on the other side of the grid subreflector (feed No. 2 in Fig. 7.15). This capability is used in the construction of two-frequency monopulse systems.

Some reduction in cross polarization can be achieved by decreasing the illumination of the edges of the reflector. This reduces the antenna gain, however, and widens the beam [21, 82].

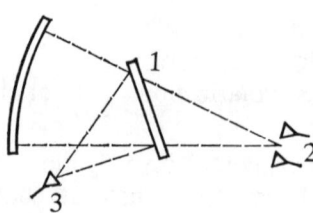

**FIGURE 7.15** Parabolic antenna with remote feed and polarized grid:

1: horizontal grid   2: feed 2   3: feed 1

Polarization filtering is also used to reduce the effects of antenna cross polarization. To this end, polarized grids, usually consisting of a network of closely spaced wires or metal plates, are placed in the antenna apertures. These grids reflect waves polarized parallel to the wires (plates), and pass waves with the orthogonal polarization. This allows a certain amount of the cross-polarized components of the signal to be filtered out, reducing their harmful effects on the angle accuracy.

With the wire radius $r_0 < 0.05\lambda$ and interwire spacing $S < 0.2\lambda$, the coefficient of attenuation for a wave polarized parallel to the wires may be calculated by the formula:

$$K_T \approx \left\{ 1 + \left[ \frac{\lambda}{2S} \frac{1}{\ln(2\pi r_0/S)} \right]^2 \right\}^{-1}$$

where $K_T$ is the ratio of the power of waves passing through the grid to the incident power.

For $r_0/\lambda = 0.005$, use of the above formula gives $K_T = 0.05$ ($-13$ dB) for $S\lambda = 0.1$, and $0.002$ ($-20$ dB) for $S/\lambda = 0.05$.

## 7.4 THE EFFECTS ON ANGLE ACCURACY OF AMPLITUDE-PHASE IMBALANCES IN MONOPULSE RECEIVER CHANNELS

Imbalances in the amplitude-phase responses of receiver channels result from difficulties in manufacturing the components with stringent tolerances, from unavoidable component aging and attendant parameter changes, from detuning upon deployment of the equipment, and from various types of mechanical and climatic interactions.

For the purposes of studying the channel imbalances, it is advisable to distinguish two portions of the receiver: the RF portion, including the antenna and waveguides up to the intermediate frequency mixer, and the IF portion. This division is advantageous, because, when examining the effect of imperfect channels at SHF, it is possible to ignore the pulsed character of the signal, due to the fact that the passbands of these channels are almost always wider than the signal spectra. This assumption is not justified for the IF channels, and the relation of the signal spectrum width to the IF amplifier passband must be considered [43].

For the purpose of simplifying the analysis of amplitude-phase response imbalances, we will first consider the case of angle tracking in the CW mode, when the limited IF amplifier passband can be ignored. Because the degree and nature of the effects of amplitude-phase imbalances depends, to a certain extent, on the design of the angle tracking system, we will apply our analysis to the most commonly employed types of monopulse systems.

### 7.4.1 The Effects of Amplitude-Phase Imbalances on the Angle Accuracy of an Amplitude-Amplitude Monopulse System

The simplified structure of an amplitude-amplitude monopulse system was shown in the block diagram in Fig. 4.1. We will assume that the target is being tracked at small off-axis angles, and that a linear pattern approximation in the neighborhood of the axis is justified.

If the phase difference between the channels is $\psi$ and the RF gains (before the IF amplifier) are $k_1$ and $k_2$, then the signals at the output of the logarithmic IF amplifier may be written as

$$\underline{u}_1(t, \theta) = k_{01} \ln[k_1 E_m F(\theta_0)](1 + \mu\theta) \exp[i(\omega_c t + \psi)] \tag{7.14}$$

$$\underline{u}_2(t, \theta) = k_{02} \ln[k_2 E_m F(\theta_0)](1 - \mu\theta) \exp(i\omega_c t) \tag{7.15}$$

where $k_{01}$ and $k_{02}$ are coefficients characterizing the slope of the amplitude responses of the first and second IF amplifiers.

At the output of the IF amplifier, the signals are linearly detected and compared by subtraction, forming an error signal given by

$$S(\theta) = \overline{|\underline{u}_1(t, \theta)|} - \overline{|\underline{u}_2(t, \theta)|}$$

Using (7.14) and (7.15), we obtain

$$S(\theta) = k_{02}\{g_0 \ln[k_1 E_0(1 + \mu\theta)] - \ln[k_2 E_0(1 - \mu\theta)]\} \tag{7.16}$$

where $g_0 = k_{01}/k_{02}$ and $E_0 = E_m F(\theta_0)$.

Making use of the assumption that the target angle is small, (7.16) may be simplified as

$$S(\theta) = k_{02}[g_0 \ln(k_1 E_0) + g_0 \ln(1 + \mu\theta)$$

$$- \ln(k_2 E_0) - \ln(1 - \mu\theta)] \tag{7.17}$$

$$= \ln[(k_1 E_0)^{g_0}/(k_2 E_0)] + (g_0 + 1)\mu\theta$$

Setting (7.17) to zero and solving for $\mu\theta$, we find the tracking condition in the form:

$$\mu\theta = -\ln[(k_1 E_0)^{g_0}/(k_2 E_0)]/(g_0 + 1) \tag{7.18}$$

With identical amplitude responses in the receiver channels ($g_0 = 1$ and $k_1 = k_2$):

$$S(\theta) = k_{02} \ln [(1 + \mu\theta)/(1 - \mu\theta)] \approx 2k_{02}\mu\theta \tag{7.19}$$

In this case, $S(\theta) = 0$ when $\theta = 0$, which corresponds to tracking with no instrumentation error.

Comparison of (7.18) and (7.19) shows that amplitude imbalance in the receiver channels of an amplitude-amplitude monopulse system leads to systematic tracking errors, the value of which depends on the degree of imbalance.

If the imbalance exists only in the RF portion of the receiver ($g_0 = 1$ and $k_1 \neq k_2$), the tracking error is given by the expression:

$$\mu\theta = -\ln(g/2) \tag{7.20}$$

where $g = k_1/k_2$.

With $k_1 = k_2 = k$ and $g_0 \neq 1$, the angular error may be calculated with the formula:

$$\mu\theta = -[\ln(kE_0)^{g_0-1}]/(g_0 + 1) \tag{7.21}$$
$$= [(g_0 - 1)/(g_0 + 1)]\ln(kE_0)$$

It follows from (7.21) that, with nonidentical logarithmic responses, the signal normalization is disturbed and the error signal is dependent on the received signal levels.

Figure 7.16 presents the generalized angular error as a function of the imbalance coefficients, as calculated by (7.20). Knowing the slope of the angle-sensing response, the absolute angular error may be calculated with the generalized error for any degree of imbalance in the receiver channels' amplitude responses. Thus, for example, if $u = 0.25$ deg$^{-1}$ and the RF amplitude imbalance is 10% ($g = 1.1$), then the instrumentation angular error will be 0.2°. It is easy to calculate the acceptable amplitude imbalance in the RF channels for any tolerable error.

The curve in Fig. 7.16 may be used to estimate the angle error for the case $g = 1$ and $g_0 \neq 1$ (*i.e.*, the channels up to the IF amplifier are balanced, but the amplifier responses are different), if the quantity $|\mu\theta|$ is interpreted as the normalized quantity $|\mu\theta|/[\ln(kE_0)]$, and $g_0$ is used in place of $g$.

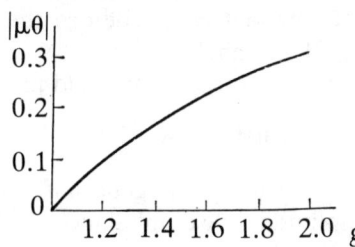

**FIGURE 7.16.** Normalized tracking and angle error as a function of disparity in the gains of amplitude-comparison monopulse receiver channels.

Analysis also shows that phase imbalance in the receiver channels has no effect on the angle accuracy of amplitude-amplitude monopulse systems.

It can be shown that the results, concerning the effect of amplitude-phase imbalances in amplitude-amplitude monopulse systems with logarithmic amplifiers, also apply to systems with linear and quadratic responses.

We have been examining the case of small target angles, when the tracking system is operating linearly. In some cases with significant amplitude imbalance in the receiver, the tracking error may exceed the limits of the linear portion of the angle-sensing response. In this case, the problem of determining the tracking error should be solved in general form, approximating the patterns with appropriate functions. Inasmuch as in the vast majority of cases, the instrumentation errors are small, this degree of complexity is unwarranted, and will not be used.

We will now examine another type of monopulse system.

### 7.4.2 The Effect of Amplitude-Phase Imbalance in the Receiver Channels on the Angle Accuracy of an Amplitude-Sum-and-Difference Monopulse System

Previous analysis [21] shows that with imbalance in the amplitude-phase responses of an amplitude-sum-and-difference monopulse receiver, the error signal may be written (to within a constant) as

$$S(\theta) = \frac{k_d\{[(1 + \mu\theta)^2 - g^2(1 - \mu\theta^2)]\cos\gamma\}}{k_s[(1 + \mu\theta)^2 + g^2(1 - \mu\theta)^2 + 2g(1 - \mu^2\theta^2)\cos\psi]}$$
$$+ \frac{k_d\{2g(1 - \mu^2\theta^2)(\sin\psi)\sin\gamma\}}{k_s[(1 + \mu\theta)^2 + g^2(1 - \mu\theta)^2 + 2g(1 - \mu^2\theta^2)\cos\psi]} \quad (7.22)$$

where $k_s$ and $k_d$ are the gains of the sum and difference channels, respectively; $\gamma$ is the phase imbalance between the sum and difference channels; $g$ is the ratio of the RF channel gains (before the mixer); and $\psi$ is the phase imbalance of the RF channels.

It follows from this that the target is tracked in accordance with

$$[(1 + \mu\theta)^2 - g^2(1 - \mu\theta)^2]\cos\gamma + 2g(1 - \mu^2\theta^2)(\sin\psi)\sin\gamma = 0$$

Solving this equation for $\mu\theta$, for the generalized angular tracking error as a function of the amplitude-phase imbalances in the receiver channels, we obtain the expression:

$$\mu\theta = \frac{-(1 + g^2) \pm 2g\sqrt{1 + (\tan^2\gamma)\sin^2\psi}}{(1 - g^2) - (2g\tan\gamma)\sin\psi} \quad (7.23)$$

The practical solution uses the $+$ sign, which gives the solution for stable tracking with comparatively small target angles. The solution using the $-$ sign gives large generalized target angles and is not justified, given the assumption that the system is operating over the linear portion of the angle-sensing response.

With identical receiver channel responses, (7.22) assumes the well-known form corresponding to the case of tracking with no instrumentation error:

$$S(\theta) = \mu\theta \tag{7.24}$$

Differentiating (7.22) by $\theta$, we find the expression for the slope of the angle-sensing response at the tracking point:

$$\left.\frac{dS(\theta)}{d\theta}\right|_{\theta=0} = \frac{k_d 4\mu g[2g\cos\gamma + \cos(\psi + \gamma) + g^2\cos(\psi - \gamma)]}{k_s(1 + g^2 + 2g\cos\psi)^2} \tag{7.25}$$

Comparing (7.22) and (7.24), we see that the amplitude-phase imbalance affects the tracking errors.

In order to elucidate the character of the dependence of the tracking error on the receiver channel imbalance, we will examine several particular cases.

**Case 1.** The amplitude-phase response of the sum and difference channels are identical, and the RF channels have identical phase responses, but different gains. In this case, $k_s = k_d = k$; $\gamma = 0$; $\psi = 0$; and $g \neq 1$.

Using these values in (7.23) and carrying out elementary transformations, we obtain

$$\mu\theta = (g - 1)/(g + 1) \tag{7.26}$$

which testifies to the fact that a difference in the amplitude responses of the RF channels of an amplitude-sum-and-difference monopulse system introduces a shift in the null axis and additional tracking errors.

Inasmuch as the RF portions of a sum-and-difference system do not usually employ active components, their similarity will be determined basically by how well they are matched, and the equality of their electrical lengths over the operating band. Therefore, particular care must be used in designing and assembling the waveguide feed lines making up the RF channels.

**Case 2.** The sum and difference channels are identical, and the RF channels have identical gains but different phase responses. In this case, $k_s = k_d = k$; $\gamma = 0$; $g = 1$; $\psi \neq 0$; and (7.23) takes the form $\mu\theta = 0$.

This indicates that there is no shift in the null axis whatever the value of the RF channel phase discrepancy. The system will not necessarily operate normally under these conditions, however; we must examine the effects on the angle-sensing slope.

Applying (7.25) to this case, we obtain

$$\frac{dS(\theta)}{d\theta}\bigg|_{\theta=0} \frac{2\mu}{1 + \cos \psi} \tag{7.27}$$

Under normal tracking conditions, $\psi = 0$ and $ds(\theta)/d\theta = \mu$.

With $\psi = \pi$, the channels are functionally reversed: the sum channel becomes the difference channel, and *vice-versa*. The tracking system is then completely inoperable. Setting $\psi = \pi$ in (7.27), we find

$$\frac{dS(\theta)}{d\theta}\bigg|_{\theta=0} = \infty$$

which is a consequence of normalizing with respect to the difference signal which is 0 for $\theta = 0$. This case is illustrated in Fig. 7.17, where the sum and difference channel patterns are shown for the partial receiving antenna patterns $F_1$ and $F_2$. The patterns for $\psi = 0$, corresponding to normal tracking operation, are shown for comparison.

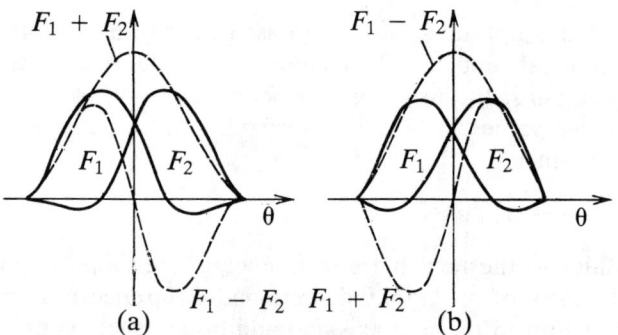

**FIGURE 7.17.** Sum and difference channel patterns:

(a)   for $\psi = 0$   (b)   for $\psi = \pi$

When $\psi < \pi$, as may be seen from (7.27), the slope of the angle-sensing response increases. If $\psi = \pi/2$, the signals may combine in quadrature in the waveguide balanced bridge, and, as a result, the signals at the output of the sum and difference branches will be equal, regardless of the amplitude relations of the signals, *i.e.*, independent of the angle error (see Fig. 7.18). The radar remains useable, however, because the angle information contained in the amplitudes of the independent sum and difference

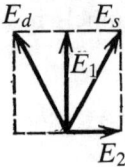

**FIGURE 7.18.** Vector diagram for clarifying formation of sum and difference signals for $\psi = \pi/2$.

signals is converted to a phase relation between the channels. Inasmuch as this phase difference on the null axis is 90°, the phase detector output will be zero in the absence of an angular error, and the tracking system remains serviceable if the target angle is estimated using the phase difference between the sum and difference channels.

It should be noted that a 90° phase shift is often inserted ahead of the sum-and-difference device for the purposes of data conversion [35].

With $\psi \neq \pi/2$ and $\psi \neq 0$, the sum and difference signals are out of phase. When there is a phase difference between the channels, the system is in an undesirable condition, leading to crosstalk between the channels and a reduction in stability.

**Case 3.** The RF channels in the receiver have identical amplitude-phase responses, and the sum and difference channels have identical gains, but different phases. In this case, $g = 1$; $k_s = k_d = k$; $\psi = 0$; $\gamma \neq 0$; and the tracking condition is given, in accordance with (7.22) and (7.25), by the expression:

$$S(\theta) = \mu\theta \cos \gamma, \qquad \left.\frac{dS(\theta)}{d\theta}\right|_{\theta=0} = \mu \cos \gamma$$

The condition $S(\theta) = 0$ is met when $\theta = 0$. This indicates that there is no shift in the null of the angle-sensing response. The phase imbalance affects the angle-sensing sensitivity and slope.

For $\gamma = \pm 90°$, the error signal in the tracking system is zero, regardless of the target angle. Auto tracking in this case is impossible. With $90° < \gamma < 270°$, the slope of the angle-sensing response is negative, which results in a loss in tracking system stability.

Thus, it is impossible to ignore a phase imbalance between the sum and difference channels. Although it does not lead to a direct shift in the null axis, any phase imbalance must be kept within acceptable limits, which may be determined by the acceptable reduction in angle-sensing sensitivity. For example, if the tolerable reduction in sensitivity is 50%, then the phase imbalance between the sum and difference channels must not exceed $\pm 60°$.

**Case 4.** The RF receiver channels have identical amplitude-phase responses, and the sum and difference channels are matched in phase, but have different gains.

In this case, $g = 1$; $\psi = 0$; $\gamma = 0$; $k_s \neq k_d$; and the tracking condition is described by the expression:

$$S(\theta) = \frac{k_s}{k_d}\mu\theta, \qquad \frac{dS(\theta)}{d\theta}\bigg|_{\theta=0} = \mu\frac{k_d}{k_s}$$

The error signal is zero when $\theta = 0$, and there is no shift in the null axis. The discrepancy in the amplitude response of the sum and difference channels affects only the angle-sensing sensitivity, which may be characterized by the formula:

$$k_d/k_s = (k_s \pm \Delta k)/k_s = 1 \pm \Delta k/k_s$$

The change in sensitivity is directly proportional to the gain imbalance in one of the channels. Thus, if the acceptable loss in sensitivity is 0.5, the gain of the difference channel should not differ from that of the sum channel by more than a factor of two.

**Case 5.** The receiver channels have identical amplitude responses, but different phase responses, both before and after the sum-and-difference converter. In this case, $g = 1$; $k_s = k_d = k$; $\psi \neq 0$; $\gamma \neq 0$; and the tracking condition, as follows from (7.23) and (7.25), is determined by the equations:

$$\mu\theta = \frac{1 + \sqrt{1 + (\sin^2\psi)\tan^2\gamma}}{(\sin\psi)\tan\gamma} \tag{7.28}$$

$$\frac{dS(\theta)}{d\theta}\bigg|_{d\theta=0} = \frac{2\mu\cos\gamma}{1 + \cos\gamma} \tag{7.29}$$

When $\psi = 0$ and $\gamma \neq 0$, which corresponds to Case 3, $\mu\theta = 0$. The effect of the phase imbalance affects only the angle-sensing sensitivity. When $\gamma = 0$ and $\psi \neq 0$ (Case 2), we again have $\mu\theta = 0$. A systematic error thus arises only when both the RF, sum, and difference channels are subject to simultaneous phase imbalances.

Generalized angular errors, calculated with (7.28), are plotted in Fig. 7.19. Using these calculated data and setting the angle-sensing sensitivity, it is possible to determine the systematic angular error as a function of the phase imbalance in the channels of an amplitude-sum-and-difference monopulse system. Thus, for example, if $\mu = 0.25$ deg$^{-1}$, $\gamma = 45°$, and $\psi = 10°$ and $45°$, the systematic phase error will be $0.35°$ and $1.27°$, respectively. If the acceptable error is one arc-minute, then the RF phase imbalance for $\gamma = 45°$ should not exceed $0.5°$.

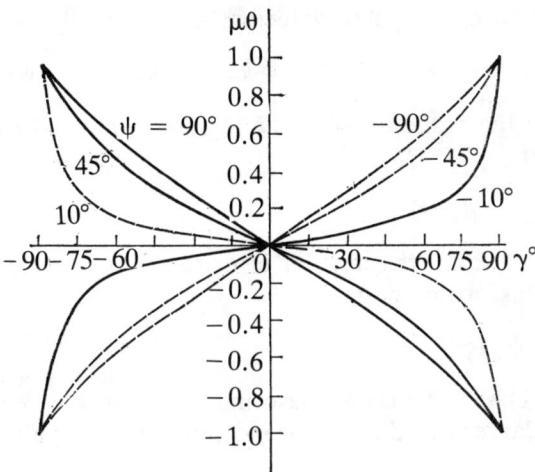

**FIGURE 7.19.** Normalized tracking error as a function of phase disparities in the receiver channels, calculated for imbalance in the RF channels.

The results of this analysis show that, in an amplitude-sum-and-difference system, as opposed to an amplitude-amplitude system, the RF receiver channels must have well matched phase responses. Considering the significant effect of RF channel phase mismatch, the radar should be designed such that the waveguide bridge, used for sum-and-difference processing, is as close to the antenna as possible, to minimize the opportunity for unequal phase responses in the RF lines.

A method for analyzing the effects of amplitude-phase imbalances on the angle accuracy with nonlinear angle-sensing responses is considered in [21], approximating the antenna amplitude pattern with a Gaussian pattern.

### 7.4.3 The Effects of Unequal Amplitude-Phase Responses on the Angle Accuracy of a Phase-Phase Monopulse Receiver

A simplified block diagram of a phase-phase monopulse system was presented in Fig. 4.2. We will assume that the gains of the receiver channels are $k_1$ and $k_2$, and that the phase difference between them is $90° + \psi$. We then obtain, at the output of the IF amplifiers:

$$\underline{u}_1(t, \theta) = u_{\lim} \exp[i(\omega_c t + \Delta\phi/2)]$$

$$\underline{u}_2(t, \theta) = u_{\lim} \exp[i(\omega_c t - \Delta\phi/2 + 90 + \psi)]$$

and at the output of the phase detector:

$$S(\theta) = \text{Re}[\underline{u}_1(t, \theta)\,\underline{u}^*{}_2(t, \theta)] = k_{pd}u^2{}_{\text{lim}} \sin(\Delta\phi + \psi) \qquad (7.30)$$

Substituting the value of $\Delta\phi$ in (7.30), and setting the error signal to zero, we obtain

$$\sin\theta = -\psi\lambda/(2\pi l)$$

With auto tracking, the angular error is comparatively small, and $\sin\theta \approx \theta$. From this, it follows that

$$\theta/\theta_3 \approx -\psi/(2\pi) \qquad (7.31)$$

Equation (7.31) shows that a phase imbalance in the receiver channels leads to angular errors directly proportional to the value of the phase mismatch.

The use of amplitude limiting prevents amplitude-response imbalances from affecting the tracking accuracy of phase-phase monopulse systems, as was shown in [43].

### 7.4.4 The Effects of Amplitude-Phase Imbalances on the Angle Accuracy of Phase-Sum-and-Difference Monopulse Systems

A simplified block diagram for a phase-sum-and-difference monopulse system was presented in Fig. 4.5. Using the same notation as was used in the analysis of amplitude-phase imbalances in amplitude-sum-and-difference systems, it can be shown that the error signal is given by the expression [21]:

$$S(\theta) = \frac{k_d\{(1 - g^2)\sin\gamma + [2g\sin(\Delta\phi + \psi)]\cos\gamma\}}{k_s[(1 + g^2) + 2g\cos(\Delta\phi + \psi)]} \qquad (7.32)$$

The tracking condition is then found to be

$$(1 - g^2)\sin\gamma + [2g\sin(\Delta\phi + \psi)]\cos\gamma = 0$$

Solving this equation for $\Delta\phi$, we find

$$\Delta\phi = \arcsin\left(\frac{g^2 - 1}{2g}\tan\gamma\right) - \psi \qquad (7.33)$$

Using the value $\Delta\phi = 2\pi l\theta/\lambda$ in (7.33), and using the approximation $\theta_3 = \lambda/l$, we obtain the following formula for the tracking error:

$$\frac{\theta}{\theta_3} = \frac{1}{2\pi}\left[\arcsin\left(\frac{g^2-1}{2g}\tan\gamma\right) - \psi\right] \tag{7.34}$$

We will now examine some practical examples.

**Case 1.** The RF channels have identical phase, but different gains, and the sum and difference channels have identical amplitude-phase responses. In this case, $\psi = 0$; $g \neq 1$; $k_s = k_d = k$; and $\gamma = 0$.

Substituting these values in (7.32) and (7.34), we find

$$S(\theta) = [2g\,\sin(\Delta\phi)]/[1 + g^2 + 2g\,\cos(\Delta\phi)], \qquad \theta/\theta_3 = 0$$

Consequently, amplitude imbalances in the RF portions of the receiver do not significantly affect the tracking accuracy: the null of the angle-sensing response is not shifted, and the angle-sensing sensitivity changes only insignificantly.

**Case 2.** The sum and difference channels have identical responses, and the RF channels have identical gains, but unequal phase shifts. In this case, $k_d = k_s = k$; $\gamma = 0$; $g = 1$; $\psi \neq 0$; and, for comparatively small angle errors:

$$S(\theta) = \frac{\sin(\Delta\phi + \psi)}{1 + \cos(\Delta\phi + \psi)} = \tan\left(\frac{\Delta\phi + \psi}{2}\right) \qquad \frac{\theta}{\theta_3} = \frac{-\psi}{2\pi} \tag{7.35}$$

As in a phase-phase monopulse system, the tracking error is directly proportional to the phase imbalance in the RF channels. The sign of the error depends on which channel has the greater phase lag.

**Case 3.** The RF channels have identical responses, and the sum and difference channels have equal phases, but different gains. In this case, $g = 1$; $\psi = 0$; $\gamma = 0$; $k_s \neq k_d$; and

$$S(\theta) = \frac{k_d}{k_s}\tan\frac{\Delta\phi}{2}, \qquad \mu' = \left.\frac{dS(\theta)}{d\theta}\right|_{\theta=0} = \frac{k_d}{k_s}\frac{\pi D}{\lambda} = \frac{k_d}{k_s}\mu$$

where $\mu$ is the slope of the angle-sensing response with identical receiver channels.

It is evident that a difference between the sum-and-difference channel gains does not result in a shift of the angle-sensing response null, and only slightly affects the angle-sensing sensitivity.

**Case 4.** The RF channels have identical responses, and the sum and difference channels have the same gains, but different phases.

In this case, $g = 1$; $\psi = 0$; $k_d = k_s = k$; $\gamma \neq 0$; and, in accordance with (7.35), $\theta/\theta_3 = 0$, which indicates that there is no shift in the null axis.

The angle-sensing response and its slope for this case, in accordance with (7.32), may be expressed through the equations:

$$S(\theta) = \frac{\sin(\Delta\phi)}{1 + \cos \Delta\phi} \cos \gamma = \left( \tan \frac{\Delta\phi}{2} \right) \cos \gamma \qquad (7.36)$$

$$\left. \frac{dS(\theta)}{d\theta} \right|_{\theta=0} = \frac{\pi}{\theta_3} \cos \gamma \qquad (7.37)$$

The phase mismatch in the sum and difference channels manifests itself, as may be seen in (7.36) and (7.37), as a change in the angle-sensing sensitivity.

With $\gamma = \pm \pi/2$, $S(\theta) = 0$ and $[dS(\theta)/d\theta]|_{\theta=0} = 0$. The error signal is zero regardless of the target angle relative to the antenna axis.

With $90° < \gamma < 270°$, the sign of the slope of the angle-sensing response and the error signal reverses. The coordinate system is now out of phase, and instability results. With a phase shift $\gamma \leq 60°$, the angle-sensing sensitivity is reduced by a factor no greater than two, in comparison with the case where the receiver channels are ideally matched.

It should be noted that amplitude and phase imbalances in the sum and difference channels have the same effect on the accuracy of both amplitude and phase systems, due to the fact that both types of systems use the same angle discriminator.

**Case 5.** The receiver channels, both up to and after the sum-and-difference device, have identical amplitude responses, but different phases. In this case, $g = 1$; $k_s = k_d = k$; $\psi \neq 0$; $\gamma \neq 0$; and the basic tracking properties of the system are determined by the following expressions:

$$S(\theta) = \frac{\sin(\Delta\phi + \psi)}{1 + \cos(\Delta\phi + \psi)} \cos \gamma = \tan \left[ \frac{\Delta\phi + \psi}{2} \right] \cos \gamma$$

$$\left. \frac{dS(\theta)}{d\theta} \right|_{\theta=\theta_t} = \frac{\pi}{\theta_3} \cos \gamma$$

where $\theta_d = -(\psi/2\pi)\theta_3$ is the operating point of the angle-sensing response, determined in accordance with the expression obtained from (7.34):

$$\theta/\theta_3 = -\psi/(2\pi)$$

These equations show that, when the channels of phase-sum-and-difference monopulse receivers have unequal phase responses both before and after the sum-and-difference device, there results a shift in the null of the angle-sensing response (a systematic error), and the angle-sensing sensitivity is reduced (the dynamic errors are increased).

Comparison of the results of the above analysis leads to the conclusion that systems using sum-and-difference angle discriminators require less precise matching of the amplitude-phase responses of the receiver channels than systems using amplitude or phase angle discriminators.

## 7.5 ADDITIONAL REQUIREMENTS ON THE EQUALITY OF THE AMPLITUDE-PHASE RESPONSES OF RECEIVER CHANNELS IN SYSTEMS USING WIDEBAND, CONTINUOUS, AND PULSED DOPPLER SIGNALS

The use of wideband transmitted signals (frequency-modulated and phase-shift keyed pulses, frequency-modulated CW signals) in monopulse systems places additional requirements on the design of these systems. This is a consequence of the fact that the signal processing for such systems is not ideal, causing additional angular errors to arise for various types of signals [35].

The antenna is usually of a sufficiently wide bandwidth so that the form of the received signal is not distorted, and, consequently, the additional demands relate primarily to the components which convert the information within the signals and the angle discriminators. We will examine these demands briefly for several transmitted waveforms.

### 7.5.1 Tracking with Frequency-Modulated Signals

When using frequency-modulated signals, the converters should perform their functions at any frequency within the spectrum of the received signal. Magic tees provide sufficient bandwidth for this purpose. The phase shifters, however, operate only with small frequency deviations.

As is demonstrated in [21, 35], the error signal at the output of the angle discriminator of a frequency-modulation system accordingly becomes a function of time, which degrades the system accuracy and interference tolerance.

If the carrier frequency of the received signals is modulated with a sinusoid:

$$\omega = \omega_0 + \Delta\omega \sin(\Omega t)$$

where $\Delta\omega$ is the frequency excursion, and $\Omega$ is the modulation frequency, then the error signal in a phase-phase monopulse system for $\theta = 0$ takes the form:

$$S(\theta) = \sin\{\omega_0\tau + \Delta\omega \sin[\Omega(t/2 + \tau)]\}$$

Correspondingly, we have for a phase-sum-and-difference system:

$$S(\theta) = \tan(\Delta\phi/2)\tan(\tfrac{1}{2}\{\omega_0\tau + \Delta\omega[\sin \Omega(t/2 + \tau)]\})$$

It follows that even with identical receiver channels, an unwanted amplitude modulation of the error signal arises.

Thus, when using frequency-modulated phase-comparison monopulse systems, the character of the angle-sensing response is altered and the angle accuracy is reduced by the unwanted amplitude modulation on the error signal. Therefore, when employing phase-comparison tracking techniques, any frequency modulation must exhibit only a slight excursion.

In amplitude-comparison monopulse systems with identical receiver channels, frequency modulation has an effect through the dependence of the antenna amplitude pattern on frequency, in addition to the parasitic amplitude modulation on the error signal, and the angle accuracy is accordingly reduced.

With unequal amplitude-phase responses in the receiver channels, the effects of frequency modulation are aggravated and the errors grow. The requirements for precise matching of the receiver channel responses, when using a frequency-modulated waveform, are thus greater than those indicated earlier. These tolerances may be estimated with the method described above for each interesting case.

### 7.5.2 Tracking by Velocity Gating with CW and Pulsed Doppler Signals

Examples of the design of monopulse systems which track by velocity were presented in Chapter 4. A common characteristic of these systems is that the target angle measurement is performed after the signals have been resolved in Doppler frequency. To this end, a Doppler filter is included in the receiver channel at the phase detector.

Inasmuch as the Doppler frequency shift in the signal depends on the velocity, trajectory, and aerodynamic properties of the target, and will, in practice, vary within rather wide limits, there are stringent requirements for maintaining identical Doppler filter responses in systems employing velocity resolution. Because these filters have narrow bandwidths and steep phase-frequency responses, any discrepancy in their responses can lead to large instrumentation errors as the Doppler frequency shift varies during the target track. In order to minimize this effect, it is necessary to stabilize the frequency of the signals passing through the Doppler filters to the phase detectors.

The demands for identical receiving channels in pulsed Doppler systems are analogous to those for CW systems.

### 7.5.3 Tracking with Pulse Compression

It is well known that improved range resolution can be obtained for given illumination energy and transmitter peak power requirements by employing intrapulse frequency (phase) modulation. This pulsed signal is passed through an optimum receiver filter and has a reduced effective duration, improving the range resolution, while the longer actual duration is sufficient to satisfy the energy requirements.

Pulse compression results from the fact that the optimum filter introduces delays for each spectral component of the signal such that the low-frequency components experience a longer delay than the high-frequency components. All of the signal components are combined in phase at a certain moment, providing a short pulse of high amplitude.

Any difference in the average tuned frequencies of the optimum filters leads to a change in the signal amplitude ratios, the appearance of an unwanted phase shift, and corresponding frequency modulation, which results in a change in the error signal and worsens the tracking accuracy.

If the mean signal frequency and disparity in the optimum filters' tuning frequencies are stable in time, then the distortion in the tracking system error signal may be cancelled by tuning the equipment. Otherwise, there is an unavoidable degradation in both accuracy and interference tolerance. Differences in the filter responses may lead to a mismatch between the rate of change of the frequency and envelope of the impulse response, and the corresponding parameters of the input signal. In this case, even if the filters are tuned identically to the center signal frequency, the parasitic phase shift and residual frequency modulation may be significant. It follows that it is extremely important to ensure that the optimum filters in monopulse systems, operating with frequency-modulated pulsed signals, are identical within strict tolerances.

### 7.6 DESIGN METHODS FOR REDUCING INSTRUMENTATION ERRORS IN ANGLE TRACKING

There are several different methods for reducing instrumentation errors in angle tracking. There are technological methods, connected with improving the technology of preparing the individual assemblies and equipment components; operating methods for preparing the system for operation, tuning, calibration and deployment; and design methods, associated with circuit solutions which allow the effects of amplitude-phase response disparities in the receiver channels to be reduced or eliminated altogether.

One widely employed operating method for reducing instrumentation tracking errors is to equalize the electrical lengths of the receiver channels with a pilot signal [112]. With this goal in mind, the receiving antenna is fitted with an auxiliary feed which is fed with signals from a special oscillator at the operating frequency. Reception of these signals allows the amplitude-phase responses of the receiver channels to be equated with the help of regulating components designed for this purpose.

Most interesting are the design approaches, including various methods for multiplexing the monopulse receiver channels, stabilizing the intermediate frequency, commutation of the receiver channels, and a number of other measures.

We will now consider the questions of multiplexing receiver channels in more detail. Multiplexing in essence replaces the two or three channels necessary for basic monopulse reception with a single channel. Inasmuch as the number of channels is thus minimized, so are the errors caused by disparities between multiple receiver channels. At the same time, the cost, size, and weight of the equipment is reduced. There are currently many methods for multiplexing the various functional channels into a single channel; we will now examine some of these.

### 7.6.1 Multiplexing the Channels at High Frequency

The block diagram in Fig. 7.20 represents the structure of one of the first amplitude-comparison monopulse radars designed for measuring coordinates in one plane [21].

The antenna system of the given system consists of two parabolic reflectors connected to one another mechanically; one receives, and the other transmits. With this design, the problems associated with switching the RF circuits to receive and transmit are avoided. The distinguishing feature of this system is the manner in which the receiver channels are multiplexed into a single channel at high frequency. This is achieved by using the time difference of the signals in the various channels, which is accomplished by delaying one of the signals by an amount exceeding the pulse length.

After multiplexing, the RF signals are converted to IF signals, amplified in a logarithmic amplifier, detected, and separated with the appropriate gating. The separated signals are compared by subtraction, for which the undelayed signal is delayed by the same amount as the other signal was delayed when they were multiplexed.

A drawback to this multiplexing method is the fact that the range resolution is worsened by a factor of two, because the system cannot resolve targets separated by less than two pulse lengths. If the targets are too close to one another, the signal from the farther target and the delayed signal

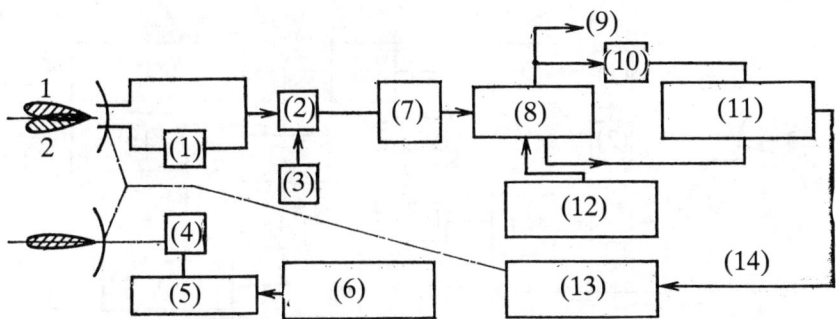

**FIGURE 7.20.** Simplified block diagram of an amplitude-amplitude radar with RF channel multiplexing.

| | | | |
|---|---|---|---|
| (1) | delay line | (8) | switching circuit |
| (2) | mixer | (9) | to range channel |
| (3) | local oscillator | (10) | delay line |
| (4) | phase detector | (11) | subtraction device |
| (5) | modulator | (12) | gating device |
| (6) | synchronization device | (13) | antenna control device |
| (7) | IF amplifier + video detector | (14) | error signal |

from the closer target will pass through the receiver simultaneously and not be separated. Another shortcoming of this system is that it is vulnerable to jamming in the form of paired pulses, separated by the same delay used for multiplexing.

One problem in implementing RF multiplexing is the bulkiness of the delay line (DL). The delay line in the system developed by Summers [21], for example, consists of a coiled waveguide measuring 30.5 m. This required compensation for the attenuation introduced in the line. With this design, it is difficult to avoid disparities between the various parts of the receiver channels.

### 7.6.2 Multiplexing Receiver Channels at Intermediate Frequency

The block diagram of a monopulse receiver, which multiplexes the channels at intermediate frequency, is shown in Fig 7.21 [21]. A phase-comparison monopulse receiver may be designed analogously.

The signals received by antennas 1 and 2 are passed through the mixers, where they are converted to IF signals with the help of a common oscillator (LO), and then amplified in IF preamplifiers. After one of the IF preamplifiers, a delay line introducing a lag exceeding one pulse length is inserted

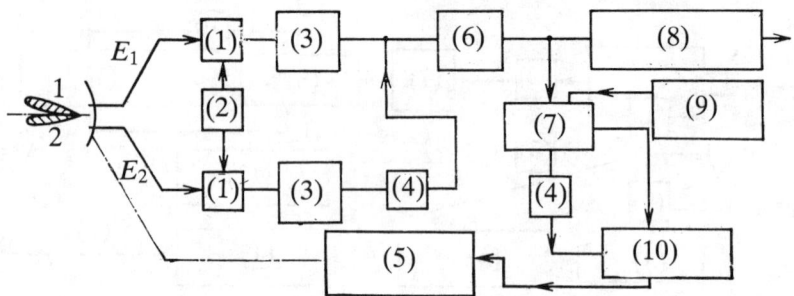

**FIGURE 7.21** Simplified block diagram for the receiver of an amplitude-amplitude monopulse radar with IF channel multiplexing.

(1)  mixer                          (6)  IF amplifier and amplitude detector
(2)  local oscillator               (7)  switching circuit
(3)  IF preamplifier                (8)  range autotracking system
(4)  delay line                     (9)  gating device
(5)  antenna control system  (10)  comparison device

into one of the signals. This displacement of the signals in time allows them both to be amplified in a common IF amplifier. After IF amplification, the signals are detected in the amplitude detector (AD), separated into two signals with the help of a gated switch, after which they reach the comparison circuitry. The diagram in Fig. 7.22 elucidates the operation of this receiver.

A device has been constructed in which the delay line is connected directly to the output of balanced mixers, and separate IF preamplifiers are unnecessary.

Inasmuch as the construction of a compact delay line is significantly simpler at IF than at RF, this circuit has an advantage over that in which the signals are multiplexed at RF (see Fig. 7.20). Another advantage to this design is that multiplexing the signals at IF permits the tracking system to be protected from the effects of pulse-pair jamming by employing a gated IF preamplifier which will pass only one of the interfering pulses.

There is a drawback to multiplexing the signals at IF, however, in that the effects of disparities in the receiver channels increase, because the length of the channels before multiplexing are longer than in the circuit of Fig. 7.20. The problem of poorer range resolution which was described for the system with RF multiplexing also exists in systems with IF multiplexing.

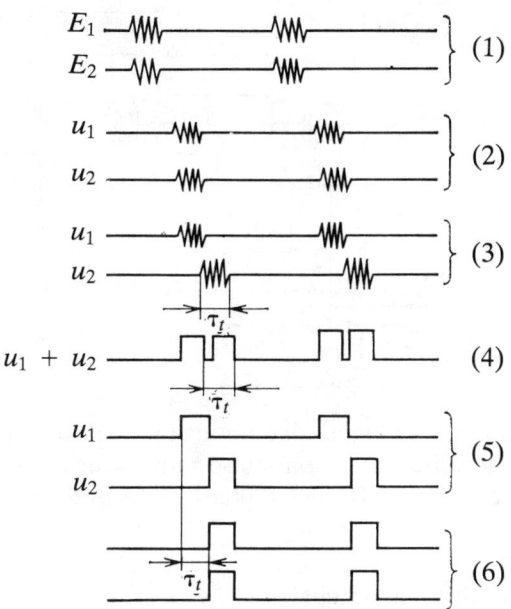

**FIGURE 7.22.** Voltage diagrams illustrating the passage of signals through the receiver of an amplitude monopulse system which multiplexes the channels at IF.

(1)   transmitted signal
(2)   received signal
(3)   signals at input to IF amplifier
(4)   signals at output of detector
(5)   signals at output of switching circuit
(6)   signals at input of comparison device

### 7.6.3 Multiplexing the Receiver Channels Using Phase Shifts and a Time Delay

The block diagram of the receiver of a monopulse radar, which multiplexes the channels using this method, is shown in Fig. 7.23 [21].

The channels are multiplexed in two stages. First, the difference channels are multiplexed by shifting the azimuth difference signal $E_{da}$ by 90° and adding it to the elevation difference signal $E_{de}$. This angle measurement signal is then down-converted, passed through an IF preamplifier, delayed

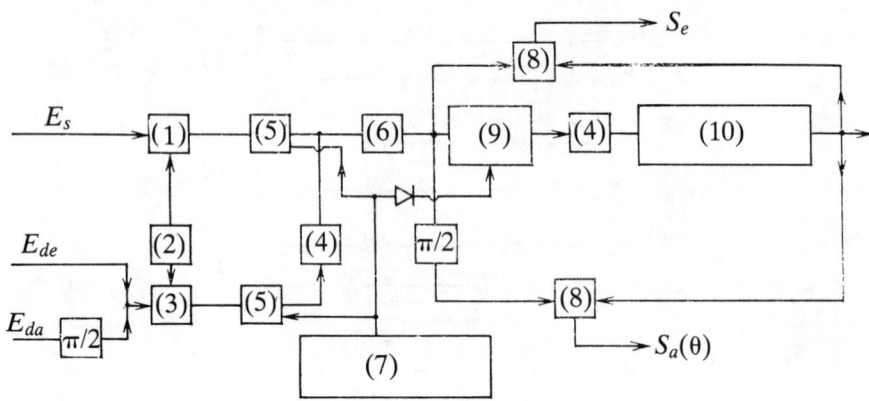

**FIGURE 7.23.** Simplified block diagram of the receiver of an amplitude-sum-and-difference monopulse radar which multiplexes the range and angle channel at IF, using phase shifts and time delays:

$E_s$    sum signal
$E_{de}$   elevation difference signal
$E_{da}$   azimuth difference signal

|  |  |  |  |
|---|---|---|---|
| (1) | mixer 1 | (6) | IF amplifier |
| (2) | local oscillator | (7) | range autotracking system |
| (3) | mixer 2 | (8) | phase detector |
| (4) | delay line | (9) | IF amplifier stage |
| (5) | IF preamplifier | (10) | 360° phase shifter |

by a time approximately equal to the pulse length, and added to the IF sum signal $E_s$, which functions as both the range signal and the reference signal for the difference channels.

The output signal of the common IF amplifier is split three ways: one part goes to the phase-sensitive detector (PD) of the elevation channel; another part is passed to the phase detector of the azimuth channel with a 90° phase shift; the third part undergoes further IF amplification, is delayed by a time equal to the delay introduced in the difference signals prior to multiplexing, passes through a controlled phase shifter to the range channel and to the difference channel phase detectors as a reference signal. The phase-sensitive detectors allow the elevation and azimuth error signals $S_e(\theta)$ and $S_a(\theta)$ to be obtained.

In obtaining the elevation error signal at the phase detector, one signal is taken directly from the common IF amplifier, while the other passes through the gated IF amplifier, delay line, and phase shifter. Due to this arrangement, the reference portion of the first signal arrives at the phase detector at the same time as the difference portion of the second (see the diagrams in Fig. 7.24), and the two combine to form an error signal proportional to the elevation offset angle of the target.

The azimuth component of this difference signal will not affect the output elevation error signal, inasmuch as it is in quadrature with the reference signal. The azimuth error signal is obtained from the other difference signal and phase detector, one input of which is taken from the common IF amplifier with a 90° phase shift, while the other input passes through the gated IF amplifier stage, delay line, and phase shifter. In this case, the difference portion of the first signal reaches the detector at the same time as the reference portion of the second signal, and produces the azimuth error signal at the output of the detector.

The elevation component of this difference signal is not passed by the phase detector, because it is in quadrature with the reference signal. Inasmuch as the circuit uses two delay lines, there may be a significant phase difference between the reference and difference signals. A phase shifter with a widely variable phase is inserted in the reference signal channel in order to compensate for these undesired phase shifts.

In order to maintain the signal-to-noise ratio through the receiver, a gated IF preamplifier is used, so that only the useful signals are amplified. Placing the delay line after the IF preamplifier is important, because it improves the signal-to-noise ratio. If the delay line is placed before the preamplifier, then the signal is attenuated prior to amplification, which reduces the signal-to-noise ratio and degrades the target angle and range tracking accuracy.

A disadvantage to this multiplexing system is its high sensitivity to disparities in the phase responses of the receiver channels. This deprives the sum-and-difference monopulse receiver of its main advantage — the insensitivity of the null axis to unwanted phase shifts. This shortcoming makes such a system inappropriate for use in radars which employ frequency agility over a wide band.

In some cases, it is possible to multiplex only the difference channels with this method, and compensate for the effects of frequency hopping by automatically adjusting the phase of the channels.

In light of the expediency of using single-channel monopulse systems and the necessity of performing certain signal manipulation in the receiver, several specialized devices in the form of functional modules have been developed for use in radar systems. One example is a device designed by an American company which comprises all the components between a four-channel antenna, transmitter and single-channel receiver. The device is meant for operation over a 10% bandwidth in the 3-cm wave band, and achieves channel operation of greater than 25 dB [95].

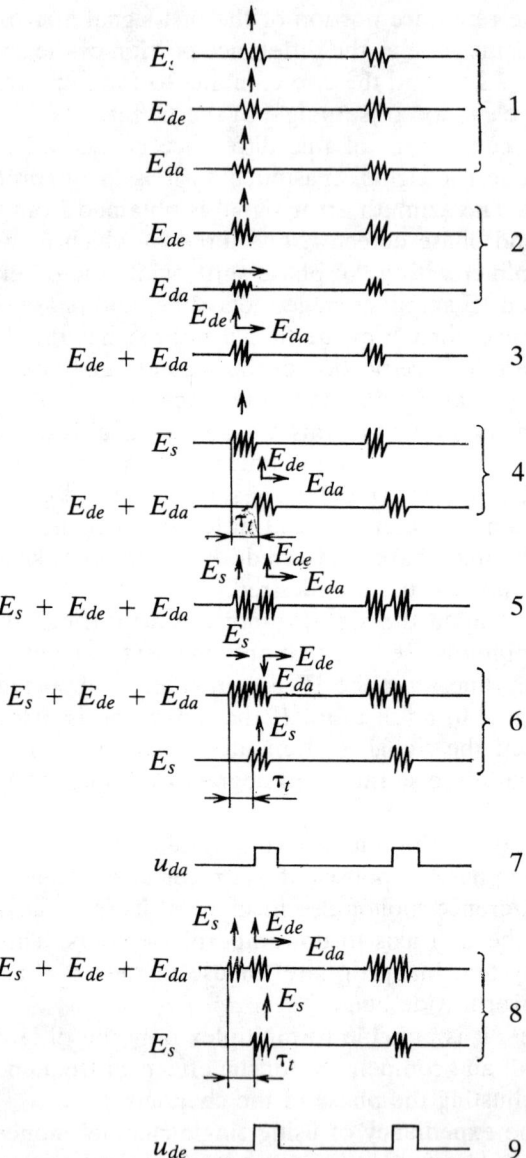

**FIGURE 7.24.** Signal voltage diagrams illustrating the operation of the amplitude-sum-and-difference receiver shown in Fig. 7.23 (the arrows indicate the signal phase relations).

1: at the output of the waveguide bridges
2: at the input to mixer 2
3: at the output of mixer 2
4: at the input of the IF amplifier
5: at the output of the IF amplifier
6: at the input to the azimuth channel phase detector
7: at the output of the azimuth channel phase detector
8: at the input of the elevation channel phase detector
9: at the output of the elevation channel phase detector

### 7.6.4 Multiplexing the Receiver Channels at Intermediate Frequency with Signal Coding

A simplified block diagram of a monopulse receiver, using a multiplexing method of a different principle than those considered above [88], is presented in Fig. 7.25. In this circuit, the signals are coded at the antenna outputs, summed at high frequency, processed in a single-channel receiver, separated in power, and finally processed in an angle processor with the usual monopulse methods.

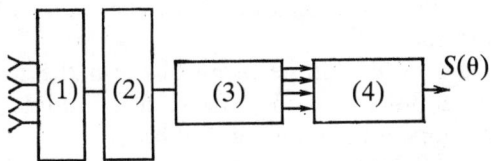

**FIGURE 7.25.** Simplified block diagram of a single channel monopulse receiver employing signal coding.

(1) active adder      (3) power divider
(2) wideband receiver (4) angle processor

The coding in effect "colors" the received signal, allowing it to be distinguished from other signals with which it is mixed and extracted from them after processing in a single-channel receiver. The signals are digitally phase-coded with the help of a double balanced mixer fed with a coded RF signal. A 15-bit code, formed in a four-stage shift register, is shown as an example in Fig. 7.26. The coded words passed to the balanced mixer (modulator) are offset from one another by a time equal to, or exceeding, the bit length, preventing interaction of the codes when the multiplexing is performed.

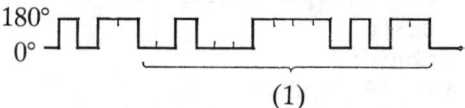

(1)

**FIGURE 7.26.** Example of a 15-bit digital code, used for multiplexing receiver channels.

(1)   15-bit code

Due to the fact that the code bit length is significantly less than that of the pulsed signal received by the antenna, the bandwidth of the coded signal is accordingly expanded. Therefore, in order to perform amplification and processing, the receiver needs to have a substantially larger passband than would a monopulse receiver which does not multiplex the channels.

The combination of IF signals, at the output of the single-channel receiver, passes through a power divider and on to a block of four dual-balanced mixers; these act as demodulators, recovering the original structure of the signals required to extract the angle information. High quality signal separation requires strict time synchronization of the modulating and demodulating codes.

Inasmuch as the coding is performed within the limits of the pulse duration with no additional time shifts in the multiplexing process, this method does not worsen the range resolution. This is a primary advantage of the method over those considered earlier.

### 7.6.5 Multiplexing the Channels Using Low-Frequency Modulation in One of the Channels

A block diagram of a monopulse receiver performing this type of multiplexing is shown in Fig. 7.27 [21].

This receiver forms the sum and difference signals in the usual way with a magic tee. The difference signal, which contains the target angle information, is amplitude-modulated at the audio frequency $\Omega$, and summed with the sum channel signal. Approximating the sum pattern with $\cos\theta$ and the difference pattern with $\sin\theta$, the resulting signal may accordingly be written as

$$\underline{E}(t, \theta) = E_m[\cos\theta + (\sin\theta)\sin\Omega t]\exp(i\omega t) \qquad (7.38)$$

With small angular errors (7.38) may be simplified to

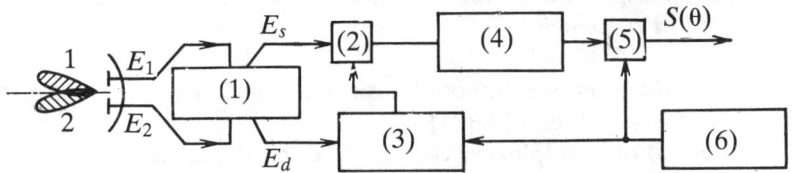

**FIGURE 7.27.** Simplified block diagram of the receiver of an amplitude-sum-and-difference monopulse radar which performs multiplexing with low-frequency modulation of the difference channel.

(1)  Magic tee            (4)  AM receiver
(2)  adder                (5)  phase detector
(3)  balanced modulator   (6)  low-frequency oscillator

$$\underline{E}(t, \theta) = E_m[1 + \theta \sin(\Omega t)]\exp(i\omega t) \tag{7.39}$$

It may be seen from (7.39) that the coefficient modulating the resultant signal is proportional to the angle offset. Amplitude detection of the signal at the receiver output thus gives a low-frequency signal of frequency $\Omega$, the amplitude of which is proportional to the magnitude of the angular error, while the phase indicates the direction of the error. As the error passes through zero, the phase of the signal changes by 180°. The low-frequency signal emerging from the AM signal receiver is compared in a phase detector with the output of a low-frequency signal generator. A dc error signal is formed, which is used to control the antenna servomechanism.

This receiver can use a wideband signal processor without reducing its sensitivity, due to which undesired phase shifts may be substantially reduced. The tuning-critical IF amplifier is no longer a source of errors, due to the fact that the sum and difference signals passing through it are identically phase-shifted.

The system is easy to tune and remains stable in a variety of operating conditions. Multiplexing the channels with low-frequency modulation, however, makes the system sensitive to amplitude fluctuations in the received signals and vulnerable to amplitude-modulated jamming. In order to generate such jamming, it is sufficient to know the modulation frequency with an accuracy equal to the tracking system bandwidth. This is a serious fault in this receiver system, which limits its application.

### 7.6.6 Multiplexing the Receiver Channels by Separating the Signals in Frequency

The block diagram of a monopulse receiver which uses this multiplexing method is shown in Fig. 7.28 [21].

This method of multiplexing converts the RF signals to signals close to IF, sums them, and clips them in a multistage amplifier-limiter. At the output of the IF amplifier, the signals reach bandpass filters, tuned to the corresponding frequencies of the multiplexed signals, and are separated into the sum and difference signals.

After the sum signal has been separated, it is converted to the frequency of the difference signal. The resultant signal is used as a reference signal in the phase-sensitive detector which affects the angle meassurement and forms the error signal.

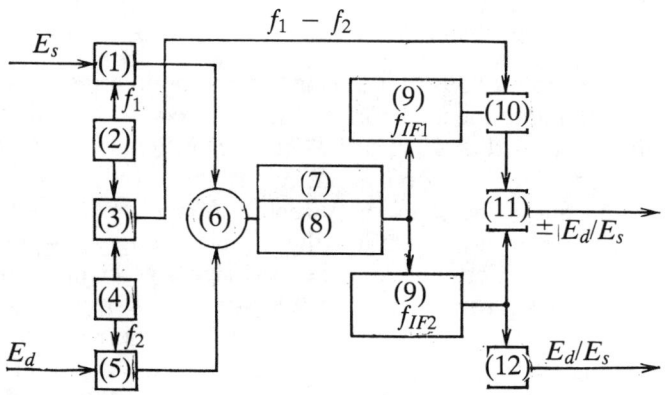

**FIGURE 7.28** Simplified block diagram of the receiver of an amplitude-sum-and-difference monopulse radar, which uses a single-channel signal processing device with frequency separation.

| | |
|---|---|
| (1) mixer 1 | (7) IF amplifier |
| (2) local oscillator 1 | (8) clipper |
| (3) mixer 2 | (9) bandpass filter |
| (4) local oscillator 2 | (10) mixer 4 |
| (5) mixer 3 | (11) phase detector |
| (6) adder | (12) amplitude detector |

This method of multiplexing can be used in systems which are designed for two-coordinate angle measurement. In this case, the multiplexed signals (one sum and two difference signals) are converted into three signals of closely spaced frequencies, such that their spectra do not overlap, allowing both difference signals to be normalized without the usual sum signal AGC systems.

Application of this sort of multiplexing is not limited to sum-and-difference systems. It may also be used in other systems—an amplitude-comparison system with logarithmic amplifiers, for example. The only difference in the system of Fig. 7.28 for such a receiver would be to replace the phase detector with an amplitude detector, as necessary for determining the target offset angle in such a radar.

The results of analyzing the signal processing for the case of an ideal wideband limiter show that if the signals at the limiter inputs have amplitudes $\underline{E}_s$ and $\underline{E}_d$ and frequencies $\omega_1$ and $\omega_2$, then the bandpass filter output signals can be expressed as

$$\overline{S_1(t)} \approx (2u_{\lim}/\pi)(E_d/E_s)\cos(\omega_1 t + \phi_1)$$
$$\overline{S_2(t)} \approx (4u_{\lim}/\pi)\cos(\omega_2 t + \phi_2).$$

The amplitudes of $\overline{S_1(t)}$ and $\overline{S_2(t)}$ are plotted as a function of the ratio of the instantaneous amplitudes of the difference and sum signals $E_d/E_s$ in Fig. 7.29. It can be seen in the drawing that when the ratio of the difference and sum signals is small, as is usually the case when automatically tracking a target, the output voltage $\overline{S_1(t)}$ is almost directly proportional to the ratio of the amplitudes $E_d/E_s$ and, consequently, directly proportional to the ratio of the amplitudes of the signals received by the radar antenna. Thus, it is possible to measure the target angle with sufficient accuracy using this signal processing method.

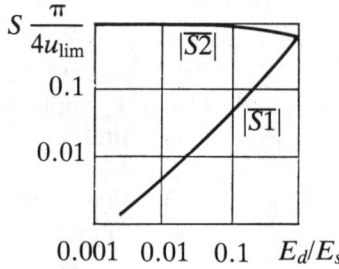

**FIGURE 7.29.** Bandpass filter output signal amplitudes as normalized functions of the ratio of the difference and sum signals at the input to the ideal limiter.

There are other methods of multiplexing the channels of a monopulse receiver by separating them in frequency. The structure of a receiver which multiplexes the channels using double frequency conversion is shown in

Fig. 7.30. The angle sensor in this system is an amplitude-phase antenna operated in two modes. Two signals of frequency $f$ are formed at the antenna output: a sum and a difference. These are then mixed at mixers 1 and 2 with signals from two oscillators of different frequencies to form IF signals with frequencies $(f - f_1)$ and $(f - f_2)$, respectively, after which they are summed and amplified in IF amplifier 1. After this amplification, the signals are detected in an amplitude detector. The detector acts as a second frequency converter, inasmuch as the output signal has a frequency $\Delta f_0 = f_1 - f_2$. This second IF signal is split to a logarithmic amplifier and IF amplifier 2.

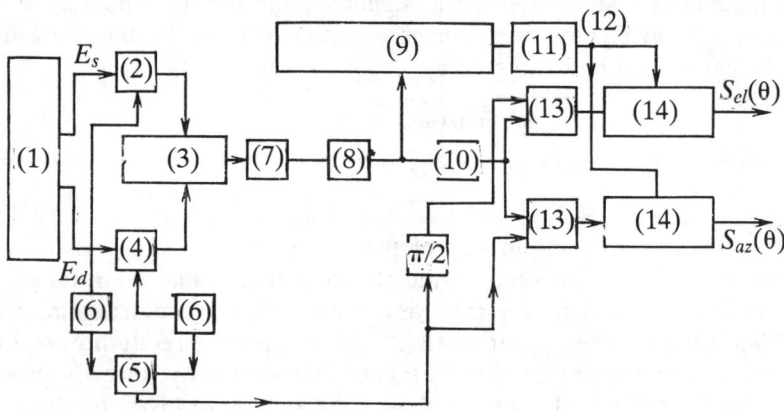

**FIGURE 7.30.** Simplified block diagram of a receiver which multiplexes the channels using dual-frequency conversion.

| | | | |
|---|---|---|---|
| (1) | dual-mode antenna | (8) | amplitude detector |
| (2) | mixer 1 | (9) | logarithmic video amplifier |
| (3) | adder | (10) | IF amplifier 2 |
| (4) | mixer 2 | (11) | filter |
| (5) | mixer 3 | (12) | AGC |
| (6) | local oscillator ($LO_{f1,f2}$) | (13) | phase detector |
| (7) | IF amplifier 1 | (14) | amplifier |

The amplified signal with frequency $\Delta f_0$ is taken from IF amplifier 2 and split to two phase detectors connected in parallel. The second inputs of both detectors are fed with reference signals at frequency $\Delta f_0$, formed directly from the local oscillator signals by mixing them in mixer 3. One of the signals is passed directly to the phase detector, while the other is shifted 90° before reaching the other phase detector. As a result, the output of one of the phase detectors is the azimuth error signal, while the output of the other is the elevation error signal. Both signals are amplified and normalized in separate amplifiers. The normalization is performed with respect to the signal passed through the logarithmic amplifier and a low-frequency bandpass filter, which ensures that the error signals are independent of the amplitudes of the signals received by the antennas.

### 7.6.7 Multiplexing the Receiver Channels with Commutation

Alternate-period channel commutation [38] is another way to compensate for unequal amplitude-phase responses in the receiver channels. The receiver channels are commutated at half the pulse repetition frequency which alternately connects the IF amplifiers first to one, and then to the other antenna. With this approach, the phase shift caused by channel response disparities will alternate between positive and negative values, so that the resultant error, due to these discrepancies, in the averaged signal will be significantly reduced. The commutation can also be performed at an arbitrary frequency unrelated to the pulse repetition frequency.

Analysis of an amplitude-amplitude monopulse system has shown [21] that with channel commutation the error signal is given by the expression:

$$S(\theta) = k_{02}(g_0 + 1)\mu\theta$$

The tracking condition $S(\theta) = 0$ is evidently independent of the unequal channel amplitude responses, but the system sensitivity has been reduced by a factor of $2/(g_0 + 1)$, where $g_0 = k_{01}/k_{02}$. Therefore, unequal gains in the receiver channels does not lead to systematic angular errors in a system employing commutation.

If the channels are commutated at a point removed from the antenna, the system will still be sensitive to disparities in the RF channel amplitude responses. Thus, when the commutation is performed at the input to the IF amplifier channels, the angular error in this system will be given by the expression [21]:

$$\mu\theta = \tan[-\ln(g/2)] \sim -\ln(g/2) \tag{7.40}$$

where $g$ is the ratio of the RF channel gains (up to the mixer).

Equation (7.40) is analogous to (7.20), which was obtained for the tracking error due to amplitude response disparities in the RF channels with no commutation.

There are other methods for multiplexing the receiver channels with commutation. One of them is based on gating the multiplexed channels, alternately enabling them for a time smaller than the pulse length by a factor of $n$ (where $n$ is the number of multiplexed channels). One of the components of the resulting signal in the receiver channel is used as a reference, and the other two for angle measurement. Owing to the time separation of the received pulses introduced by this process, signal conversion and amplification may be accomplished with a single-channel receiver. Appropriate delay lines are used to compensate for the phase lags resulting in the signals, and the signals are separated at the output of the receiver by gating synchronized with the gating of the input signals.

There is a method for multiplexing the receiver channels with alternating antenna connections without splitting the pulses. The block diagram of an original system for multiplexing the channels with commutation in a monopulse receiver is shown in Fig. 7.31 [21]. The commutating elements in this receiver are ferrite phase shifters placed in the antenna waveguide channels, which are controlled from a special source with a specially programmed pulsed voltage.

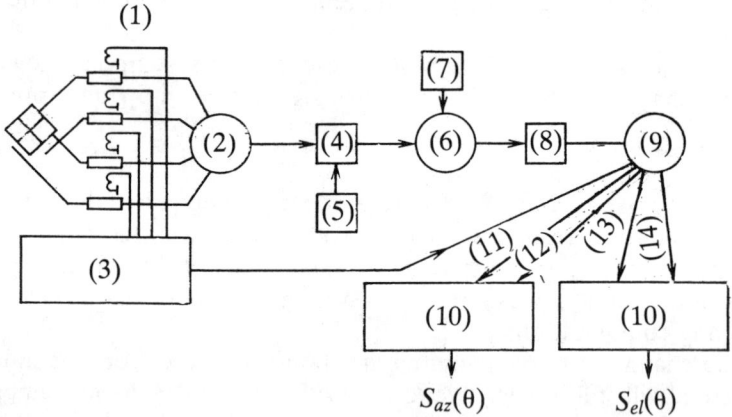

**FIGURE 7.31.** Simplified block diagram of a monopulse radar employing commutation of the receiving/transmitting RE channels.

| | | | |
|---|---|---|---|
| (1) | ferrite phase shifters | (8) | IF amplifier |
| (2) | adder | (9) | receiver |
| (3) | switch control device | (10) | comparison device |
| (4) | receiving/transmitting switch | (11) | channel "left" |
| (5) | transmitter | (12) | channel "right" |
| (6) | mixer | (13) | channel "lower" |
| (7) | local oscillator | (14) | channel "upper" |

When there is no magnetizing voltage across the ferrite phase shifters, all of the channels are in phase, and the resulting antenna pattern has a maximum in the direction of the antenna boresight. The application of a calibrated voltage pulse to any of the phase shifters introduces an additional phase shift in the corresponding channel, leading to a shift in the axis of the resulting pattern from the boresight. Depending on which channel is affected, the beam is deflected downward, upward, to the left, or to the right.

The sequential beam positioning which results leads to signals separated in time at the receiver input, which allows the ensuing conversion and amplification to be performed in a single-channel receiver. The signals are separated at the output of the receiver with the help of a switch controlled in synchronization with the phase shifters, and are then passed in pairs to comparison devices where the azimuth and elevation error signals are formed.

The beam direction may be controlled with a speed approaching several kilohertz. Combined with coding, such a commutation system is fairly insensitive to coherent jamming and the low-frequency amplitude fluctuations intrinsic to reflected signals.

## 7.7 REDUCING THE EFFECTS OF FREQUENCY-DEPENDENT DISPARITIES IN THE AMPLITUDE-PHASE RESPONSES OF RECEIVER CHANNELS ON THE ACCURACY OF MONOPULSE RADARS

The ability to vary the operating frequency over a wide range is a characteristic of modern radars. This is mainly a result of the efforts to improve the noise tolerance of the radar, and ensure that it will continue to operate normally under the difficult conditions of jamming. The frequency changes can lead to errors resulting from disparities in the amplitude-phase responses of the receiver channels, which, in turn, degrade the system accuracy.

To control this source of angular errors, careful experimentation with the equipment is usually performed over the entire operating frequency range. The values of the angle errors, arising as the frequency is varied, are determined and noted, and methods are developed for compensating for these errors during radar operation [48].

# Chapter 8

# Interference Tolerance of Monopulse Radars

## 8.1 THE SIGNIFICANCE OF RADAR INTERFERENCE TOLERANCE

Interference tolerance is usually understood to mean the ability of a radar system to generate a useful signal and perform its assigned detection, coordinate measurement, and tracking operations with sufficient accuracy under conditions of interference.

The significance of interference tolerance follows directly from the key role played by radars in modern weapon systems. Owing to their reliability, long operating ranges, ability to obtain reliable target information at any time and in any meteorological conditions, and the ease with which they may be incorporated into automatic systems, radars determine, to a large extent, the effectiveness of the weapon systems using them. Therefore, radar countermeasures employing jamming are very important in reducing the effectiveness of modern weapon systems. The basic objective of a jamming system is to distort (mask) the information in the receiving channels of the given radars, and to create redundant (spurious) information, making it difficult to detect targets, measure their coordinates, and organize a defensive response to dangerous targets.

The capabilities of radar countermeasures are based on the same principles as radar operation. As is known, radar operates by establishing radio contact with an object (target); the target becomes an element of a closed loop within which information is obtained on the position of the target and the parameters of its motion. Changes in the corresponding characteristics of the target affected with electronic countermeasures (ECM) lead to an unavoidable loss of radar performance. The comparative vulnerability of the majority of radar systems, including monopulse systems, is due mainly to the fact that they cannot work without external connections closing the loop through the target. Thus, radars give themselves away with their radiation, and give the enemy initial information for organizing their ECM.

Ever since the development of the first radars the possibility of radar countermeasures has been known, and they have been used, as necessary, for military goals. Radars and radar countermeasures have accordingly been developed in parallel, as two areas of new technology. As radars have been improved, radar countermeasures have been improved in pace, and the methods for applying the new techniques have been perfected as new jamming methods are developed. The development of radar countermeasures has, in turn, stimulated further improvements in radar and antijamming techniques. Radar and radar countermeasure devices, together, constitute the means for conducting electronic warfare.

As has been reported in the US press, the military operations in Vietnam and the Near East have shown that electronic warfare capabilities are necessary factors in supporting the military operations of all forms of armed forces. A US analysis [50, 92] of experience with electronic warfare showed that the use of electronic warfare techniques significantly increased the "survivability" of a plane in evading antiaircraft defenses, and in Vietnam prevented damage to planes, each costing approximately a million dollars. Without the use of electronic warfare techniques, the loss of aircraft in Vietnam would have been five times greater [50].

Considering the role of radar in modern weapon systems, and the ability to take effective countermeasures against them, radar designers must devote much attention to the interference immunity of the radar. One of the fundamental causes of the development of the monopulse method, and its intense introduction into radar technology, was the effort to increase the interference immunity of radars.

Before the introduction of the monopulse method, single-channel angle measurement and autotracking systems were widely used. These employ conical-scanning or sequential lobing techniques, are vulnerable to jamming, and are incapable of performing their military missions when modern jammers are used against them. For example, these systems are vulnerable to repeater jamming, which is amplitude-modulated at the scanning frequency [92, 109].

In estimating the interference immunity of monopulse radars, relative to modern jamming methods, it must be remembered that the monopulse method is applied only to measure angular coordinates. The methods of detection, determining range and speed, and tracking by range and speed, do not differ in principle from the methods used in the usual single-channel tracking radars. Therefore, there is continuity in the forms of jamming, and the means of countering it, applied to single-channel and dual-channel angle measurement systems. Inasmuch as the measurement of angular coordinates (target tracking by direction) occurs after target search, detection and selection, the interference immunity of the angle channel also depends on the interference immunity of the target detection and selection channels. It is, therefore, advisable to examine the possible means of jamming all the basic channels in monopulse radars.

## 8.2 THE GENERAL CHARACTERISTICS
## OF RADAR INTERFERENCE

Interference is any radio signal which, upon reception by a radar, worsens the radar operation and destroys the reliability and accuracy of the target information provided by the radar. Interference may be either natural or man-made [5, 21, 30]. Natural interference can result from storm activity in the atmosphere, various forms of precipitation (snow, hail, rain), the operation of poorly shielded electrical equipment, thermal noises, cosmic noise, and so on. Man-made interference is often called intentional interference or jamming. In assessing the interference immunity of radars, we will consider only jamming.

Jamming may be categorized as either active or passive. Active jammers create interference with special signal generators, producing radio emissions of varying structure over the operating band of the radar. Such jamming may be continuous or pulsed high-frequency signals with various modulations (amplitude, phase, frequency, pulse length, pulse repetition interval, polarization, *etcetera*). Passive jammers reradiate the signals transmitted by the radar itself, without using any special signal generators. Dipole reflectors are often used for passive jamming, usually constructed as metallized strips with a length equal to half the operating wavelength of the radar. Various corner and lens reflectors are also employed in passive jamming.

All jammers may also be categorized as either of the masking or deception type. Masking jammers are designed to prevent or significantly hinder target detection, and worsen the ability to determine its position and the parameters of its motion. The most common active method for creating this type of interference is use of a jamming transmitter; the most common passive device is chaff.

Deception jamming is designed to confuse the radar operator or automatic system by creating redundant information or spurious signals whose parameters mimic those of real targets, thus making it more difficult to select the real targets. Therefore, this type of jamming is sometimes referred to as simulation jamming. Deception jamming may be either active (with special transmitters) or passive (decoy targets).

There is a substantial difference between deception jamming and masking jamming in terms of their operational security. The radar operator is usually unable to determine that his system is being subject to deception, whereas masking jamming is easily distinguished on radar display screens. It should also be noted that active deception jamming is favorable over active masking jamming from the point of view of power considerations, inasmuch as with active deception practically all of the jammer transmitter power may be concentrated in the receiving band of the victim radar. Active deception jamming, however, is more complex from a technology point of view, because it requires rather advanced specialized equipment.

Jammers may also be characterized by the target of their interference: the target detection channel, the target selection channel (by velocity or range), the angle measurement channel, or a remote control channel.

Interference from high-altitude nuclear explosions falls outside the categories above, and includes the whole series of phenomena resulting from nuclear explosions, including disturbances in radio wave propagation and effects on the radio equipment.

Regardless of its character and implementation, jamming is ultimately designed to degrade the tactical and technical capabilities of the radar through the following: disturbing the process of target detection (decreasing the probability of detection and increasing the false alarm rate); confusing the operator; delaying target detection or track initiation; overloading the data processing system with an excessive number of signals with various parameters; inhibiting the ability to measure target range, speed, and direction; interrupting automatic target tracking; creating errors in the measurements of target range, speed, angular coordinates, and so on.

In order to assess a radar's resistance to jamming, it is necessary to know the particular effects which the interference has on the operation of the various portions of the radar. We will accordingly examine, in more detail, the effects of various types of jamming on the operation of radar channels, and possible protection against these effects.

## 8.3 TARGET DETECTION CHANNEL JAMMING

The job of the detection channel in a radar is to detect targets and determine the actual situation in the observation zone prior to making the appropriate decisions. Thus, the operation of this channel is always associated with a search for true target signals in the zone of observation, and target detection is a probabilistic process during which both correct and incorrect decisions are made. The goal in jamming the detection channel is to complicate the process of detecting real targets, to increase the false alarm rate, or to eliminate the ability to detect targets altogether.

Inasmuch as the process of target detection is characterized by operation of the receiver with maximum sensitivity and without the use of target selection devices which limit the capabilities of jammers, the radar is more vulnerable to jamming in the detection mode than during target tracking.

Jamming during the process of target detection forces the operator to expend additional time analyzing interfering signals, which reduces the time available to search for real targets, lowers the probability of timely target detection, and reduces the operating range of the radar. Jamming during the target search and detection stage is accordingly an important factor in the defense of various objects, naval ships in particular [60]. In this case, jamming confuses the enemy in regard to the number, type, and location of interesting objects, and makes the process of target indication and acquisition for automatic tracking more difficult. Jamming can cause enemy systems to lock on to false targets, which reduces the time available to engage the real target. Since it is much harder to interrupt an automatic target track with jamming than it is to prevent the initial lock-on, jamming during this stage of the action is extremely important.

Both masking and deception jamming, capable of inhibiting target detection during the search stage, are used to jam the detection channel.

### 8.3.1 Active Masking Jamming of the Target Detection Channel

The most universal form of masking jamming, among those now known, is noise jamming [5, 109]. With sufficient power, this jamming causes the radar indicator screen to be partially or completely illuminated by interference, so that real targets are masked (see Fig. 8.1). The nature of this noise is such that useful information can be fully extracted without distortion only when the characteristics of the noise at the receiving point are completely known. Such a situation, however, is unrealistic for radar countermeasures. Noise jamming affects not only the detection channel, but also the range and velocity resolution channels, and also hinders angular resolution when the jamming power is sufficient to affect reception through the sidelobes.

Noise jamming is characterized as either spot or barrage jamming, depending on the spectrum width. Spot noise jamming is characterized by a narrow spectrum commensurate with the passband of the radar receiver, and in this respect has a power advantage over barrage jamming. Spot noise jamming, however, can suppress only one radar, and if it is not sufficiently tunable, it may be avoided by changing the radar's operating frequency. Barrage noise jamming is characterized by a wide spectrum, and in principle may operate simultaneously against several radars, including any radars with limited frequency agility. This capability is obtained at the price of power loss, however, since to create an identical effect (for any given radar) a barrage noise jammer transmitter must transmit significantly more power than a spot jammer, and much power is transmitted uselessly to cover portions of the spectrum which are unused by the radars. One advantage of barrage jamming is that a radar may not, as a rule, avoid its effects by changing its frequency.

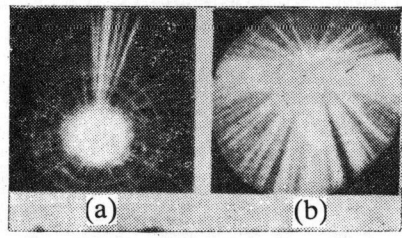

**FIGURE 8.1.** The appearance of noise jamming on a PPI indicator for (a) moderate and (b) high jamming power.

To create spot noise jamming it is possible to use a modulation method in which the carrier frequency is modulated in amplitude, frequency, or phase by a noise signal. The structure of the resulting noise jamming is determined by the type of modulation and the characteristics of the modulating noise signal which is used. Transmitters which are built to perform such jamming should be tunable over a wide band, so that they may react to changes in the electronic environment.

Barrage noise jamming, formed by direct amplification of the noise from a low-power wideband source with uniform spectrum, is called direct noise jamming, and, in principle, is the most effective type of masking jamming. It is extremely difficult to protect a radar against this sort of jamming.

In addition to spot and barrage noise jamming, frequency-swept noise jamming, which is narrowband noise tunable over a wide band, is also used [83]. Such jammers allow the benefits of both spot and barrage jamming to be retained to some extent, because, for a certain amount of time, a high-power density is maintained in all covered radar channels. The masking effect may also be produced by sweeping the transmitter frequency randomly without employing narrowband noise. The effect of jamming based on random frequency variation is somewhat different from the effect of interference amplitude-modulated by noise. If the variation of the carrier frequency significantly exceeds the receiver passband, and is sufficiently narrow in comparison with the spectrum of the carrier frequency of the modulating noises, this interference will produce, at the receiver output, pulses of almost constant amplitude whose shape is similar to the frequency response of the receiver. These pulses will be obtained every time the jammer is swept through the receiver passband, and the interval between them will change in a random fashion.

Masking jamming may also be achieved with asynchronous pulsed interference with low- and high-pulse repetition rates. Such jamming causes coarse- or fine-grained patterns on an intensity-modulated display, or a series of ordered or random pulsed blips on amplitude displays, creating excess information and making it difficult to detect and automatically track the real target.

These active masking jammers fall in the category of *power* jammers, inasmuch as they achieve suppression of the radar capabilities at the cost of high jammer power. Active deception jamming, as will be shown later, has lower power requirements. Power jamming also includes jamming in the form of an unmodulated carrier at the same frequency as the radar carrier. Without enough power, such a jammer may overload the radar receiver to such an extent that the entire radar appears to be malfunctioning.

Dual-frequency jammers may also achieve covert masking jamming [59]. This jammer functions by transmitting at two frequencies within the operating band of the radar, with the two frequencies offset from one another by the intermediate frequency of the radar receiver. In operation, such a jammer will produce an IF signal at the output of the receiver mixer which overloads the receiver and prevents the reception of useful information. In a pulsed radar, this power suppression results primarily from the video detector being saturated by the jamming signal. In a Doppler system, dual-frequency jamming saturates the AGC system. In either case, the radar operator will not even know that he is being jammed. He will simply lose the target information, and not know the reason for this disappearance.

An analogous effect may be achieved by amplitude-modulating an RF signal at the IF frequency of the radar receiver. This type of jamming is especially effective against receivers which do not have balanced mixers or RF filters which pass only the transmitted band of frequencies. The frequency tuning accuracy of such a jammer is determined, as with the dual-frequency jammer, by the frequency-selection properties of the radar receiver preselector.

Besides the methods just enumerated, it is also possible to create active masking jamming with miniature emitters which are jettisoned. These emitters are usually of solid state design, and generate interference at a particular frequency for an amount of time sufficient to complete the combat operation [96].

### 8.3.2 Passive Masking Jamming of the Target Detection Channel

One of the methods of producing passive masking jamming, which first found practical application, is chaff, which usually consists of metallic strips with a length equal to half the radar's operating wavelength. Ejected from planes, chaff packs are dispersed by the wind and form extended reflecting clouds. The display of an unprotected radar is brightly illuminated as a result, masking any targets within the chaff cloud. Since most radars operate over a band of frequencies, the chaff strips are prepared in various lengths in accordance with the known and probable operating frequencies of the victim radar.

An effective means for countering this sort of passive jamming is to use Doppler filtering based on the frequency difference between the target and chaff signals which is caused by their different velocities. The speed of the chaff is determined by the mass of the strips, their aerodynamic properties, and the wind speed. In comparison with the plane, the chaff will be a slowly moving target, which allows it to be resolved in velocity.

To increase the effectiveness of the chaff against the possible methods of countering it, the chaff is sometimes illuminated by RF signals generated by special transmitters on the object being defended [65, 109]. The illuminating signal to be transmitted is shifted either in frequency or time, in order to simulate false chaff velocities or ranges, making the characteristics of the chaff signals more similar to those of real moving targets and thus harder to resolve.

Because the chaff length corresponds to the radar's wavelength and increases with the latter, changing radars to the upper portion of the decimeter and meter wavebands has been examined abroad as one possible way of weakening chaff interference [109].

### 8.3.3 Active and Passive Deception Jamming of the Target Detection Channel

Deception jamming of the detection channel is distinguished from masking jamming in that excessive information is produced mainly by generating signals with spurious parameters which differ little from the parameters of real reflected signals. Such jamming does not act to mask the real target signals, but deludes the operator and adds false information to an estimate of the electronic situations produced by the automatic radar devices.

An example of active deception jamming of the detection channel is multiple repeater jamming. A multiple repeater jammer reacts to received radar pulses by transmitting a series of analogous pulses with the same shape on the same frequency, distributed in range and azimuth, and imitating a large number of targets [65, 109]. The existence of such interference makes it difficult for the radar operator to chose the true target and reduces the effectiveness of the detection and indication systems. Many spurious targets may be intercepted in place of the real target. This leads to dispersion of the defense forces, which increases the probability of the combat airplanes reaching their objective.

The effectiveness of multiple repeater jamming is increased when it is combined with noise jamming (see Fig. 8.2), by which means targets covered by the jammer are simulated. Jammer performance may also be improved by generating spurious target responses in the sidelobes, as opposed to the main lobe, creating false angular coordinate information. This may be achieved by synchronizing the jammer transmitter with the scan period of the radar [21].

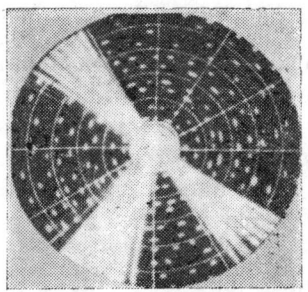

**FIGURE 8.2.** The appearance of multiple repeater jamming combined with noise jamming on a PPI radar indicator.

Spurious targets may also be generated with the help of miniature emitters (retranslators) ejected from the defended object. Such miniature repeaters may be of solid state design and use, for example, tunnel diodes, to "capture" the radar signals and retransmit them with some time delay. This approach is considered to be promising for overcoming antiballistic missile systems [96].

Passive deception jamming of the detection channel may also be accomplished with decoy targets which are very similar to the actual target. If the radar cannot distinguish the decoy from the real target, the operator or automatic systems may take it for a dangerous target and designate it for interception and destruction, thus increasing the probability that the real target will not be attacked.

The simplest decoys are corner reflectors, which, even with small dimensions, are capable of reflecting signals with levels comparable to those from large targets; chaff is also a simple decoy. Chaff may be ejected automatically in any direction from the defended target. Chaff ejected in the forward direction can sometimes affect the range and velocity tracking capabilities of the radar, and even cause the radar to lock on the chaff [67, 80, 109].

## 8.4 TARGET SELECTION CHANNEL JAMMING

Range or velocity target selection channels are employed in almost all foreign monopulse radar systems and perform the following:

- they increase the selectivity of guidance and autotracking systems, ensuring that the system works only against the selected target;
- they increase the antijamming capabilities of the system by reducing the time the receiver devices are open and narrowing the system passband;
- they affect measurement of the target range and velocity.

Jamming is applied against the selection channels in order to interrupt the target selection process or to introduce errors into the range and velocity measurements. It should be noted that jamming directed against the selection channels also affects the angle measurement channels, inasmuch as breaking the automatic target track in range or velocity necessitates repeated returns to search mode in the corresponding parameters. If there are multiple targets in the search zone, then the original target may be lost and the radar may lock onto a new target in a different direction. This entails redirecting the angle-sensing system and increases the angle errors. Interfering with the target selection reduces the power requirements when jamming an automatic angle tracking system [109].

Range and velocity are also used in systems in the West to calculate the lead angle in rocket launches with command guidance and in firing antiaircraft artillery. If these parameters are fed to the fire control computer with significant errors, the intercept point will also be calculated incorrectly, which affects the firing accuracy. In launching interceptor missiles, errors in calculating the intercept point can increase the missile movement overloads and reduce the probability of destroying the target [5].

In accordance with the problems described above, providing the target selection with an antijamming capability is an important matter, although not of the most urgency, because the primary channel in autotracking radar systems is the angle measurement channel. In turn, Western ECM specialists believe that designers should place appropriate stress on developing techniques for jamming selection channels [109].

Repeater jammers are currently used the most for jamming selection channels, allowing the normal operation of these channels to be disrupted with relatively simple technology [109]. Repeater jammers are widely used in foreign radar countermeasure systems for disorientation and the creation of false situations on radar indicators. The method for creating repeater jamming is based on the synchronous radiation of interfering signals which either imitate false targets with real motion parameters, or distort the real signal [109].

In the opinion of Western specialists, range (or velocity) deception jamming is the most interesting form of selection channel jamming. The transmitters in such jammers emit pulse signals with parameters analogous to those of reflected signals, but with a continuously varying time delay which imitates the motion of a target with a different velocity, thus creating interference. If the jammer power exceeds that of the reflected signals, the autotracking range gate will be pulled off the real target to the jamming pulse. This will result in an erroneous target range measurement, significantly disrupting the process of locating the target. The jamming signal delay may be increased or decreased until there is a sufficient range error. The jammer may be switched off once the range gate has been shifted far enough; the real target will be lost and the system will return to search mode [103].

Velocity deception jamming is performed the same way with continuous or pulsed Doppler transmission, the only difference being that now the frequency is continuously varied relative to the radar's operating frequency [5, 76, 109]. This frequency variation, in turn, produces a continuous variation in the Doppler frequency which mimics a moving target. With sufficient jamming power, the velocity gate of the tracking radar will shift in accordance with the jamming frequency, and the tracking system will produce false target velocity information. If the jammer is switched off once the jamming frequency has been shifted by an amount exceeding the passband of the narrowband tracking filter (velocity gate), the target will be lost and the radar will have to switch to search mode, just as with range deception jamming. Velocity deception jamming makes it possible to divert the range gate to a decoy, thus redirecting the aiming system to a false target. In general, it is possible to imitate a large number of false targets, and, although they will be in the same direction as the real target and spurious angle information is not generated, the radar operator may be unable to determine the number of targets properly, and may conclude that there are many targets where there is really only one.

In US range or velocity deception jammers, delayed feedback oscillators are usually used; the signals are instantaneously stored, and programmed jamming signals are then directed to the subject radar. To perform the storage of the frequency and structure of the signal an RF delay line with digitally controlled delay is used; the fine structure of the radar signal may then be reproduced directly in the microwave band. Thus, it is possible to maintain coherency in the jamming signal and to generate jamming signals for radars with complex signals [109].

Jamming the target selection channel of a radar which uses complex signals is, on the whole, a difficult problem, inasmuch as it requires detailed information on the signal structure, correlation processing methods and pulse compression schemes used by the radar receiver. The problem becomes even more complex if there is a high density of long pulse radars using FM or phase code modulation, in which case the signal selection, recognition, and processing is made more difficult. In this situation, angle jamming assumes a greater importance.

Disruption of a system performing automatic tracking in range or velocity may be disrupted and the radar redirected to false targets with the help of chaff, which may be ejected by special devices, for example, into the hemisphere in front of the aircraft being defended [109].

## 8.5 TECHNIQUES FOR JAMMING THE ANGLE CHANNEL OF AUTOTRACKING MONOPULSE RADARS

The parameters determined by radar systems are not all of equal importance. Range, velocity, and acceleration, for example, are of secondary importance in some cases. These parameters allow targets to be resolved in range and velocity, which is useful, of course, especially in multiple target situations. However, if these parameters are absent for any reason, the system can continue with some probability of acquiring and destroying the target. There are systems, for example, which can aim at a source of radio or jamming radiation using only target angle information; range and velocity data is not needed [109]. With missile systems employing on-board homing devices, range information is necessary only to determine the time of launch.

The situation is completely different if the radar is an autonomous system and is, for some reason, incapable of determining the direction to the target. In this case, the system is unable to perform the aiming necessary to acquire and destroy the target. Therefore, the operation of the angle channel in a radar, when jamming is directed against it, is of the highest importance. Electronic countermeasure designers thus place the most emphasis on jamming angle measurement channels, and foremost among these, the angle channel of a system which performs automatic angle tracking.

The improved jamming immunity of the angle channel in monopulse systems results from the principles by which they operate. As has been established, monopulse systems may, in principle, determine the target direction accurately with just a single pulse. Inasmuch as the reflecting surface of the target is invariant for the duration of a pulse, monopulse systems are largely insensitive to amplitude fluctuations in the reflected signal and are more accurate than single-channel systems. Due to the effective normalization performed with the help of the reference pattern, any internal amplitude modulation is destroyed. It follows that the angle channel of a monopulse system is not very sensitive to amplitude-modulated jamming. In fact, AM jamming, in which the jamming power exceeds the radar signal power, can improve the operation of a monopulse angle channel, because it increases the effective target cross section and the range region over which the target may be tracked normally. This is true, not just for AM jamming, but for jamming with any sort of modulation (frequency or phase modulation, for example) which originates from a single point in space, as long as the receiver channels have identical amplitude-phase responses.

In practice, however, it is difficult to build a perfect monopulse receiver, and there are various design and manufacturing shortcomings. These deficiencies can reduce the immunity of radar systems to active jamming, including jamming radiated from a single point.

All possible forms of monopulse angle channel jamming may be divided into two groups. In the first group are those jamming techniques which are made possible by the design and construction faults in the monopulse system. The second group comprises those jamming methods which are feasible even for ideal monopulse systems, including cross-polarized, dual frequency, coherent, blinking, and ground bounce jamming [109].

### 8.5.1 Jamming Exploiting Design and Manufacturing Faults in Angle Channels

In addition to receiver channels possessing unequal amplitude-phase responses, monopulse systems may have other faults, including nonlinear amplitude responses in the receiver channels, the AGC time constant, different antenna channel gains, the appearance of an image receiving channel, and so on.

Jamming of sufficient power can disrupt the normal operation of a receiver channel with a nonlinear amplitude response by overloading the channel, because saturation of any element of the tracking loop destroys the amplitude variations in the target signal, in turn partially or totally preventing the formation of the correct error signal in an amplitude-comparison system. This effectively disables the tracking system and leads to an unavoidable increase in the angle errors. The errors will be increased if the jamming is blinked (switched on and off); in this case, it is insufficient for the receiver channels to possess identical static responses—they must also have identical dynamic responses. With limitations on the speed and dynamic range of the AGC system, this balance of the dynamic responses cannot always be maintained.

The use of powerful jamming which is "blinked" at the proper frequency can cause the antenna to undergo increasing oscillation on account of the resonance properties and inertia of the tracking system. The possibilities for blinking jamming are significantly reduced if logarithmic receivers or fast AGC systems are used in the radar.

When a narrowband AGC is used in the radar receiver, switching off a powerful jamming signal can break the tracking loop for the time necessary for the receiver sensitivity to return to the level of the reflected target signal. During this time the antenna will either not move or will continue moving by inertia; either condition increases the tracking errors. With a logarithmic receiver or receiver with fast AGC, the inertial or uniform antenna motion will be minimized, since the system will almost instantly begin tracking the target signal as soon as the jammer is switched off.

The tracking accuracy in some monopulse receivers is reduced when the receiver is imprecisely tuned to the received signal. This is a result of unequal intermediate frequency responses in the receiver channels. This effect may be exploited in jamming a monopulse system by transmitting the jamming signal with a carrier frequency which differs from the radar frequency by half the radar receiver's intermediate frequency. Reception of this jamming signal will cause increased angular errors.

Image channel interference may be produced by jamming with an RF signal whose frequency is offset from the radar receiver's local oscillator frequency by the IF, but on the opposite side from the radar's RF [76]. This type of jamming produces an error signal out of phase with the true error signal, thus making the tracking system unstable in the region of the axis. In order to use this method, however, it is necessary to know the radar's IF and the position of the local oscillator frequency relative to the RF. If this information is unavailable, this technique may still be used by transmitting on two image frequencies with appropriately wide spectra to make up for the lack of exact data on the intermediate frequency.

An effective means of countering image channel jamming is to employ preselectors which attenuate signals in the image channel by 60 dB or more. Image channel jamming may then be created only with practically unrealizable jammer power [76].

Monopulse systems which employ commutation in order to diminish the requirements for equal receiver channel responses present a different situation with respect to antijamming capabilities. Under certian conditions, the use of commutation makes these systems vulnerable to amplitude-modulated jamming transmitted from a single point.

It has been shown that when the jamming signal is amplitude-modulated at the commutation frequency, systematic tracking errors arise in a logarithmic receiver which may be described by the following expression [21]:

$$\mu\theta = 2\frac{[(1 - m_p)/(1 + m_p)]^x - 1}{[(1 - m_p)/(1 + m_p)]^x + 1}$$

where $m_p$ is the percentage of modulation.

As can be seen in Fig. 8.3, when the receiver channels are identical ($g_0 = 1$), the tracking error is zero, and jamming amplitude-modulated at the commutation frequency has no adverse effects.

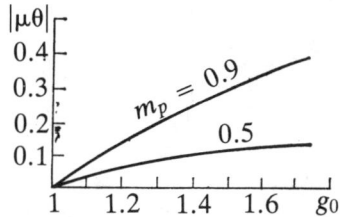

**FIGURE 8.3.** The normalized systematic tracking error as a function of the discrepancy in the gains of the receiver channels, in the presence of jamming amplitude-modulated at the commutation frequency.

Using the calculated dependencies, it is possible to determine the absolute tracking errors for a known angle-sensing sensitivity. Thus, with $m_p = 0.9$ and $g_0 = 1.3$, the systematic tracking error will be 0.8° for $\mu = 0.25$ deg$^{-1}$; 0.33° for $\mu = 0.6$ deg$^{-1}$ and 0.27° for $\mu = 0.75$ deg$^{-1}$.

The tracking errors resulting from jamming synchronously amplitude-modulated at the commutation frequency are close in vaiue to those resulting from unequal receiver channel amplitude responses in an analogous system without commutation. Jamming with a signal synchronously amplitude-modulated at the commutation frequency thus counteracts the advantages of employing commutation.

The case just considered was under the ideal condition that the jammer signal was modulated in phase with the receiver channel commutation. If, in reality, this is not the case and the jamming signal is generated with only approximate knowledge of the commutation frequency, then the effectiveness of the jammer will depend not only on the degree of modulation, but also on the modulation frequency and the phase relationship between the modulation and the jamming signal. If this phase difference is assumed to be distributed uniformly between $-\pi$ and $+\pi$ during the course of jamming, then the rms tracking error in this case will be

$$\sigma = \theta_{max}/\sqrt{3}$$

Considering the unknown value of the commutation frequency, it is improbable that the jammer will operate synchronously with the commutation. This is even more true if the commutation is implemented with a special coding scheme. This is only true, however, if the commutation device is ideally matched and does not affect the transmitter channel. If this condition is not satisfied, then the transmitted signal will contain information on the frequency and phase of the commutation process, and if appropriate equipment is deployed along with the jammer, then this information may be extracted and used to generate an effectively matched jamming signal modulated at the radar commutation frequency [83, 102].

### 8.5.2 Cross-Polarized Jamming

Cross-polarized jamming is based on illuminating the radar antenna with a signal at the radar frequency and with a polarization orthogonal to that at which the radar operates [21]. As was shown in Chapter 6, the existence of cross-polarization radiation in reflector antennas causes the antenna pattern, and, consequently, the angle-sensing response, to be dependent on the polarization of the received signals. This not only causes a reduction in the target tracking accuracy, but can lead to total disruption of the angle-sensing system. It is accordingly of interest to consider the effect of the polarization characteristics of the received signals on monopulse systems. To this end, a theoretical analysis of the complex antenna patterns and angle-sensing response for reception of signals with varying polarization is presented below. Chosen for this analysis were the two most common monopulse systems, namely, the amplitude- and phase-sum-and-difference systems.

*Amplitude-sum-and-difference monopulse system.* We will consider a two-plane angle-sensing system with a square parabolic receiving antenna which forms four partial patterns, the maxima of which lie in the four quadrants 1, 2, 3, 4 and are squinted from the axis by the angle $\theta_0$ (see Fig. 8.4). The field distribution over the aperture is given by the function:

$$\underline{\psi}(x, y) = \underline{\psi}_{px}(x, y)\mathbf{e}_x + \underline{\psi}_{py}(x, y)\mathbf{e}_y$$

where $\psi_{px}(x, y)$ and $\psi_{py}(x, y)$ are functions describing the field excitation at the operating polarization and cross polarization; $p$ is the lobe number ($p = 1, 2, 3, 4$); and $\mathbf{e}_x$ and $\mathbf{e}_y$ are the unit vectors.

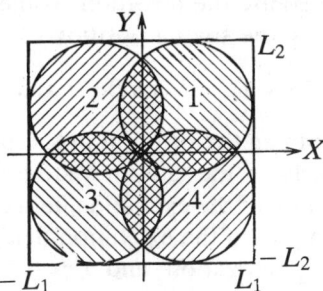

**FIGURE 8.4.** Beam positions at operating polarization in an amplitude-sum-and-difference monopulse system for tracking in two planes.

The excitation fields at operating polarization in this case may be determined by the following functions:

$$\underline{\psi}_{1x,3x}(x, y) = \exp[\pm ik_\lambda(x \cos \alpha_0 + y \cos \beta_0)]f_{1,3}(x, y)$$

$$\underline{\psi}_{2x,4x}(x, y) = \exp[\pm ik_\lambda(-x \cos \alpha_0 + y \cos \beta_0)]f_{2,\,4}(x, y)$$
$$\text{for } -L_2 \leqslant y \leqslant L_2, \ -L_1 \leqslant x \leqslant L_1$$

Considering that the cross-polarized patterns in neighboring quadrants are directly out of phase, the cross-polarized excitation fields may be expressed as follows:

$$\underline{\psi}_{1y,3y}(x, y) = a_c \exp\{\pm i[k_\lambda(x \cos \alpha_0 + y \cos \beta_0) + \phi_{ce}]\}f_{1,3}(x, y)$$
$$\text{for } 0 \leqslant y \leqslant L_2, \ 0 \leqslant x \leqslant L_1$$

$$\underline{\psi}_{1y,3y}(x, y) = -a_c \exp\{\pm i[k_\lambda(x \cos \alpha_0 + y \cos \beta_0) + \phi_{ce}]\}f_{1,3}(x, y)$$
$$\text{for } 0 \leqslant y \leqslant L_2, \ -L_1 \leqslant x \leqslant 0$$

$$\underline{\psi}_{1y,3y}(x, y) = a_c \exp\{\pm i[k_\lambda(x \cos \alpha_0 + y \cos \beta_0) + \phi_{ce}]\}f_{1,3}(x, y)$$
$$\text{for } -L_2, \leqslant y \leqslant 0, \ -L_1 \leqslant x \leqslant 0$$

$$\underline{\psi}_{1y,3y}(x, y) = -a_c \exp\{\pm i[k_\lambda(x \cos \alpha_0 + y \cos \beta_0) + \phi_{ce}]\}f_{1,3}(x, y)$$
$$\text{for } -L_2 \leqslant y \leqslant 0, \ 0 \leqslant x \leqslant L_1$$

$$\underline{\psi}_{2y,4y}(x, y) = a_c \exp\{\pm i[k_\lambda(-x \cos \alpha_0 + y \cos \beta_0) + \phi_{ce}]\}f_{2,4}(x, y)$$
$$\text{for } 0 \leqslant y \leqslant L_2, \ 0 \leqslant x \leqslant L_1$$

$$\underline{\psi}_{2y,4y}(x, y) = -a_c \exp\{\pm i[k_\lambda(-x \cos \alpha_0 + y \cos \beta_0) + \phi_{ce}]\}$$
$$\times f_{2,4}(x, y) \qquad \text{for } 0 \leqslant y \leqslant L_2, \ -L_1 \leqslant x \leqslant 0$$

$$\underline{\psi}_{2y,4y}(x, y) = a_c \exp\{\pm i[k_\lambda(x \cos \alpha_0 + y \cos \beta_0) + \phi_{ce}]\}f_{2,4}(x, y)$$
$$\text{for } -L_2 \leqslant y \leqslant 0, \ -L_1 \leqslant x \leqslant 0$$
$$\underline{\psi}_{2y,4y}(x, y) = -a_c \exp\{\pm i[k_\lambda(-x \cos \alpha_0 + y \cos \beta_0) + \phi_{ce}]\}$$
$$\times f_{2,4}(x, y) \qquad \text{for } -L_2, \leqslant y \leqslant 0, \ 0 \leqslant x \leqslant L_1$$

where $\cos \alpha_0$ and $\cos \beta_0$ are the direction cosines in a spherical system of coordinates determining the spatial position of the pattern lobes:

$$\cos \alpha_0 = (\sin \theta_0)\cos \vartheta_0, \qquad \cos \beta_0 = (\sin \theta_0)\sin \vartheta_0$$

$\vartheta_0$ is the angle determining the position of the plane in which the beam is deflected by an angle $\theta_0$ in the chosen system of coordinates; $f_p(x, y)$ is the amplitude distribution of the excitation field over the antenna aperture; $\phi_{ce}$ is the phase shift between the excitation field components at operating polarization and cross polarization; and $a_c$ is the attenuation coefficient of the cross-polarized excitation, equal to the ratio of the orthogonally polarized excitation field components ($+i$ corresponds to the function with indices $p = 1, 2$). In order to simplify the expressions we will make the assumption that

$$f_1(x, y) = f_2(x, y) = f_3(x, y) = f_4(x, y) = \tfrac{1}{2}L$$

If the received signal is given by the expression:

$$\underline{E}_s = \underline{E}_{sx}(\theta, \vartheta)e_x + \underline{E}_{sy}(\theta, \vartheta)e_y$$

where $\underline{E}_{sx}(\theta, \vartheta) = E_s \exp[ik_\lambda(x \cos \alpha + y \cos \beta)]$ is the signal component at the operating polarization;

$$\underline{E}_{sy}(\theta, \vartheta) = bE_s \exp[ik_\lambda(x \cos \alpha + y \cos \beta) + \phi_{cs}]$$

is the cross-polarized signal component; $E_s$ is the signal amplitude; $\phi_{cs}$ is the phase shift between the orthogonally polarized components of the received signal; $b$ is a coefficient determining the relation of the orthogonally polarized component amplitudes; and $\cos \alpha$ and $\cos \beta$ are the direction cosines determining the spatial position of the signal sources:

$$\cos \alpha = (\sin \theta)\cos \vartheta, \qquad \cos \beta = (\sin \theta)\sin \vartheta$$

Here $\theta$ and $\vartheta$ are elements of the spherical system of coordinates shown in Fig. 8.5 ($\theta$ is the angle of the source relative to the radar axis).

Under the given conditions, the signals at the output of the antenna receiver channels are given by the expression:

$$\underline{E}_p(\theta, \vartheta) = \int_s \underline{\psi}_p(x, y)\underline{E}_s \, ds$$

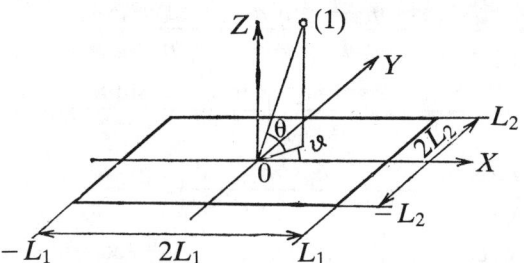

**FIGURE 8.5.** Coordinate system for finding the direction to a signal source.

(1) signal source

The angle-sensing response in an amplitude-sum-and-difference system will be formed in accordance with the expression:

$$S(\theta, \vartheta) = \frac{\mathrm{Re}[\underline{u}_d(\theta, \vartheta)\underline{u}_s^*(\theta, \vartheta)]}{\underline{u}_s(\theta, \vartheta)\underline{u}_s^*(\theta, \vartheta)}$$

where $\underline{u}(\theta, \vartheta)$ and $\underline{u}_d(\theta, \vartheta)$ are the complex envelopes of the sum and difference signals in the receiver at the output of the phase detector.

Carrying out the necessary mathematical operations to determine the signals and transform them in accordance with the operating principles of an amplitude-sum-and-difference system laid out in [21], we obtain expressions for the angle-sensing responses in the azimuth and elevation channels in the form:

$$S_{az}(\theta, \vartheta) = \frac{A_1 - A_2 + A_3}{B_1 - B_2 + B_3}, \qquad S_{el}(\theta, \vartheta) = \frac{A_4 - A_5 + A_6}{B_1 - B_2 + B_3}$$

where

$$A_1 = \left(\frac{\sin^2 m_1}{m_1^2} - \frac{\sin^2 m_2}{m_2^2}\right)\left(\frac{\sin^2 n_1}{n_1} + \frac{\sin^2 n_2}{n_2}\right)$$

$$A_2 = 2a_c b\left(\frac{\sin n_1}{n_1} + \frac{\sin n_2}{n_2}\right)\left(\frac{\cos n_1 - 1}{n_1} + \frac{\cos n_2 - 1}{n_2}\right)$$

$$\times \left[\frac{\sin m_1}{m_1}\left(\frac{\cos m_1 - 1}{m_1}\right) - \frac{\sin m_2}{m_2}\left(\frac{\cos m_2 - 1}{m_2}\right)\right]\cos \Phi$$

$$A_3 = a_c^2 b^2 \left[\left(\frac{\cos m_1 - 1}{m_1}\right)^2 - \left(\frac{\cos m_2 - 1}{m_2}\right)^2\right]\left[\frac{\cos n_1 - 1}{n_1} + \frac{\cos n_2 - 1}{n_2}\right]^2$$

$$B_1 = \left(\frac{\sin m_1}{m_1} + \frac{\sin m_2}{m_2}\right)^2 \left(\frac{\sin n_1}{n_1} + \frac{\sin n_2}{n_2}\right)^2$$

$$B_2 = 2a_c b \left(\frac{\sin m_1}{m_1} + \frac{\sin m_2}{m_2}\right)\left(\frac{\sin n_1}{n_1} + \frac{\sin n_2}{n_2}\right)$$

$$\times \left(\frac{\cos m_1 - 1}{m_1} + \frac{\cos m_2 - 1}{m_2}\right)\left(\frac{\cos n_1 - 1}{n_1} + \frac{\cos n_2 - 1}{n_2}\right)\cos \Phi$$

$$B_3 = a_c^2 b^2 \left(\frac{\cos m_1 - 1}{m_1} + \frac{\cos m^2 - 1}{m_2}\right)^2 \left(\frac{\cos n_1 - 1}{n_1} + \frac{\cos n_2 - 1}{n_2}\right)^2$$

$$A_4 = \left(\frac{\sin^2 n_1}{n_1^2} - \frac{\sin^2 n_2}{n_2^2}\right)\left(\frac{\sin m_1}{m_1} + \frac{\sin m_2}{m_2}\right)^2$$

$$A_5 = 2a_c b \left(\frac{\sin m_1}{m_1} + \frac{\sin m_2}{m_2}\right)\left(\frac{\cos m_1 - 1}{m_1} + \frac{\cos m_2 - 1}{m_2}\right)$$

$$\times \left[\frac{\sin n_1}{n_1}\left(\frac{\cos n_1 - 1}{n_1}\right) - \frac{\sin n_2}{n_2}\left(\frac{\cos n_2 - 1}{n_2}\right)\right]\cos \Phi$$

$$A_6 = a_c^2 b^2 \left[\left(\frac{\cos m_1 - 1}{m_1}\right) + \left(\frac{\cos m_2 - 1}{m_2}\right)\right]^2$$

$$\times \left[\left(\frac{\cos n_1 - 1}{n_1}\right)^2 - \left(\frac{\cos n_2 - 1}{n_2}\right)^2\right]$$

$$m_1 = k_\lambda L(\cos \alpha + \cos \alpha_0)$$
$$n_1 = k_\lambda L(\cos \beta + \cos \beta_0)$$
$$\Phi = \phi_{ce} - \phi_{cs}$$
$$m_2 = k_\lambda L(\cos \alpha - \cos \alpha_0)$$
$$n_2 = k_\lambda L(\cos \beta - \cos \beta_0)$$

With $a_c b = 0$ (reception at operating polarization):

$$S_{az}^0(\theta, \vartheta) = \frac{(\sin m_1)/m_1 - (\sin m_2)/m_2}{(\sin m_1)/m_1 + (\sin m_2)/m_2}$$

$$S_{el}^0(\theta, \vartheta) = \frac{(\sin n_1)/n_1 - (\sin n_2)/n_2}{(\sin n_1)/n_1 + (\sin n_2)/n_2}$$

Because we are assuming that the beams are asymmetrical about the optical antenna axis, $\theta = \theta_3/2$, $\vartheta = 45°$, and

$$m_{1,2} = \pi[\Delta \cos(\vartheta \pm 0.35)],$$
$$n_{1,2} = \pi[\Delta \sin(\vartheta \pm 0.35)], \qquad \Delta = \theta/\theta_3$$

If the tracked target lies in the azimuth plane ($\vartheta = 0$), then

$$S_{az}^0(\Delta_x) = \frac{\dfrac{\sin[\pi(\Delta_x + 0.35)]}{\pi(\Delta_x + 0.35)} - \dfrac{\sin[\pi(\Delta_x - 0.35)]}{\pi(\Delta_x - 0.35)}}{\dfrac{\sin[\pi(\Delta_x + 0.35)]}{\pi(\Delta_x + 0.35)} + \dfrac{\sin[\pi(\Delta_x - 0.35)]}{\pi(\Delta_x - 0.35)}}$$

$$S_{el}^0(\Delta_x) = \frac{(\sin 0.35\pi)/0.35\pi - (\sin 0.35\pi)/0.35\pi}{(\sin 0.35\pi)/0.35\pi + (\sin 0.35\pi)/0.35\pi} = 0$$

These expressions show that when tracking a signal which is matched with the operating polarization, there is no crosstalk between the angle-sensing channels.

We will now consider the other extreme case, when $a_c b = \infty$ and signals are received only at cross-polarization. In this case:

$$S_{az}^c(\theta, \vartheta) = \left[\frac{\sin^2(m_1/2)}{m_1/2} - \frac{\sin^2(m_2/2)}{m_2/2}\right] \Big/ \left[\frac{\sin^2(m_1/2)}{m_1/2} + \frac{\sin^2(m_2/2)}{m_2/2}\right] \quad (8.1)$$

$$S_{el}^c(\theta, \vartheta) = \left[\frac{\sin^2(n_1/2)}{n_1/2} - \frac{\sin^2(n_2/2)}{n_2/2}\right] \Big/ \left[\frac{\sin^2(n_1/2)}{n_1/2} + \frac{\sin^2(n_2/2)}{n_2/2}\right]$$

After substituting the values for $m_{1,2}$ and $n_{1,2}$ when $\vartheta = 0$ and the appropriate transformations, we obtain

$$S_{az}^c(\theta, \vartheta) = \infty, \qquad S_{el}^c(\theta, \vartheta) = \infty$$

Despite the fact that the tracked signal source is located in the azimuth plane ($\vartheta = 0$), the elevation channel error signal is not zero, but attains an infinite value. This is a manifestation of strong coupling between the channels, which is independent of the offset angle.

The error signal in the azimuth channel will also be infinite when the signal is completely cross polarized. It follows that when receiving cross-polarized signals, the angle-sensing system will not operate normally.

Equating the numerator of (8.1) to zero, we find the equilibrium points:

$$\frac{\sin^2[(\pi/2)(\Delta_x + 0.35)]}{(\pi/2)(\Delta_x + 0.35)} - \frac{\sin^2[(\pi/2)(\Delta_x - 0.35)]}{(\pi/2)(\Delta_x - 0.35)} = 0$$

Solving the transcendental equation, we obtain $\Delta_x = \pm 0.75$, and it can be shown that these are points of stable equilibrium. The angle-sensing response is distorted when operating on cross-polarized signals because the antenna patterns are distorted. To illustrate this effect, the sum and difference pattern lobes are diagrammed in the plane of the page, where + and − denote the relative phases of the lobes, which differ by 180°.

Typical spatial arrangements of the difference channel lobes of the system under consideration are shown in Figs. 8.7 and 8.8. Analysis shows that when receiving cross-polarized signals, the sum and difference patterns have deep minima along the axis. The sum pattern essentially becomes the difference pattern in planes other than the main angle-sensing planes (see Fig. 8.6(a)), and instead of difference patterns, sum patterns are observed to be formed in the orthogonal planes. Thus, the cross-polarized azimuth difference pattern is characterized by the formation of sum lobes in the elevation plane (along the $Y$-axis in Fig. 8.6(b)), and the elevation pattern by sum lobes in the azimuth plane (along the $X$-axis in Fig. 8.6(c)). Moreover, these sum patterns lying on opposite sides of the angle-sensing plane are 180° out of phase. As a result, a pattern typical of elevation angle sensing is formed in the azimuth channel. In other words, when receiving cross-polarized signals there is a functional reversal of the angle-sensing channels: the elevation channel becomes the azimuth channel, and *vice versa*.

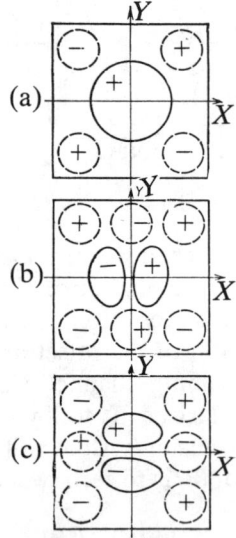

**FIGURE 8.6.** Typical positions of the lobes of an amplitude-sum-and-difference monopulse system at opposite polarization (————) and cross polarization (— — —):
    (a) sum
    (b) azimuth difference
    (c) elevation difference

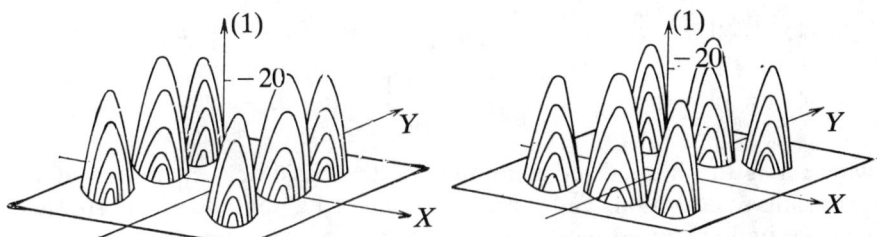

**FIGURE 8.7.** Typical spatial structure of lobes of amplitude-sum-and-difference monopulse azimuth pattern at cross polarization.

(1) $F_{d\ el}(\theta)$, dB

**FIGURE 8.8.** Typical spatial structure of lobes of amplitude-sum-and-difference monopulse elevation pattern at cross polarization.

$F_{d\ az}(\theta)$, dB

This functional reversal of the angle-sensing channels and transformation of the sum channel into the difference channel, in the case being considered, makes the amplitude-sum-and-difference system completely inoperable. This result is expressed in the angle-sensing responses found above for $a_c b = \infty$. The effect of polarization mismatch is further aggravated in a sum-and-difference monopulse system, because the difference signals are no longer properly normalized by the sum signal, inasmuch as the latter has become the difference signal and is no longer the reference signal. Analysis also shows that the influence of the phase shift $\Phi$ (for $a_c b = \infty$) affects the positions of the pattern maxima and minima. As the initial angular errors increase, the shifts of the maxima and minima, as a function of $\Phi$, also increase.

The distortion of the angle-sensing response in the plane of the quadrants 3 and 4, when receiving cross-polarized signals, is shown in Fig. 8.9 in comparison with the normal response (for matched reception).

*Phase-sum-and-difference tracking monopulse system.* Phase-comparison systems typically form a pair of parallel beams in each coordinate plane (see Fig. 1.1). This may be achieved in a parabolic reflector with a square aperture by the following aperture excitations, in phase:

$$\psi_{1x}(x,\ y) = \begin{cases} u, & \text{if } -L_2 \leqslant y \leqslant 0 \\ 1, & \text{if } 0 \leqslant y \leqslant L_2 \end{cases} \quad \text{for } -L_1 \leqslant x \leqslant L_1$$

$$\psi_{2x}(x,\ y) = \begin{cases} 1, & \text{if } -L_2 \leqslant y \leqslant 0 \\ u, & \text{if } 0 \leqslant y \leqslant L_2 \end{cases} \quad \text{for } -L_1 \leqslant x \leqslant L_1$$

$$\psi_{3x}(x, y) = \begin{cases} u, & \text{if } 0 \leqslant x \leqslant L_1 \\ 1, & \text{if } -L \leqslant x \leqslant 0 \end{cases} \quad \text{for } -L_2 \leqslant y \leqslant L_2$$

$$\psi_{4x}(x, y) = \begin{cases} 1, & \text{if } 0 \leqslant x \leqslant L_1 \\ u, & \text{if } -L_1 \leqslant x \leqslant L_1 \end{cases} \quad \text{for } -L_2 \leqslant y \leqslant L_2$$

where 1 is the normalized amplitude of the basic field excitation at the operating polarization; and $u$ is the normalized amplitude of the spurious field excitation caused by the unwanted leakage currents.

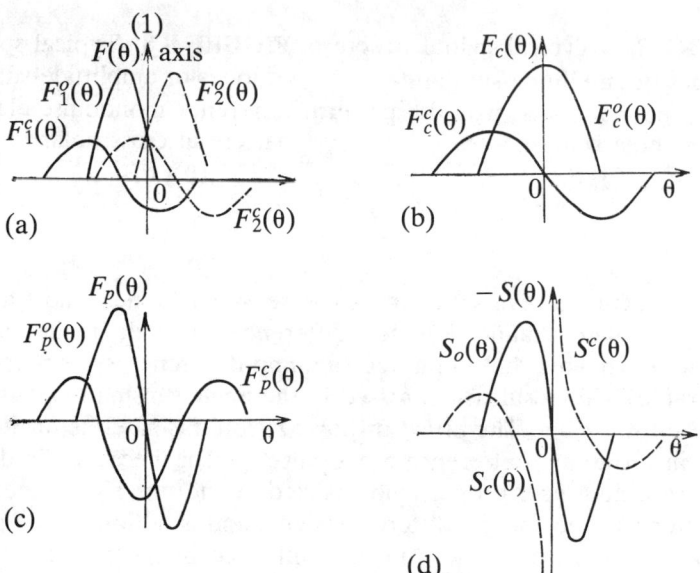

**FIGURE 8.9.** Typical antenna pattern and angle-sensing response of an amplitude-sum-and-difference monopulse system when receiving signals at operating polarization (index "*o*") and cross polarization (index "*c*"):

    (a)   partial patterns        (c)   difference pattern

    (b)   sum pattern           (d)   angle-sensing response

As a result, four parallel beams are formed as shown in Fig. 8.10, where the operating polarization is along the $X$-axis.

Considering that they are out of phase in neighboring quadrants, the cross-polarized components of the field excitation (along the $Y$-axis) may be expressed as follows:

$$\left. \begin{array}{l} \psi_{1y,4y}(x, y) = a_c \\ \psi_{2y,3y}(x, y) = a_c u \end{array} \right\} \quad \text{for } 0 \leqslant y \leqslant L_2, \, 0 \leqslant x \leqslant L_1$$

$$\left.\begin{array}{l}\psi_{1y,3y}(x,\ y)\ =\ -a_c\\\psi_{2y,4y}(x,\ y)\ =\ -a_cu\end{array}\right\}\quad\text{for }0\leqslant y\leqslant L_2,\ -L_1\leqslant x\leqslant 0$$

$$\left.\begin{array}{l}\psi_{1y,4y}(x,\ y)\ =\ a_cu\\\psi_{2y,3y}(x,\ y)\ =\ a_c\end{array}\right\}\quad\text{for }-L_2\leqslant y\leqslant 0,\ -L_1\leqslant x\leqslant 0$$

$$\left.\begin{array}{l}\psi_{1y,3y}(x,\ y)\ =\ -a_cu\\\psi_{2y,4y}(x,\ y)\ =\ -a_c\end{array}\right\}\quad\text{for }-L_2\leqslant y\leqslant 0,\ 0\leqslant x\leqslant L_1$$

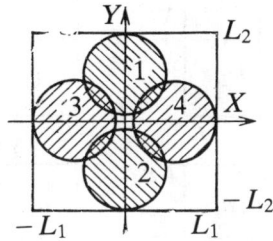

**FIGURE 8.10.** The antenna pattern of a phase monopulse system tracking in two planes at basic polarization.

where $a_c$ is the excitation field attenuation coefficient at cross polarization, equal to the ratio of the amplitudes of orthogonally polarized field components.

The excitation patterns for the paraboloid are shown in Fig. 8.11 at both operating polarization and cross polarization.

A simplified block diagram of a phase-sum-and-difference monopulse system for tracking in one coordinate was presented in Fig. 4.5. Carrying out all of the necessary operations to form the signals and process them in accordance with the methods laid out above, the angle-sensing responses for the system under consideration may be obtained in the following form:

$$S_{az}(\theta,\ \vartheta)\ =\ \frac{1\ -\ u}{1\ +\ u}\ \frac{A_1A_4(A_3^2\ -\ 4a_c^2b^2A_5^2)\ +\ a_cbA_3A_5(A_1^2\ -\ 4A_4^2)\cos\ \Phi}{A_1^2A_3^2\ -\ 8a_cbA_1A_3A_4A_5\ \cos\ \Phi\ +\ 16a_c^2b^2A_4^2A_5^2}$$

$$(8.2)$$

$$S_{el}(\theta,\ \vartheta)\ =\ \frac{1\ -\ u}{1\ +\ u}\ \frac{A_3A_5(A_1^2\ -\ 4a_c^2b^2A_4^2)\ +\ a_cbA_1A_4(A_3^2\ -\ 4A_5^2)\cos\ \Phi}{A_1^2A_3^2\ -\ 8a_cbA_1A_3A_4A_5\ \cos\ \Phi\ +\ 16a_c^2b^2A_4^2A_5^2}$$

$$(8.3)$$

where

$$A_1\ =\ (sin\ m)/m,\qquad A_2\ =\ [(2\ sin\ n)/2]/n,\qquad A_3\ =\ (sin\ n)/n,$$

$$A_4 = [(\sin^2 m)/2]/m, \quad A_5 = [(\sin^2 n)/2]/n;$$
$$m = k_\lambda L \cos \alpha, \, n = k_\lambda L \cos \beta$$

$\Phi$ is the phase shift of the orthogonally polarized signal components taking into account their phase shift at the antenna and their spatial orthogonality.

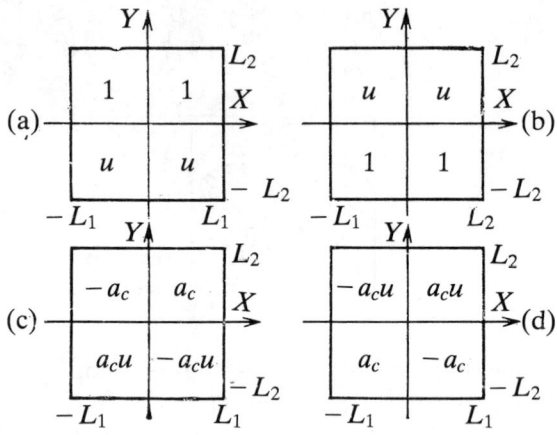

**FIGURE 8.11.** Paraboloid excitation for forming the (a) upper and (b) lower patterns at operating polarization, and the (c) upper and (d) lower pattern at cross polarization.

We·will consider the extreme cases.

1. $a_c b = 0$. Then

$$S_{az}^0(\theta, \vartheta) = \frac{1}{2} \frac{1-u}{1+u} \tan \frac{m}{2}, \quad S_{el}^0(\theta, \vartheta) = \frac{1}{2} \frac{1-u}{1+u} \tan \frac{n}{2}$$

The slope of the angle-sensing responses at the tracking point ($\theta = 0$) for $a_c b = 0$ are, accordingly,

$$\mu_{az}^0 = \frac{dS_{az}(\theta, \vartheta)}{d\theta} \bigg|_{\theta=0} = \frac{1-u}{4(1+u)} k_\lambda L \cos \vartheta$$

$$\mu_{el}^0 = \frac{dS_{el}(\theta, \vartheta)}{d\theta} \bigg|_{\theta=0} = \frac{1-u}{4(1+u)} k_\lambda L \sin \vartheta$$

2. $a_c b = \infty$. We obtain

$$S_{az}^c(\theta, \vartheta) = -\frac{1}{2} \frac{1-u}{1+u} \cot \frac{m}{2}, \quad S_{el}^c(\theta, \vartheta) = -\frac{1}{2} \frac{1-u}{1+u} \cot \frac{n}{2},$$

$$\mu_{az}^c = \left.\frac{dS_{az}^c(\theta,\vartheta)}{d\theta}\right|_{\theta=0} = \left.\frac{1-u}{2(1+u)}\frac{k_\lambda L\cos\vartheta}{2}\frac{1}{\sin^2(m/2)}\right|_{\theta=0} = \infty,$$

$$\mu_{el}^c = \left.\frac{dS_{el}^c(\theta,\vartheta)}{d\theta}\right|_{\theta=0} = \left.\frac{1-u}{2(1+u)}\frac{k_\lambda L\sin\vartheta}{2}\frac{1}{\sin^2(n/2)}\right|_{\theta=0} = \infty$$

The expressions which have been derived show that there is a substantial transformation of the angle-sensing responses in a phase-sum-and-difference monopulse system when the signal polarization changes. The approximate form of the angle-sensing responses for the cases considered are shown in Fig. 8.12, using the azimuth plane as an example.

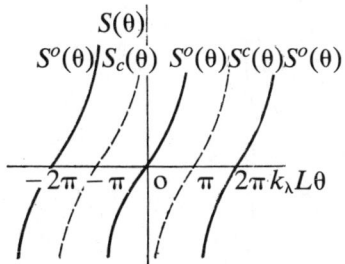

**FIGURE 8.12.** The typical angle-sensing response of a phase-sum-and-difference monopulse system when tracking a signal source with matched polarization ($S^o$) and cross polarization ($S^c$).

When $a_c b = \infty$ there are discontinuities in the angle-sensing response in the region $\theta = 0$, and tracking becomes unstable for signal sources on the axis. There are also two points at which $S(\theta) = 0$, where the response has positive slope:

$$m = n = \pm\pi$$

Inserting the values $m = k_\lambda L\theta\cos\vartheta$ and $n = k_\lambda L\theta\sin\vartheta$ and setting $\theta_3 = \lambda/L$, we obtain the points of stable equilibrium: $\theta_{az} = \pm\theta_3/2$ for azimuth angle sensing, and $\theta_{el} = \pm\theta_3/2$ for elevation angle sensing.

The presence of the factor $(1-u)/(1+u)$ in the previous expressions is a consequence of the reduction in the angle-sensing slope due to the coupling of the channels by the excitation field. For $u = 1$, when the antenna fields of the angle-sensing channels overlap completely, angle sensing becomes impossible. When $u = 0$, there is no coupling, and the angle-sensing slope is a maximum.

Calculations using (8.2) and (8.3) show that if the source of a linearly polarized signal ($\Phi$) is offset from the axis in the azimuth plane ($\Delta_y = 0$), the azimuth angle-sensing response will not depend on the quantity $a_c b$ (as long as $a_c b \neq \infty$). There is coupling between the channels in this case (see Fig. 8.13), and it increases with increasing $a_c b$. This results in the generation of an elevation channel error signal even though there is no elevation offset between the target and the axis. When there is, in fact, an elevation offset in addition to the crosstalk, the azimuth angle-sensing response is distorted, and there is a shift in the null axis which increases with the quantity $a_c b$ (see Fig. 8.14).

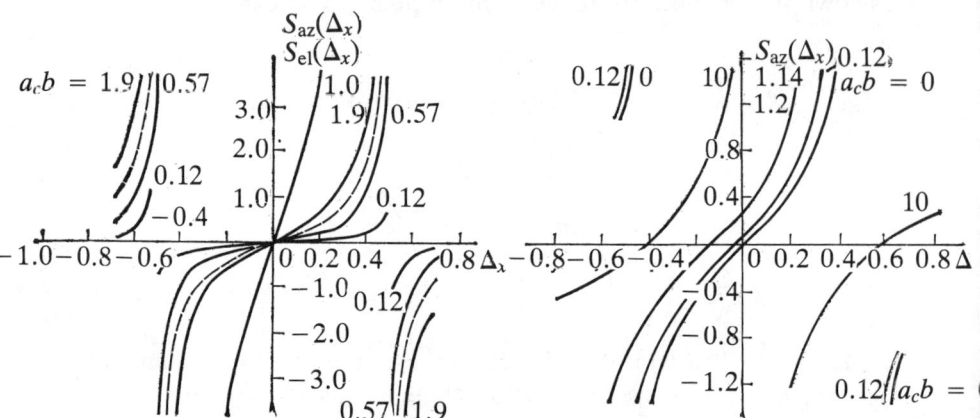

**FIGURE 8.13.** The angle-sensing response calculated for a phase-sum-and-difference monopulse system as a function of $a_c b$ for $\phi = 0$; $\vartheta = 0$.

**FIGURE 8.14.** The angle-sensing response calculated for a phase-sum-and-difference monopulse system as a function of $a_c b$ for $\Delta_y = 0.3$ and $\phi = 0$.

The angle-sensing responses which result when receiving elliptically polarized signals from a source offset diagonally from the axis are shown in Fig. 8.15. In this figure each value of $a_c b$ corresponds to a definite elliptical polarization coefficient, which may be determined from the known quantity $a_c$.

Analysis of the antenna patterns confirms their dependence on the polarization of the received signals. In a phase-sum-and-difference monopulse system, just as in the amplitude-comparison case, the patterns are functionally reversed when receiving cross-polarized signals: the sum channel becomes the difference pattern with a four-lobed structure, and the difference pattern becomes the sum pattern. The functions of the difference channels are exchanged as a result: the azimuth channel becomes the elevation channel and *vice versa.* Proper signal normalization, which is normally performed with the sum signal, is disrupted at the same time. All of this supports the conclusion reached earlier that, in the region of the axis, the angle-sensing system becomes inoperable and the antenna system will be steered towards spurious nulls when attempting to track cross-polarized signal sources.

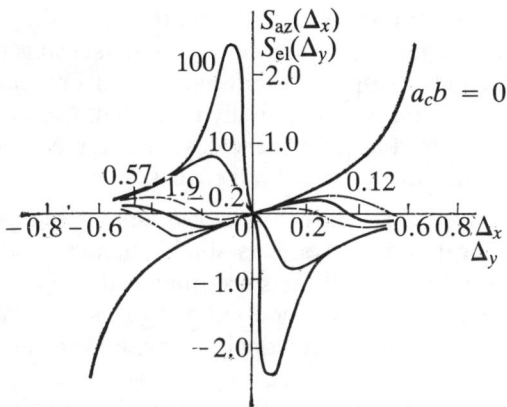

**FIGURE 8.15.** The angle-sensing response calculated for a phase-sum-and-difference monopulse system as a function of $a_c b$ for $\phi = 90°$ and $\vartheta = 45°$ $[S_{az}(\Delta_x) = S_{el}(\Delta_y)]$.

Analysis of the performance of various types of monopulse systems when receiving signals of varying polarization has thus led to the following results:

1. Cross-polarized interference affects all the types of monopulse systems which were considered. The effects of this interference possess identical characteristics, but have quantitative differences relating to the particular signal processing which is employed and the structure of the antenna patterns.
2. The fundamental cause of cross-polarized interference is distortion of the amplitude-phase distribution in the electromagnetic fields exciting the aperture, which causes distortion of the antenna patterns and, as a result, distortion of the system angle-sensing responses.
3. As a result of the transformations of the partial patterns resulting from cross-polarized interference, the sum pattern becomes the difference pattern and *vice versa,* in all angle measurement planes. This

leads to a reversal of the angle-sensing channels and complete tracking system inoperability due to the dephasing of the coordinate system and improper signal normalization.

4. The positions of the pattern minima and maxima do not change in the main angle-sensing planes, those in which the antenna is steered, as a result of cross-polarized interference. The interference in this case causes "leakage" of the minima and the appearance of troughs in the maximum to an extent determined by the accuracy of the orthogonally polarized jamming; in the limiting case, this leads to a reversal of the sum and difference patterns.

5. Cross-polarized jamming distorts the angle-sensing responses, causing a change in the slope and shift in the points of stable equilibrium. When the jamming is orthogonally polarized, the slope of the angle-sensing response in all cases reverses near the axis, and the tracking system becomes unstable in both planes.

It should be noted that inasmuch as cross-polarized jamming distorts the radar receiving pattern, it affects single-channel angle measurement systems, and in this sense its effects are universal.

In order to create cross-polarized jamming it is necessary for the jamming power to exceed the radar signal power, and to direct the jamming to orthogonal polarization with the required accuracy. If the radar to be jammed uses a single antenna for both reception and transmission, the jammer may maintain orthogonal polarization automatically according to the polarization of the signal illuminating the jammer platform. The transmitted radar signal is received and analyzed to determine its polarization, and jamming polarization is automatically oriented to be orthogonal to the radar's operating polarization [21]. The accuracy with which the jamming polarization is kept orthogonal to the radar signal depends on the antenna's cross-polarized radiation level, and must be rather high (on the order of several degrees). This is because, when the radar receives jamming signals which are not perfectly orthogonally polarized, a useful (for the radar) signal is formed at the antenna output. This signal, derived from the matching polarization component, limits the jamming-to-signal ratio which can be achieved in the radar receiver, and, consequently, limits the effectiveness of the cross-polarized jamming. It has been established that if the relative cross polarization level of the antenna is $-15$ dB, and the jamming polarization plane is $10°$ or more off the orthogonal orientation, the jamming has little effect on the stability of an automatic angle tracker, and the angle errors caused by the jamming are insignificant.

The necessary ratio of jamming power to signal power is also determined by the antenna's cross-polarized pattern levels, and can be in the range of 20–40 dB. If cross polarized jamming is combined with range (velocity) deception jamming, however, the jamming power only needs to be 6 dB higher than the signal power [86]. Use of a phased array antenna in the jammer greatly increases the effective jamming power. It has been shown that in the band 4.8–9.6 GHz it is possible to obtain an effective power of 112 kW, and a jamming-to-signal power ratio in the range of 2–40 dB. There are promising proposals for designing wideband jammers with phased array antennas and polarization control [57].

According to foreign sources, cross-polarized jamming is considered to be a promising means for jamming monopulse radars, and is receiving steady attention from both developers and customers [109].

### 8.5.3 Two-Frequency Jamming

In addition to disrupting the detection channel, two-frequency jamming may be employed to suppress the angle channel of monopulse radars [59]. We will illustrate this using a sum-and-difference monopulse system as an example. When operating normally, such systems are characterized by angle-sensing responses which are symmetric about the axis with a change of sign, which allow stable target tracking.

The two-frequency jamming signal may be expressed as

$$\underline{E}(t) = [2E_m \cos(\omega_d t/2)] \exp[i(\omega_1 + \omega_d/2)t]$$

where $\omega_d$ is the frequency difference between the two high-frequency carriers, and $\omega_1$ is the lower carrier frequency.

The signals resulting at the output of the antenna system of an amplitude-sum-and-difference monopulse system accordingly will be given by the expressions:

$$\underline{E}_1(t, \theta) = [2E_m \cos(\omega_d t/2)] F(\theta_0 - \theta) \exp[i(\omega_1 + \omega_d/2)t]$$
$$\underline{E}_2(t, \theta) = [2E_m \cos(\omega_d t/2)] F(\theta_0 + \theta) \exp[i(\omega_1 + \omega_d/2)t]$$

and the sum and difference signals will be

$$\underline{E}_s(t, \theta)$$
$$= [2E_m \cos(\omega_d t/2)][F(\theta_0 - \theta) + F(\theta_0 + \theta)]\exp[i(\omega_1 + \omega_d/2)t]$$
$$\underline{E}_d(t, \theta)$$
$$= [2E_m \cos(\omega_d t/2)][F(\theta_0 - \theta) - F(\theta_0 + \theta)]\exp[i(\omega_1 + \omega_d/2)t]$$

If the receiver mixer has a response in the form:

$$u = k(\alpha + \beta u^2_{in})$$

then direct detection will result in intermediate-frequency signals at the mixer outputs with amplitudes proportional to the square of the sum and difference patterns, i.e.,

$$\underline{u}_s(t, \theta) = [F(\theta_0 - \theta) + F(\theta_0 + \theta)]^2 \exp(i\omega_d t)$$

$$\underline{u}_s(t, \theta) = [F(\theta_0 - \theta) - F(\theta_0 + \theta)]^2 \exp(i\omega_d t)$$

The signal at the output of the phase detector is thus found to take the form:

$$S(\theta) = \frac{\mathrm{Re}[\underline{u}_d(t, \theta)\underline{u}_s^*(t, \theta)]}{\underline{u}_s(t, \theta)\underline{u}_s^*(t, \theta)} = \frac{[F(\theta_0 - \theta) - F(\theta_0 + \theta)]^2}{[F(\theta_0 - \theta) + F(\theta_0 + \theta)]^2}$$

In the case being considered, the angle-sensing response evidently is an even function.

We may obtain the angle-sensing response for a phase-sum-and-difference monopulse system analogously. In this case, upon reception of two-frequency jamming the antenna output signals will be

$$\underline{E}_1(t, \theta) = [2E_m \cos(\omega_d t/2)]F(\theta)\exp[i(\omega_1 + \omega_d/2)t + \Delta\phi_1/2]$$

$$\underline{E}_2(t, \theta) = [2E_m \cos(\omega_d t/2)]F(\theta)\exp[i(\omega_1 + \omega_d/2)t - \Delta\phi_1/2]$$

The sum and difference signals emerging from the difference channel will be described by the expressions:

$$\underline{E}_s(t, \theta) = [\sqrt{2}E_m \cos(\omega_d t/2)]F(\theta)$$
$$\times [\exp(i\Delta\phi_1/2 + \exp(-i\Delta\phi_1/2)]\exp[i(\omega_1 + \omega_d/2)t]$$

$$\underline{E}_d(t, \theta) = [\sqrt{2}E_m \cos(\omega_d t/2)]F(\theta)$$
$$\times [\exp(i\Delta\phi_1/2 - \exp(-i\Delta\phi_1/2)]\exp[i(\omega_1 + \omega_d/2)t]$$

The mixer output signals will be proportional to

$$\underline{u}_s(t, \theta) = [F^2(\theta)\cos^2(\Delta\phi_1/2)]\exp(i\omega_d t)$$

$$\underline{u}_d(t, \theta) = [-F^2(\theta)\sin^2(\Delta\phi_1/2)]\exp(i\omega_d t)$$

The phase detector output is then

$$S(\theta) = \frac{\mathrm{Re}[\underline{u}_d(t, \theta)\underline{u}_s^*(t, \theta)]}{\underline{u}_s(t, \theta)\underline{u}_s^*(t, \theta)} = -\tan^2(\Delta\phi_1/2) = -\tan^2(k_\lambda l\theta/2)$$

The angle-sensing response of a phase-sum-and-difference monopulse system is also an even function of the offset angle. In both of these cases, then, the two-frequency jamming leads to the formation of angle-sensing responses which have no stable nulls, owing to which autotracking of the source of such jamming is completely impossible. The null axis of the radar antenna system will stray from the direction to the jamming source and the automatic track will be interrupted.

It should be noted that the effects of two-frequency jamming are not the same for other types of monopulse systems.

We will now consider an amplitude-amplitude monopulse system. In this case the signals resulting at the output of the first and second channels due to two-frequency jamming will be given by the expressions:

$$\underline{E}_1(t, \theta) = [2E_m \cos(\omega_d t/2)]F(\theta_0 - \theta)\exp[i(\omega_1 + \omega_d/2)t]$$
$$\underline{E}_2(t, \theta) = [2E_m \cos(\omega_d t/2)]F(\theta_0 + \theta)\exp[i(\omega_1 + \omega_d/2)t]$$

after direct square-law detection in the mixers, these signals will be proportional to

$$\underline{u}_1(t, \theta) = F^2(\theta_0 - \theta)\exp(i\omega_d t)$$
$$\underline{u}(t, \theta) = F^2(\theta_0 + \theta)\exp(i\omega_d t)$$

If the channels have identical amplitude-phase responses, the signals at the input of the subtraction device will be

$$u_1(t, \theta) = 2\ln[F(\theta_0 - \theta)], \qquad u_2(t, \theta) = 2\ln[F(\theta_0 + \theta)$$

and the output response will be

$$S(\theta) = 2\ln[F(\theta_0 - \theta)/F(\theta_0 + \theta)] \approx 4\mu\theta$$

Comparison with (4.4) shows that the effect of two-frequency jamming on an amplitude-amplitude monopulse system is only to change the slope of the angle-sensing response by a factor of two. The system will automatically track a two-frequency jammer almost as well as if it were experiencing no jamming.

We will now consider a phase-phase monopulse system. When receiving two-frequency jamming, the system may be described (to within a constant) by the following group of equations.

● Antenna output signals:

$$\underline{E}_1(t, \theta) = [2E_m F(\theta)\cos(\omega_d t/2)]\exp\{i[(\omega_1 + \omega_d/2)t + \Delta\phi_1/2]\}$$
$$\underline{E}_2(t, \theta) = [2E_m F(\theta)\cos(\omega_d t/2)]\exp\{i[(\omega_1 + \omega_d/2)t + \pi/2 - \Delta\phi_1/2]\}$$

● Signals after quadratic detection in the mixers:

$$\underline{u}_1(t, \theta) = [2E_m F(\theta)\cos(\omega_d/2)]^2 \exp[i(\omega_d t + \Delta\phi_1)]$$

$$\underline{u}_2(t, \theta) = [2E_mF(\theta)\cos(\omega_d/2)]^2 \exp[i(\omega_d t + \pi - \Delta\phi_1)]$$

- Signals at the inputs of the phase detector, with amplitude limiting:

$$\underline{u}_1'(t, \theta) = u_{\lim} \exp[i(\omega_d t + \Delta\phi_1)]$$

$$\underline{u}_2'(t, \theta) = u_{\lim} \exp[i(\omega_d t + \pi - \Delta\phi_1)]$$

- Signal at the output of the phase detector:

$$S(\theta) = k_{pd}\text{Re}[\underline{u}_1'(t, \theta)\underline{u}_2'^*(t, \theta)] = k_{pd}u_{\lim}^2 \cos(2\Delta\phi_1)$$

The angle-sensing response is again an even function of the offset angle, making a phase-phase monopulse system inoperable in the examined case.

Thus, out of the monopulse systems considered, only the amplitude-amplitude system can maintain an accurate track when subject to two-frequency jamming. The other systems lose the ability to maintain automatic target tracks.

### 8.5.4 Coherent Jamming from Two Points in Space

The basis of coherent jamming is the generation of phase irregularities in the receiving antenna aperture by illuminating it with coherent signals from two spatially separated sources [5, 21, 60, 92, 109].

The fundamental physics behind this type of jamming were laid out in sufficient detail in Chapter 6 while discussing the phase front of a two-point target signal. It was shown that if a target has two signal sources, then in comparison with the case of a single-point target, the phase front will be distorted according to the amplitude and phase relations between these two signals. This results in a degradation of the angle accuracy in various radars, including monopulse systems.

In order to elucidate the phenomena associated with receiving signals from a two-point target and the associated possibilities for ECM, the reception and processing of signals from two-point sources in an amplitude-sum-and-difference system were studied [21]. A mathematical expression describing the error in tracking a two-point target was obtained as a function of the amplitude and phase relations:

$$\frac{\theta}{\psi_s} = \frac{1 - a^2}{2(1 + a^2 + 2a \cos \alpha)} \tag{8.4}$$

where $\theta$ is the angular error relative to the center of the baseline (halfway between the sources).

Equation (8.4) is analogous to (6.4), which describes the deflection of the phase front from a two-point source relative to the phase front which would be obtained from a single-point signal source.

This supports the fact that when tracking point sources with small angle errors, a radar system seeks the direction normal to the phase front of the waves reflected from the target, and distortion of the phase front due to interference phenomena associated with multiple-scatterer targets or jamming signals unavoidably increases the tracking errors. The value of the angle error, as may be seen from (8.4), depends on the distance between the radiating sources, the phase difference between the signals and the ratio of their amplitudes at the input of the angle-sensing system.

Figure 8.16 presents the angle error in units of the source separation as a function of the amplitude and phase relations between the jamming signals. These thoeretical curves show that the error does not depend on the direction in which the axis deviates relative to the sources.

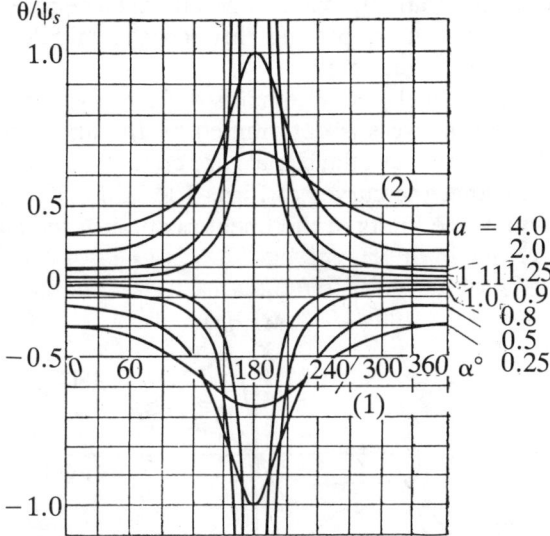

**FIGURE 8.16.** The angle errors when tracking a two-point target, calculated as a function of the amplitude and phase relations of the signals radiated by the target.

    (1) position of target 1      (2) position of target 2

It may be seen that the angle error relative to source 1 is approximately 0.6 times the source separation for the case $a = 1.25$. Thus, even if the radar is pointed accurately at one of the sources, the null axis indicates a point located approximately half-way between the targets. As the phase shift between the jamming signals increases, so does the error, which reaches a maximum value when the phase difference is 180°. The error in this case theoretically may attain extremely large values. In practice, the error is limited by the antenna pattern, and may not exceed the width of the receiving pattern of the jammed radar.

It should be noted, however, that with $\alpha = 180°$ and $a \approx 1$, the normal single-lobed receiving pattern is transformed into a two-lobed pattern, and the angle-sensing responses of the radar are significantly distorted. In sum-and-difference monopulse radars, for example, the sum and difference channels are functionally reversed under this condition. The signal normalization which is performed with respect to the sum signal is destroyed, and, as a result the tracking system becomes unstable and inoperable, and may be forced out of the autotracking mode. In this case, the angle error may exceed the receiving beamwidth.

When the received signals are in phase ($\alpha = 0$), the error decreases to $(1 - a)/[2(1 + a)]$, which corresponds to the *energy center* of the two sources. If the signal amplitudes are equal, the energy center coincides with the geometric center, passing through the middle of the baseline between the two coherent sources.

The direction in which the antenna is deflected when receiving signals from two coherent sources is determined by the amplitude ratio, and is reversed at the point where the signals are equal (see Fig. 8.17). It should be kept in mind that the curves of Figs. 8.16 and 8.17 are valid only for errors lying within the limits of the linear portion of the antenna pattern.

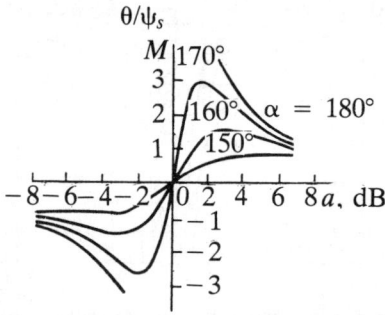

**FIGURE 8.17.** The relative angle error when tracking a two-point target as a function of the signal ratio for specific phase shifts.

These phenomena, which are observed when tracking two-point coherent signal sources, represent the basis for creating coherent jamming of monopulse radars. Two separated sources are used to generate coherent signals, the amplitude and phase relations of which are adjusted to produce the maximum tracking error in the jammed radar. This coherent jamming acts to deflect the null axis of the radar from the direction to the target, which is the jamming source.

One limitation of this type of jamming is that it is applicable only at short ranges [21, 109]. This is a consequence of the fact that the angle error is proportional to the angle subtended by the jamming sources. At long ranges, the subtended angle of the sources is necessarily small, because they must both lie within the geometric limits of the jamming platform; the induced angle errors are insignificant and the jammer is, therefore, expected to be ineffective at long ranges. Placing the sources on different aircraft in order to extend the baseline is plainly unworkable, due to the difficulty in maintaining the coherency of unconnected sources.

Coherent jamming is also limited due to power considerations, inasmuch as large jamming power levels are required. This is due to the fact that the maximum jamming effect is obtained with signals of opposite phase, which then cancel each other to a large extent as they are radiated through the separate antennas. Large jamming power levels are, therefore, required for the residual jamming signal level to exceed the reflected radar signal level; these levels must, in practice, exceed those of known jammers radiating from a single point in space. Thus, for example, if the induced error is to exceed the baseline between the transmitting antennas, the jamming signal must exceed the reflected signal by 20 dB [83]. If coherent jamming is combined with range (velocity) deception jamming, however, the jamming signal need exceed the radar signal by only 6 dB [86].

Despite these limitations, coherent jamming is considered to be a promising ECM method by specialists abroad [109], and therefore, the nature of its operation needs to be considered. It should be kept in mind that coherent jamming affects both single-channel and multiple-channel angle measuring systems equally, and monopulse systems therefore do not possess any advantage over the usual single-channel systems in immunity from coherent jamming.

## 8.5.5 Blinking Jamming from Two Points in Space

All known single-target radars are placed in an almost hopeless situation when they are confronted with two or more spatially unresolved targets [92]. The fundamental cause of this effect is the random nature of the instantaneous reflectivities associated with the targets located in the radar beam; the radar axis wanders from one target to another, and, as a result, large angle errors arise. The degradation in radar performance can be increased if the targets are supplied with programmed jammers, which, in the simplest approach, alternately switch on and off. In this situation, the radar observes the various targets in the same sequence as the jammers switch on and off, or as the jamming frequencies pass through the radar passband. The jammers may be switched on and off in a random fashion, denying the radar the ability to track an individual target for the time necessary to perform an accurate measurement of its location. The result is blinking jamming, which is employed in the group defense of aircraft.

The jamming effect of this scheme is based on the limited resolving capability of angle measurement systems. As has been shown in [21], when there are two jamming sources which are unresolved in angle, the radar will track the energy center, as described by the expression:

$$\frac{\Delta\theta}{\psi_s} = \frac{P_{j1}(t) - P_{j2}(t)}{2[P_{j1}(t) + P_{j2}(t) + 2P_s(t)]} \tag{8.5}$$

where $P_{j1}(t)$ and $P_{j2}(t)$ are the first and second jamming powers, and $P_s(t)$ is the signal power reflected from each jamming platform.

It follows from (8.5) that the position of the emission energy center is determined basically by the ratio of the jammer source powers and the character of the jammer power variation with time. As the jammers are switched on and off alternately, the jammed radar will attempt to track first one, then the other target, causing the radar antenna to oscillate in angle in time with the jammer switching. This will make angular resolution and measurement of the targets more difficult. The separation necessary for the targets to be resolved will be increased with blinking jamming, which will lead to an unavoidable increase in the distance by which the interceptor missile misses, inasmuch as this miss distance and the critical resolution angle are connected through the dependence [5]:

$$\Delta l = \frac{l_b}{2} - \frac{1}{2}\frac{ngl_b^2}{V_{cl}^2\theta_c^2}$$

where $\Delta l$ is the miss distance (in linear units); $l_b$ is the projection of the baseline between the targets onto the plane normal to the line of sight; $ng$ is the maximum acceleration of the rocket; $V_{cl}$ is the closing velocity of the missile to the target; and $\theta_c$ is the critical resolution angle.

Blinking jamming is accordingly considered to be an effective means of jamming missile guidance radars [21, 109].

It is clear that if the tracking system is to be able to track the jumping energy center created by blinking jamming, the frequency with which the jammers are switched should be matched to the tracking system passband $\Delta F_{ts}$ in accordance with the condition

$$F_s \leq \Delta F_{ts}/2$$

If the switching rate is too high, the tracking system will average the angle errors and track the direction to the energy center of the jammer sources. It is also clear that if the switching rate is too low, the jamming will fail since the attacking missile will be successfully guided to the emissions from one of the jammer sources. The optimum blinking frequency is usually between 0.5 and 10 Hz [76].

Inasmuch as blinking jamming operates by alternately redirecting the jammed radar's antenna to a stronger signal source in another direction, a parameter which is no less important than the blinking rate is the jamming power, or more correctly, the amount by which the jamming power needs to exceed the signal power to achieve the desired effect. If the linearized angle-sensing response is used, then the amount by which the jamming power must exceed the signal power in order to redirect the radar antenna may be determined from (8.5). Calculations show that in order to induce antenna jumps in an angular sector $0.8\psi_s$ ($\psi_s$ is the target separation), the jamming power must exceed the signal power by 9 dB [21].

If the angular target separation exceeds the linear portion of the angle-sensing response, then the calculations should be carried out taking account of the actual antenna patterns.

Like two-source coherent jamming, blinking jamming may be considered universal, in that, under certain conditions, it may be used against angle trackers of different types. This is explained by the fact that its operation is fundamentally connected with a change in the slope of the phase front impinging on the radar's receiving antenna.

Blinking jamming, which is generated from several aircraft, is considered to be an effective means of group defense using individual jammer transmitters. It should be noted that the degradation of the radar's angular resolving capability caused by blinking jamming is achieved only if the targets are also unresolved in range and velocity. Foreign specialists [92, 109] accordingly feel that the simplest and most effective means of creating blinking jamming is to use spot-barrage noise transmitters, asynchronously switched at low frequencies.

Because aircraft tend to be close together in formation missions, the most realistic case will be that of jamming through the jammed radar's sidelobes with the targets being resolved in range, velocity, and direction. Future improvement in blinking jamming will, therefore, entail control of the individual airborne jammers through communications channels.

To provide mutual defense, continuously transmitting noise jammers may be employed. If the power of the noise jamming is sufficient to mask the reflected radar signals, the radar will track the direction to the energy center of the sources, as in the case of blinking jamming with a high blinking rate. If the distance between the targets exceeds the radar beamwidth, the radar will track one of the targets and the mutual defense will be ineffective.

## 8.5.6 Frequency-Swept Jamming Generated from Two or More Points

It was pointed out earlier that transmitters with continuously swept frequency may be employed as sources of masking jamming. As with blinking jamming, the radar will seem to see many targets, each occurring as the jamming transmitters pass through the radar receiver's passband. The interference thus created in the angle channel is similar to that caused by a group of targets, being different only in that the targets cannot be resolved in range, because the jammers transmit continuously [21, 109].

If planes flying in tight formation are equipped with transmitters whose sweep rates are adjusted to meet the requirements for jamming the angle channel, frequency-swept jamming can, in principle, produce the effects of blinking jamming, and is considered to be a means for suppressing the angle channel of radars, including monopulse systems.

The blinking effect produced by frequency-swept jamming is explained by the fact that the radar will redirect itself to track the jamming source whose frequency enters the receiver passband. When the jamming signal leaves the radar receiver's passband, the radar will continue to track the same target, but with the reflected radar signal. As soon as the jammer on another target is swept into the receiver passband, the radar's range channel will be jammed, preventing target selection by range, and the radar will automatically begin to track the new target by its jamming signal. Since this new target will, in general, be in a different direction, the antenna system will redirect itself to this new direction. This sequence will be repeated as the jamming signal from the various targets enter the radar's passband. As a result of the alternating influence of the spatially separated frequency-swept jammers, the antenna system will track first one target, and then another, and will consequently attempt to jump in accordance with the programmed jammer transmissions.

The acceptable sweep rates are fairly low, which is a limitation on this type of jamming; if the jammers are swept over wide frequency bands, the intervals between the occurrences of jamming will be long, significantly reducing the intended blinking effect of the tracked target. If the sweep rates are high and the time between occurrences of jamming is comparable to the receiver time constant, the expected effect upon the angle channel is similar to that obtained with a formation target, regardless of the different individual target ranges, inasmuch as range resolution is suppressed by the jamming.

The advantages of frequency-swept jamming include its effectiveness against many radars whose operating frequencies lie within the sweep band of the jammers, and the relative simplicity of the jamming equipment.

### 8.5.7 Bistatic Jamming

Foreign specialists consider bistatic jamming to be an effective means for disrupting automatic target tracking and missile guidance. The principle behind this type of jamming involves redirecting the victim radar from the defended object to a false target. There are now many different bistatic jamming methods, including redirection to chaff clouds, to active and passive towed decoys, to expendable jammers, and to the ground (or water) [109].

*Chaff.* To defend aircraft, chaff clouds may be ejected with automatic devices and illuminated with special signals [109]. The use of automatic ejection devices removes the possibility of the chaff burning in the atmosphere, and the selection of special signals prevents the chaff from being resolved in velocity. As a result, the chaff clouds imitate actual moving targets and are an effective means of deception jamming, creating the conditions under which the radar will be diverted from the defended target to the chaff. The effectiveness of this redirection will be determined largely by the timing of the chaff ejection, which in a complex defense system may be controlled with onboard computers, modern electronic surveillance equipment, and fast acting automated chaff ejectors. It should be noted that because the chaff cloud will have a large angular extent, if it is ejected close to the radar, it will cause serious difficulties in angle tracking, as the radar axis will wander around the chaff cloud and lose the target in an automatic tracking mode.

When defending relatively slow moving objects such as ships or helicopters, it is possible to use modulated coherent signals to illuminate different chaff clouds (or parts of a single cloud). This will produce an effect on missile guidance systems similar to that produced by spatially separated blinking jammer sources, and will cause the missile trajectory to miss the target.

*Towed decoys.* Towed decoys differ from ejected decoys in that they may be reused, inasmuch as after the threat presented by a given missile has been eliminated, the decoys may be hauled back into the plane and stored in special containers until they are needed again. This redirection method consists of reflecting off the decoy a signal whose level greatly exceeds that of the radar signal reflected by the target [109]. The radar or missile seeker will begin to track the decoy, and the probability of hitting the plane decreases. The decoy can be one of many objects, such as balloons covered with reflective material, Luneberg lenses, corner reflectors, dielectric rod antennas, Van Atta arrays, and so on. With sufficient jamming power, the effectiveness of towed decoys is very high, because it is practically impossible to distinguish them from the actual targets. Unlike ejected decoys, towed decoys have a velocity equal to that of the target, so that they cannot be resolved in velocity. In addition to the passive decoys just listed, active decoys may also be towed. The decoy may be outfitted with its own transmitter, or the decoy antenna may be fed at the appropriate frequency through a cable or waveguide directly from the defended object, giving rise to the ability to modulate the decoy signal to increase its effectiveness. In order to improve the performance against missiles equipped with homing devices, several active and passive decoys may be towed at various distances.

In foreign systems employing towed decoy defenses, the decoys are put into operation automatically when the warning equipment indicates that the aircraft is faced with a dangerous situation [109].

***Redirection to jettisoned decoys.*** One shortcoming of towed decoys is that they slow down the aircraft and reduce its maneuverability. Therefore, it is often preferable to use jettisoned decoys illuminated with RF signals [109]. Illuminating the decoys with RF signals enables passive homing missiles to be diverted to the radiation source, along with systems supplementing passive with active homing. Foreign specialists believe that the ejection method is particularly important, and that it is desirable for the decoy speed to be several times that of the aircraft, but not to exceed velocities achieved during actual evasive maneuvers (otherwise the impression of an interrupted trajectory will be given, and the missile will not go in the direction of the decoy) [109].

As the jettisoned decoy moves farther from the aircraft, the signal which is reflected off it will grow smaller and approach zero. In some cases, the signal may be lost altogether at a certain stage in the process of diverting the radar. If the interceptor missile employs an active-passive guidance system, when the jamming signal is lost the seeker will switch to the active mode and acquire the actual target. It is advisable to eject chaff along with the decoys in order to prevent this reacquisition of the target by the missile [109].

Corner reflectors are often used as decoys. If they are ejected at the moment of attack, it is possible in some cases to divert the interceptor from the defended object to the false target. More developed decoys may possess their own propulsion systems so that they can imitate the speed and motion of the target in addition to its reflective properties. A typical example of such a decoy is a small drone missile carried under the wings of a bomber and ejected into free flight at the proper time [21, 30]. Active transponders are sometime placed on these drones in order to increase the intensity of the signals which would otherwise simply be reflected off the small objects [109].

A great deal of attention is being given abroad to decoys as a means for protecting warheads from an antimissile defense system. Considering the specific nature of the warhead flight, light objects such as inflated spheres or balloons are used in the middle portions of the trajectory, while as the warhead enters the dense atmosphere, the decoys will be heavier objects, which can move through the atmosphere without burning.

*Ground bounce jamming.* Among the available bistatic jamming methods, and according to the foreign press, the most important, is ground bounce jamming. This is partly explained by the recent attention which has been devoted to the development of methods for aircraft to break through defense systems at low and very low altitudes. Such tactics enhance the surprise nature of the attack and lower aircraft losses [92, 109].

The low flight path of the aircraft in itself reduces the accuracy of the radar tracker, due to interfering reflections and multipath effects. If the underlying surface is also illuminated with jamming signals from a narrow-beam antenna, then the radar task is further complicated [92, 109]. In this case, the surface becomes a source of jamming signals acting against the radar through its sidelobes. The direction to a given source point differs from that to the jammer platform, creating a situation equivalent to that which would be created if the jammer were remotely deployed at the location of the image below the surface (see Fig. 6.1). In some cases, the jamming power from the image source may exceed the radar power reflected from the target, in which case the radar will primarily track the direction to the image. Foreign specialists believe that this leads to the diversion of semiactive and active homing missiles, and also interceptor aircraft.

In order to generate a false target in a direction different from that to the actual target, the power of the jamming signal reflected off the surface must exceed the radar power reflected directly from the airborne jammer platform back to the radar. The reflected jamming signal should also have approximately the same delay and frequency as the directly received radar signal. If the aircraft is flying at high altitudes, it may be shown that the target will be tracked by the main radar beam, and the reflected jamming signal will enter the antenna sidelobes. This requires a further increase in the effective jamming power, which may be obtained by increasing the jammer transmitter power and the directivity of the antenna through which the signals are beamed off the surface. The reflected jamming power may also be increased by taking measures to reduce the direct signal at the radar antenna, for example, by using coherent signals radiated from two points, or transmitting the jamming signals at orthogonal polarizations. As has been established, both methods form a null (or close to a null) in the jamming pattern on the radar's null axis [83].

*Sidelobe jamming.* The existence of sidelobes in a radar's antenna pattern gives rise to possibilities for jamming from a remote point, and especially for jamming the angle channel. Jamming signals which are received by the sidelobes originate from directions different from that to the tracked targets, and produce an unavoidable decrease in tracking accuracy, under certain circumstances even causing an automatic angle track to be broken.

Until recently, the creation of jamming through the sidelobes was held back by high jammer power requirements in conjunction with the power limitations of jammer transmitters. As has been indicated in foreign publications, the current intense development of phased array antennas and their integration into electronic warfare systems has significantly increased the potential of jamming sites, and, as a result sidelobe jamming is now considered to be a realistic and very effective means of jamming [5, 109]. These developments have also led to a requirement for sensitive receivers to be deployed on board the defended objects in order to detect the radar through its sidelobes. It should be expected that drones will be widely employed to jam radars through the sidelobes with rather powerful onboard jammers.

*Expendable jamming transmitters.* Expendable jamming transmitters have assumed an important role in foreign radar jamming systems [109]. Such jammers may be ejected with missiles, guns, mortars, or chaff ejectors, or they may be transported by pilotless drones, dropped from aircraft, and so on. Foreign specialists believe that such transmitters, when deployed at a distance from the defended aircraft, will significantly interfere with the enemy's defense and protect the aircraft from missiles which will be guided to the jamming source. Such jammers may produce spurious targets and cause premature operation of the proximity fuse on antiaircraft missiles. They may be ejected from missiles prior to impact, and their effectiveness will be determined by their proximity to the radar and their capability for being tuned and steered. The spectral power density of such devices is typically below 10 W/Mhz, which is thought to be sufficient to jam radars 1 to 2 km away. In addition to jammers transmitting continuous noise, pulsed jammers may be used to interrupt automatic target tracking.

Interest in expendable jammers increased in the US after the appearance of several pilotless delivery systems in various forms, which become low vulnerability targets to an antiaircraft system. Drones with autopilots which eject jammers with their own power supplies at short time intervals (every five minutes, for example) can ensure that there are always four to five such jammers in the air at all times, substantially lowering the effectiveness of the antiaircraft system. Special parachutes may be used to decrease the speed of the jammers' descent [109].

## 8.6 METHODS FOR PROTECTING MONOPULSE RADARS FROM SEVERAL TYPES OF JAMMING

The deployment of radar jamming systems poses serious problems for radar designers. One of these problems is to develop antijamming techniques which do not have an adverse effect on the system's tactical and technical characteristics or on the operator's conditions. Investigations have established that there are no universal antijamming techniques, just as there are no universal jamming methods. The radar designer must respond to various forms of jamming with an ensemble of various antijamming approaches to counter them. Standard antijamming methods are usually not simply combined mechanically to provide such a coordinated defense; rather, the various components and blocks of the radar are modified while making sure that the various antijamming adjustments are compatible.

In addition to these system design methods for defending against jamming, which are associated directly with the system engineering, there are also tactical methods connected with the manner in which the radar is deployed. There are currently many engineering antijamming techniques, many of which are described in the literature. In [87], for example, there is a bibliography containing a description of approximately 150 methods for defending against jamming. Many antijamming techniques are described in [13]. The goals of these methods are: preventing the radar receiver from being overloaded (saturated); maintaining a constant false alarm rate; maintaining the signal-to-noise ratio; discriminating the jamming by its electronic and directional parameters; eliminating spurious targets; and maintaining a stable track on targets while being jammed in various ways. We will examine those methods which are of the most interest in protecting monopulse radars from jamming.

### 8.6.1 Antijamming Techniques for Monopulse Detection Channels

The following are the basic methods for protecting the detection channel from jamming [13, 84, 85, 87]:

- adaptively programming the radar's operating frequency;
- cancelling and blanking any jamming received through the sidelobes;
- employing a moving target indicator (MTI);
- employing double frequency conversion in the radar receiver;
- using STC to maintain a constant false alarm rate;
- using a sweep-to-sweep correlator;
- incorporating Dicke fix circuitry;

- using wobbulated and staggered pulse repetition intervals (PRIs);
- changing the operating polarization.

Adaptive frequency tuning allows the radar to operate in the portion of the band least affected by barrage noise jamming or free of spot noise jamming. Such frequency diversity makes it necessary for the jammer to distribute its energy over the entire band in which the radar can operate, thus decreasing the jamming spectral density and reducing the effectiveness of the jammer. The radar may be retuned when the interference level in the operating channel reaches some threshold, or when the radar's performance has noticeably worsened. An especially effective method for protecting the detection channel is to change the frequency pulse to pulse. Even when a constant pulse repetition frequency is used, the jammer can only affect the radar performance at ranges longer than that to the jammer itself, so that the jammer cannot keep itself from being detected.

Sidelobe cancellation reduces the sensitivity of the radar to jamming which originates in a direction away from the main beam, and in the ideal case will render sidelobe jamming completely ineffective. Coherent sidelobe jamming cancellation corrects the phase and amplitude of the jamming signal received in an auxiliary channel in order to match it with the signal from the main antenna, from which the corrected jamming signal is subtracted.

A coherent suppression device may be combined with a sidelobe jamming blanking device to counter deception jamming and jamming from pulse transponders with low duty factors [84]. The blanking device is attached to the same auxiliary antenna and receiving channel as the coherent suppression system. It extracts the signal, makes use of internal logic circuits to determine whether the same signals are being received through the main antenna's sidelobes, and gates the unneeded signals. When jamming from several different directions must be dealt with, a corresponding number of cancellation loops is necessary. The use of phased array antenna significantly increases the capability of cancelling a large number of jamming sources.

The sidelobes may also be reduced by optimizing the amplitude-phase distribution of the aperture excitation, and by placing radio absorbing components around the feed.

The use of a moving target indicator produces good results in the presence of such passive interference as chaff, local objects, and weather. A moving target indicator usually consists of a filter which is fixed to reject the zero Dopper frequency, in order to suppress stationary interference, and an adaptive filter to suppress nonstationary interference [84].

It should be noted that a moving target indicator will only be effective against chaff which is not illuminated by special jamming signals which give the chaff an apparent velocity equal to that of the actual targets. Interference from nearby objects may also be controlled through temporal control of the receiver sensitivity.

Double frequency conversion in the radar receiver protects the system from two-frequency jamming, inasmuch as it is impossible in this case for the jammer to form a signal at the intermediate frequency [84]. Other measures which may be taken against two-frequency jamming are [12, 28, 59]:

- placing preselectors in the RF lines ahead of the mixers so as to narrow the system passband and prevent the simultaneous reception of two signals separated by the intermediate frequency;
- using balanced mixers to lower the cross-product levels and decrease the generation of the difference frequency;
- avoiding the use of standard intermediate frequencies in the receiver, making it more difficult for the enemy to form the proper two-frequency jamming signal.

The incorporation of constant false alarm rate (CFAR) circuitry in combination with a sweep-to-sweep correlator is an effective technique against noise and asynchronous pulse jamming. Under normal circumstances, the false alarm rate is proportional to the mean internal receiver noise level in a range resolution cell. In the presence of powerful noise jamming, if the threshold is fixed, the false alarm rate increases sharply, with an attending decrease in true target detection.

If the threshold is set adaptively, as in automatic detection systems, a constant false alarm rate may be maintained, thus lowering the number of false alarms resulting from noise jamming. This does not improve the visibility of the signal against the interference; however, it serves only to extract the signal from the interference as much as possible before comparing it with a threshold in a target detection device, and to improve the results of subsequent signal processing. The use of digital signal processing in CFAR systems improves their protection of the detection channel from jamming.

The *Dicke fix* is widely used abroad in protecting detection channels, and consists of three sequential elements in the receiver: a wideband filter, a limiter, and a narrow-band filter. This method has been proven against barrage and spot noise jamming, fast frequency-swept jamming, and pulsed jamming [85, 87].

Frequency diversity and polarization switching give good results against many forms of jamming. The latter technique attempts to reduce the interference-to-signal ratio by adjusting the receiving antenna away from the jamming signal in polarization. Engineering solutions are theoretically possible, which can completely cancel the jamming in polarization.

Employing a wobbulated or staggered pulse repetition frequency is a good defense against multiple pulse jamming. This causes the screen to display spurious pulses spread in both range and angle. If the PRF is stable, then the jamming echoes on the screen will be no different from actual target echoes, surrounding the actual target. If the PRF is wobbulated, then jamming induced at ranges shorter than the range to the jammer will become blurred (defocused) on the screen, and the jammer may be detected on the boundary between blurred and defined echoes. Sharp jumping of the PRF produces an even larger effect. In this case, the brightness of the interfering pulses at close ranges is substantially reduced, and the boundary at the position of the jammer platform becomes even more well defined [85, 87].

### 8.6.2 Antijamming Methods for Monopulse Target Selection Channels

Resolving targets in range or velocity is an important means for protecting monopulse angle channels against jamming, and protecting the selection channels themselves is, therefore, of great importance. Inasmuch as automatic target tracking is the ordinary mode, countering the effects of deception jamming is the most critical antijamming problem in the selection channels.

To counter range and velocity deception jamming, it is possible to use [85, 87]:

- a wobbulated PRF;
- leading-edge tracking;
- acceleration limiting;
- comparison of the Doppler target velocity measurement with that obtained from range channel data;
- guard gating or inertial gating.

Use of a wobbulated PRF disturbs the smoothness of the shift of the jamming pulse in range and velocity, and lowers the probability that the range (velocity) gate will be diverted from the target echo.

Leading-edge tracking allows the deception jamming to be rejected in range on the basis of its delay, relative to the target signal.

Acceleration limiting allows the deception jamming signal to be gated if the apparent acceleration it creates exceeds the target's acceleration or a predetermined threshold.

Comparing the velocity measurements obtained from the Doppler channel and range channel data allows deception jamming to be rejected because it is created only in the velocity channel.

Comparison of the signals in a guard range (velocity) gate with those in the target gate permits rejection of deception jamming, such as is created, for example, by forward-ejected chaff [85].

Manual tracking is another option for dealing with deception jamming. In this case the operator observes the target on the screen and can distinguish the false targets created by deception jamming, and keep the range and velocity gates on the true target [86]. The effectiveness of this method is lessened if repeater pulse jamming matched to the tracking pulse period is used [61]. If a stable PRF is being used in this case, the jamming can completely cover the target signal and the existing time delay of the jamming pulse relative to the target pulse will not be observed by the operator.

### 8.6.3 Antijamming Methods for the Angle Channel

The monopulse method itself is an angle channel antijamming technique in that it makes the system insensitive to the majority of jammers, which are effective against single-channel radars and which transmit from a single point in space. There are other methods, such as were outlined above, against which the angle channel is not protected by the monopulse technique. We will now examine some of the additional antijamming measures which are then necessary.

*Adaptive beam patterns.* Rejecting jamming sources in the spatial domain has, in recent years, become one of the most important approaches to increasing the jamming immunity of radars. The basic technique is to control the receiving beam pattern and put "notches" in it in the jammer directions. This produces a corresponding loss in receiver sensitivity to signals from these directions, thus lowering the effectiveness of the jammers.

Adaptive beam forming has undergone extensive development thanks to the introduction of phased array antennas, which permit control of the beam forming process and the pattern structure. Almost all phased array radars recently developed in the US have the capability of performing such cancellation. Monopulse systems, which are characterized by having several independent receiver channels, may also notch out the directions from which jamming signals originate.

The use of a monopulse system in this case may be considered to be a special case of an antenna array which forms a beam with the number of nulls being one fewer than the number of elements. To suppress signals from a jammer lying in a different direction than the target, the parameters of the signals received through the various beams are automatically adjusted so that the difference channel output signal is statistically zero. It is possible to control both the amplitude and phase relations of the signals in this approach.

A system which suppresses jamming by adjusting the phase relations of the signals is shown as an example in Fig. 8.18 [58]. The target and jamming signals received by the system undergo sum-and-difference processing in two balanced bridges, connected in parallel to the output of the receiving antenna. The sum and difference signals from the output of the first bridge go to the computer, where the angle error signal is formed. This signal is also used to control a phase shifter inserted in one of the second bridge's input channels, which is used to create a null in the difference pattern in the direction of the jammer. We will demonstrate the feasibility of this approach mathematically. The target signal and jamming arrive from the directions $\theta_s$ and $\theta_j$ relative to the null axis, their amplitudes at the antenna input are $E_s$ and $E_j$, the relative phase shift is $\alpha$, and frequency $\omega$.

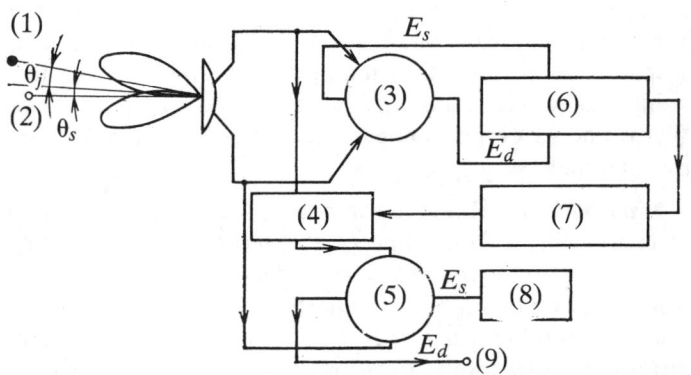

**FIGURE 8.18.** Block diagram of an amplitude-sum-and-difference monopulse system with angular gating of jamming sources.

| | |
|---|---|
| (1) jamming source | (6) computer |
| (2) target | (7) phase shifter control |
| (3) balanced bridge 1 | (8) load |
| (4) phase shifter | (9) output |
| (5) balanced bridge 2 | |

Assuming that the angular separation between the jammer and target does not exceed the linear portion of the pattern, and using (5.2), the error signal may be obtained after elementary transformations in the form:

$$S(\theta) = u \, \frac{E_s^2 \theta_s + E_j^2 \theta_j + E_j E_s (\theta_s + \theta_j) \cos \alpha}{E_s^2 + E_j^2 + 2E_j E_s \cos \alpha} \tag{8.6}$$

It is evident that if there is no jamming present, the tracking condition will be given by the expression:

$$S(\theta) = \mu\theta_s$$

In this case the null direction of the difference pattern will correspond to the direction to the target, inasmuch as $S(\theta) = 0$ when $\theta_s = 0$.

When two signal sources are being measured, the direction of the difference null will be determined by the expression:

$$\theta_{sj} = \frac{E_s^2\theta_s + E_j^2\theta_j + E_jE_s(\theta_s + \theta_j)\cos\alpha}{E_s^2 + E_j^2 + 2E_jE_s\cos\alpha} \tag{8.7}$$

When the null is directed to the angle $\theta_{sj}$, the error signal at the output of the computer will have the form:

$$S(\theta_{sj}) = \mu(\theta_s - \theta_{sj})$$

and the jamming-to-signal ratio at the output of the system will be

$$\frac{P_j}{P_s} = \left|\frac{E_j\mu(\theta_j - \theta_{sj})}{E_s\mu(\theta_s - \theta_{sj})}\right|^2$$

Inserting the value of $\theta_{sj}$ from (8.7), after elementary transformations we obtain

$$\frac{P_j}{P_s} = \left|\frac{E_s + E_j\cos\alpha}{E_j + E_s\cos\alpha}\right|^2 = \left|\frac{1 + a\cos\alpha}{a + \cos\alpha}\right|^2, \quad \text{where } a = \frac{E_j}{E_s} \tag{8.8}$$

From this equation, it is clear that in order to notch the pattern in the jammer direction to obtain $P_j/P_s = 0$, it is necessary to produce the condition $1 + a\cos\alpha = 0$, or

$$\alpha = \arccos(-1/a)$$

This operation is performed by the phase shifter, which is controlled by the error signal formed in the computer. If the phase difference between the signals fluctuates, then the pattern null must be formed using the average value of $\theta_{sj}$. If the phase is uniformly distributed from $-\pi$ to $+\pi$, the root mean square value of $\theta_{sj}$ may be found as follows:

$$\bar{\theta}_{sj} = \frac{1}{2}\frac{E_s^2\theta_s + E_j^2\theta_j + E_jE_s(\theta_c + \theta_j)\cos\alpha}{E_s^2 + E_j^2 + 2E_jE_s\cos\alpha}\,d\alpha$$

$$= \begin{cases} \theta_s & \text{for } E_s > E_j \\ (\theta_s + \theta_j)/2 & \text{for } E_s = E_j \\ \theta_j & \text{for } E_s < E_j \end{cases}$$

Usually $E_j \gg E_s$ and the difference pattern will be formed exactly in the direction to the jammer, as a result of which the jamming signal is completely cancelled.

With rapid changes in the signal directions, as with rapid phase difference changes, a fast-acting null direction control is needed. If the control is almost instantaneous, the jamming-to-signal power ratio at any given moment will be given by (8.8). If the phase difference $\alpha$ fluctuates randomly and is uniformly distributed between $-\pi$ and $+\pi$, the average value of the jamming-to-signal power ratio at the system output will be given by the expression:

$$\left| \frac{\overline{P_j}}{P_s} \right| = \frac{1}{2\pi} \int_{-\pi}^{\pi} \left( \frac{1 + a \cos \alpha}{a + \cos \alpha} \right)^2 d\alpha = a(a - \sqrt{a^2 - 1}) \quad \text{for } a > 1$$

It may be shown that with input jamming-to-signal ratios exceeding 6 dB, the jamming power to signal power ratio at the output of the system will be close to $-3$ dB. Consequently, the method just described will always weaken the effect of strong jamming and is an effective means of improving the jamming immunity of a monopulse radar.

In systems which suppress jamming by adjusting the signal amplitudes, the difference pattern null is produced using an AGC system [66]. With the help of the AGC, the jamming signal received by one of the antenna pattern beams is adjusted to the level of the same signal received by a different beam. As a result, when the signals are subtracted, the jamming will be completely or almost completely cancelled at the output of the system. The system operates in a tracking mode, in which the output signal formed from the filtered jamming signal is used as an error signal to control an RF attenuator to balance the noise jamming in the receiver channels.

It is possible to perform target detection in this case using the difference signal taken from the output of the balanced bridge if the system is of the sum-and-difference type. There will be a loss in sensitivity to the useful signal in this case, inasmuch as the target, which is displaced by some angle from the jammer, will be detected with a deformed antenna pattern.

When using a phased array radar, it is possible to suppress jamming from several angularly separated sources simultaneously. In this case, the pattern is notched adaptively in all directions from which jamming is being received. Electronically scanned phased arrays are usually used in these systems, the beams being formed through phase variations alone.

### 8.6.4 Methods for Countering Multiple Blinking Jammers

Inasmuch as blinking jamming, as has been shown, attempts to worsen target resolution, one of the more important means for countering it is to increase the resolving capabilities of the radar when tracking formation targets. Therefore, the monopulse system with improved resolution, examined in Chapter 5 (see Fig. 5.13), possesses an enhanced capability against blinking jamming.

In the generation of blinking jamming, the jamming transmitters placed on board the formation targets are switched either periodically or randomly, making it difficult for the radar to track any individual target long enough to determine its position and engage it. In these conditions, the monopulse system described above permits either the left or right target to be tracked, according to the operator's choice.

If, for example, the left-hand target is being tracked, and the signal from this target suddenly disappears and is replaced by another signal coming from the right, then the resulting negative error signal may be gated in the selection channel, and the tracking system will not be diverted to the new target. Thus the basic goal of blinking jamming, interruption of target tracks or degradation of target resolution, is not achieved.

Swept-frequency jamming transmitted from several platforms operates in much the same fashion as blinking jamming, and may thus also be countered by the method just described.

To further improve the radar's performance against blinking and frequency-swept jamming, the beam may be narrowed and the sidelobes lowered, inasmuch as this increases the jamming power requirement and improves the angular resolving capability of the radar.

### Countering Cross-Polarized Jamming

Inasmuch as jamming transmitted at cross polarization differs substantially in polarization from the radar's operating signal, one obvious method to protect the radar from it is to select signals by polarization. One such method involves placing polarized grids (filters) in the radar's antenna apertures. As was described in Chapter 6, these grids pass signals at the operating polarization with little attenuation, and strongly block signals at a polarization orthogonal to the operating polarization. Cross-polarized jamming is thus also strongly rejected and its effectiveness substantially reduced. The antenna reflector itself is sometimes used as the polarization filter, for which purpose it is constructed out of parallel plates (wires) [21]. In this case the cross-polarized components of the signals pass through the reflector and are not reflected to or from the feeds, which significantly weakens the cross-polarized component in the transmitted and received signals.

The measures which may be taken to reduce antenna cross-polarization levels, which were partially described in Chapter 6, may also be considered protection against cross-polarized jamming. The development of radars with polarization diversity has great significance for this sort of antijamming. These are multichannel systems which may operate on whichever receiving channel is closest to the signal in polarization. This permits significant cancellation of the effects of signal depolarization caused by the target, and also increases the immunity of the radar to cross-polarized jamming.

## Methods for Countering Ground-Bounce Jamming

Illumination of the underlying surface with jamming signals leads to the formation of a false signal source lying in a direction different from that to the actual target. As a result, in place of a single target, the radar channel must now contend with a paired target and all of the attending consequences (the tracking accuracy is reduced, and it is possible to divert the radar to the false target and interrupt the automatic track mode). To reduce the influence of the specularly reflected signal components, two basic methods are employed abroad: off-axis tracking, breaking the closed tracking loop and using an additional bias signal, and processing the quadrature components [62].

In off-axis tracking, the beam is positioned at a constant elevation angle $\epsilon$ on the order of $0.7\theta_3$ above the horizon, and the error signal is converted to an angle and added to $\epsilon$. The signal from the image jamming source is weakened as a result, and the track is not interrupted. The typical errors have been shown experimentally to be reduced by a factor of three or four with this technique.

It has been shown in Western studies [62], that the presence of signals reflected from the ground may be established using the quadrature components of the received signal. When there is no ground-bounce signal, the sum and difference signals in the monopulse receiver are ideally either in phase or directly out of phase, and there will be no quadrature components. When there is a signal which has been reflected by the surface in addition to the direct target signal, quadrature components will arise. These components may be extracted if an additional phase detector and 90° phase shifter are placed in the standard monopulse radar in the circuitry for each measured coordinate, as shown in Fig. 8.19. With such a modified system, it is possible to obtain, at the output, four independent equations, the solution of which allows the directions to both the target and jammer to be determined.

Ground-bounce jamming may also be cancelled by rotating the antenna until the quadrature components are zeroed. The position of the antenna at which the quadrature components are zeroed determines the direction to the target and excludes the influence of ground-bounce jamming.

When the tracking system can form an angle-sensing response with two symmetric nulls, the direction of one of which may be changed, then the influence of ground-bounce jamming may be lowered if the second null is directed to the image signal source. Tracking may then be performed on the real target.

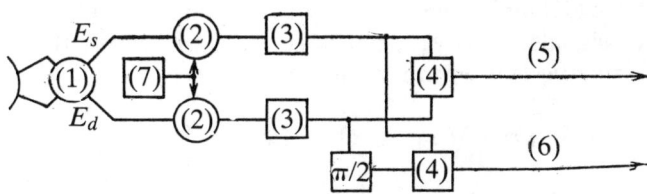

**FIGURE 8.19.** Block diagram of a sum-and-difference monopulse system with quadrature channels.

(1) bridge      (5) in phase component
(2) mixer      (6) quadrature component
(3) IF amplifier      (7) oscillator
(4) phase detector

*Use of Wideband Signals*

Wideband signal techniques include the use of long frequency- or phase-modulated pulses. The antijamming potential of radars using these signals results from the corresponding signal processing which is performed, and which usually consists of pulse compression. Pulse compression reduces the intensity of noise jamming at the output by a factor equal to the compression ratio.

In an optimum pulse compression receiver, coherency between the input and reference signals is maintained. These systems are, therefore, extremely difficult to jam, even in a "one on one" situation, inasmuch as it is extremely difficult to form a wideband jamming signal which imitates the structure of the radar signal to the level of the phase variation. If many such radars must be jammed simultaneously, the difficulty confronted by the jammer increases many times [108, 109]. Hopping the operating frequency in a random or pseudorandom fashion is another approach utilizing wideband signals which improves the jamming immunity of the radar. Narrowband filters are used to extract the useful signal, whose frequency at the antenna input is known, and to suppress any jamming. The local oscillator signal is provided by a frequency generator which is synchronized with the operating frequency variation. If the jamming signal does not follow this exact pattern of frequency hopping, then it will be severely attenuated in the narrowband filter. The jamming immunity of the radar in this case is dependent upon all the necessary transmitting and receiving information being available to the radar, and not to the jammer [108].

The signal-to-jamming advantage offered by such wideband processing is lost if the jammer imitates the radar waveform perfectly. But the creation of successful wideband jamming depends not only on information about the types of signals employed, but also about the correlation and pulse compression methods used in the radar receivers.

*Maintaining the Security of Radar Operations*

Jamming is generated using information contained in the transmitted signals themselves. Receiving these signals, the enemy measures the most important parameters and generates his jamming signals based on this information. Clearly, if the enemy is deprived of the ability to glean the necessary information from the radar signals, than the jamming task is made much more difficult, if not impossible.

Of particular interest is the work of an American company in developing a target recognition and tracking radar for use in a complex combat environment, which includes the use by the enemy of various ECM techniques including antiradar missiles [64, 97].

One characteristic of this radar is its high operational security, along with its ability to track targets "on the fly." Unlike traditional radars, which employ one or several beams to scan the search volume, this radar forms many (several thousand) narrow beams, which transmit in a pseudorandom sequence for short time intervals. Each beam is formed at its own frequency, which is constantly changing. This radar uses a relative low power continuous transmission with pseudorandom biphase modulation. The radar performs search, target detection, and target track functions simultaneously. The objective was to minimize the ability of antiradiation missiles and reconnaisance instruments to detect and parametrize the radar emissions [64, 97].

It is clear that these principles of maintaining the security of the radar emissions may be applied to monopulse radars of various types.

Another method for protecting the various waveform parameters is to employ active masking; additional signals are radiated at spurious frequencies which are close to the operating frequency [90]. To jam such a radar it should be expected that jamming signals will be transmitted at all of the frequencies, including the spurious ones, resulting in a reduction of jamming power spectral density at the radar frequency and a corresponding degradation in its effectiveness. In some cases the jammer may be completely ineffectual, because it may jam only a spurious frequency. Such a situation is realistic if the radar transmits spurious signals at frequencies both higher and lower than the operating frequency, and frequency storage is used to control the jamming signal.

*Use of a Higher Frequency Band*

The development of millimeter wave radars has received great attention abroad since the latter half of the 1970s. This is due, in part, to the greater jamming immunity afforded millimeter wave systems by the narrower beams and lower sidelobes, resulting from transmission at millimeter wavelengths, and the greater range, velocity, and angular resolution which may be achieved in such systems [113].

# Chapter 9

# Estimating the Characteristics of Monopulse Radars With Simulations

## 9.1 GENERAL PROBLEMS AND AREAS OF SIMULATION

When developing and testing monopulse radars, an analytical prediction of the radar characteristics is often impossible due to intractable mathematical formulations. In addition, difficulties arise when bringing modern radars into production due to the impossibility of estimating the system characteristics from full-scale testing. Computers are generally used to automate the operation of complex modern radars, and there are often problems in developing the algorithms and programs during preparation for testing and actual testing. To develop software without the benefit of mathematical simulation is inadvisable and often practically impossible. All of these motives steer the developers to the use of digital simulation during radar system development and production.

Modeling systems on a digital computer has several advantages:

- computer simulation offers the ability to study extremely complex processes with a large number of parameters;
- experiments are easy to design, take a comparatively short time to perform, and are relatively inexpensive;
- it is possible to control all of the factors and results of computer experiments;
- the results of computer simulation may be repeated, and dependent tests may be used.

A model which describes a radar's formalized functional processes may only include its basic operating characteristics. Despite the designer's attempt to minimize quantitative errors in the simulated radar processes, the model nonetheless is usually not identical to the actual radar. Therefore, a calibration stage is necessary in the modeling process.

The performance of any complex system usually may be assessed with a finite number of characteristics. There are two general trends in the development of simulations: a general model may be developed in order to estimate all of the radar characteristics, or partial models may be designed to estimate one or a small number of the characteristics. A comprehensive simulation is extremely complex, and may not always be implemented or even designed. The smaller-scale partial simulation, therefore, is usually used.

It is advisable to employ a modular design in developing a simulation. The model may be greatly simplified by excluding certain system components or finding equivalent descriptive elements, and finding closely connected components of the system to combine in a block structure. Such a modular design simplifies control of the simulation and division of the programming labor, and individual modules may be left out or used depending on the given problem.

It must be remembered that increasing the detail and expected accuracy of the simulation greatly increases its complexity. The complexity of the model should be chosen on the basis of the available computer capacity, and the relations between the accuracy of the system component models, the initial data, and the statistics to be obtained from the simulation. On the basis of these considerations, a compromise must be struck between simulation accuracy and complexity.

Stand-alone and integrated "hardware in the loop" simulations may be designed. In the first case, the model is implemented on a general purpose computer, is not executed in real time, and models the flight trajectories and characteristics of the targets, basic components, and algorithms of the radar. In the second type of simulation the target characteristics, trajectories, and basic radar operations are modeled, while the actual radar algorithms developed on a standard computer are used. This sort of simulation is usually implemented on the actual radar system computer, and the simulation runs in real time.

An estimate of the technical approaches and radar characteristics may be obtained at the design stage with computer simulations; the effects of individual parameters may be studied and these parameters may be optimized. The simulation will provide an estimate of the radar's operation in complex scenarios against real targets, and full-scale testing methods may be designed (the most characteristic scenarios may be determined, along with the effectiveness of various testing approaches and the expected test results). With a computer simulation calibrated against the results of limited full-scale tests, it is possible to verify the radar algorithms. A number of statistical characteristics of the radar may be estimated and compared with the requirements, leading to recommendations for system modifications.

The selection of the extent and conditions of experimentation is driven by the requirement for performance estimates of a certain accuracy and reliability. Some of the goals of full-up testing are:

- obtaining the necessary number and type of initial data to be used in the computer simulation;
- obtaining experimental results (on the radar system and its components) to verify the performance of the simulation and indicate corrections;
- obtaining the necessary number of statistical control results to exclude methodological errors.

These goals should be achieved with a minimum of actual full-scale testing in light of the difficulties with conducting such experiments. Statistics, which are in accord with the accuracy of the input data, are all that should be obtained. Several simulation methods are discussed below, along with some monopulse radar models.

## 9.2 MODELING ANTENNA-FEED DEVICES

Modeling the antenna-feed system of a radar begins with an examination of the problem [25]:

- what is the objective of the model?
- what signal parameters are used by the modeled radar to measure the target characteristics?
- what are the requirements on the form of and errors in the output data of the antenna model?

The most interesting antenna characteristics to be included in the model are the antenna pattern and directivity.

For the purposes of modeling, it is convenient to distinguish between the various types of antenna systems as follows:

- in one type, the pattern remains constant while measuring the target coordinates;
- in the other type, the pattern changes while measuring the expected target direction.

An approximation to the antenna pattern is modeled on the basis of the relation between the pattern and the aperture excitation function, which is expressed through the Fourier transform. Transforms of some common aperture excitation distributions are contained in [14]. In the case of a planar aperture, the pattern is a product of two functions, each of which may be found in the tables or by using such standard functions as

$$[(\sin \theta)/\theta]^n, \quad \cos^n \theta, \quad \exp(-|\theta|^n), \quad 1 - |\theta|^n, \quad n = 1, 2, 3, \ldots$$

If tabulated data is used, the pattern is approximated within half of the radar's search sector, and the argument of the tables is the absolute value of the angular coordinate ($|\theta|$). The sampling interval $\Delta\theta$ is chosen so that the quantization error which results does not produce a significant distortion in the modeled signal fluctuations.

To describe the pattern with standard functions, the following combination is sufficiently universal:

$$F(\theta) = k_1(\sin \theta_1)/\theta_1 + k_2 \exp(|-\theta_2|^n) \qquad (9.1)$$

where $\theta_i = \xi_i\theta$, $i = 1, 2$; and $k_i$, $\xi_i$ and $n$ are constants chosen to match the modeled beamwidth and first sidelobe levels, with the values obtained from system testing or exact calculations.

For an antenna system of the second (variable) type, the system by which the excitation fields are varied must be specified. In a phased array antenna, for example, formation and steering of the beam is controlled by phase shifters which create discrete phase shifts at the elements. In order to simplify the design, development, and operation of the antenna, and to allow more flexible beam control, a phased array antenna is usually comprised of individual subarrays. Each of these modules is usually itself a phased array, and has its own transmitting and receiving amplifiers, phase shifters and beamforming control.

The subarray aperture takes on varied forms, and is determined by many factors, including the ease of construction, ability to produce the desired patterns, and so on. The subarray elements are often placed at the nodes of a triangular grid, thus allowing the number of elements to be reduced by a factor of 1.8 while maintaining the desired directivity characteristics over a scan sector of $\pm 40°$ [2].

The overall phased array aperture may take on different forms, and is typically rectangular or circular. In order to reduce distortion of the pattern (with scanning over the given sector) and to make more full use of the potential of the monopulse technique, the receiving array should have a circular or near-circular aperture.

When filling a circular phased array aperture, the modules are usually placed in rings to form a fairly thinly populated space-tapered array. When necessary, it is possible to switch out some of the module rings. A modular phased array design allows the radar to form several individual or grouped beams, which significantly extends the capabilities of a monopulse radar.

Two levels of phase shifters are used to form and steer the beam in a subarrayed antenna. The first level, usually located directly at the elements, produces *coarse* steering within each module, while *fine* steering is provided by the second level of shifters which supply bias phases to each subarray.

The pattern of a subarrayed phased array antenna may be expressed as

$$F^2_{\Phi(\theta)} = F^2_M(\theta)F^2_a(\theta) \tag{9.2}$$

where $F_M(\theta)$ is the pattern of an individual subarray, and $F_a(\theta)$ is the pattern of the subarray system; these factors are analogous to the element factor and array factor of a simple array antenna.

If the aperture is hexagonal, as is usually the case, and the elements are positioned at the corners of a triangular grid, then the power pattern may be calculated with the following algorithm:

$$F^2_M(\theta) = \left[ \sum_{p=1}^{L_c} \sum_{g=1}^{M^p} \cos(\xi^p_g \, \psi^p_g + \kappa^p_g \, \Delta\psi^p_{\Phi g} + \Delta\psi^p_g) \right]^2$$

$$+ \left[ \sum_{p=1}^{L_c} \sum_{g=1}^{M^p} \sin(\xi^p_g \psi^p_g + \kappa^p_g \, \Delta\psi^n_{\Phi g} + \Delta\psi^p_g) \right]^2 \tag{9.3}$$

where $L_c$ is the number of element columns; $M^p$ is the number of elements in the $p$th column; $\psi^p_g = \psi^p_{Ng} - \psi^p_{0g}$ is a generalized angle;

$$\psi^p_{hg} = k\left[ l_c\left(p - 1 - \frac{L_c - 1}{2}\right)\sin\beta' + l_e\left(g - 1 - \frac{M - 1}{2}\right)\sin\epsilon' \right]$$

$$\psi^p_{\Phi g} = v^p_g\Delta_{\Phi M}, \quad v^p_g = \left[ \frac{\psi^p_{0g} - \psi^p_{\text{in } g}}{\Delta_{\Phi M}} + \frac{1}{2} \right], \quad [\cdots] - \text{integral part}$$

$$\psi^p_{0g} = k\left[ l_c\left(p - 1 - \frac{L_c - 1}{2}\right)\sin\beta'_0 \right.$$

$$\left. + l_e\left(g - 1 - \frac{M^p - 1}{2}\sin\epsilon'_0\right) \right]$$

where $g$ and $p$ are the indices of the element within a column and the column within a module; $l_e$ and $l_c$ are the distances between elements within a column and between columns, respectively; $k = 2\pi/\lambda$ is the wave number; $\epsilon'$, $\epsilon'_0$, $\beta'$, and $\beta'_0$ are the angles relative to the normal to the array for the observation angle and the beam angle (subscript 0); $\psi^p_{\text{in } g} = \eta_\psi\Delta_{\Phi M}$ is the initial phase shift of the $p g$th element; $\eta_\psi$ is a random number, uniformly distributed in the interval $(-1, +1)$; $\Delta_{\Phi M}$ is the unit of phase shift provided by the coarse phase shifters (element shifters); $\Delta\psi^p_{\Phi g} = \eta_\psi\rho_{\Phi M}$ is the phase error resulting from the random phase distribution and inaccurate phasing of the $p g$th element; $\rho_{\psi M}$ is the spread of the phase, determined by the initial distribution and the errors in setting the coarse phase shifters:

$$\rho_{\Phi M} = \sqrt{\Delta_{\Phi M}^2 + \Delta_s^2}$$

$\Delta_s$ is the error in setting the coarse phase shifters: $\Delta \psi_g^p$ is the phase error caused by instability in the element feed lines, mutual coupling, and so on; $\eta_g^p$ is the operating state of the $pg$th element shifter;

$$\xi_g^p = \begin{cases} 1 & \text{for } \eta_g^p \geq P_\Phi \\ \mu_g^p & \text{for } \eta_g^p < P_\Phi \end{cases}$$

$n_g^p$ is a random number, uniformly distributed in the interval $(0, 1)$; $\mu_{pg} = 2\pi\eta_\psi$ is random phase, uniformly distributed in the interval $(-2\pi, +2\pi)$

$$x_g^p = \begin{cases} 1 & \text{for } \xi_g^p = 1 \\ 0 & \text{for } \xi_g^p = \mu_g^p \end{cases}$$

and, finally, $P_\Phi$ is the probability of a given shifter-element being out of commission.

The random numbers are taken from a random number generator prior to calculating the pattern of each module. The array factor for a circular aperture may be calculated with the following formula:

$$F_\alpha^2(\theta) = \left[ \sum_{n=1}^{N^k} \sum_{m=1}^{M^n} t_m^n A^n \delta_m^n \cos(\xi_m^n \psi_m^n + \kappa_m^n \Delta\psi_{\Phi m}^n + \Delta\psi_m^n) \right]^2$$
$$+ \left[ \sum_{n=1}^{N^k} \sum_{m=1}^{M^n} t_m^n A^n \delta_m^n \sin(\xi_m^n \psi_m^n + \kappa_m^n \Delta\psi_{\Phi m}^n + \Delta\psi_m^n) \right]^2 \qquad (9.4)$$

where $N^k$ is the number of columns in the phased array aperture; $M^n$ is the number of modules in $n$th ring;

$$\psi_m^n = \psi_{nm}^n - \psi_{\Phi m}^n, \qquad \psi_{nm}^n = \kappa r_k^n[(\cos \phi_m^n)\sin \epsilon' + (\sin \phi_m^n)\sin \beta']$$

$r_k^n$ is the radius of the $n$th ring; $\phi_m^n$ is the angle giving the position of the first module of the $n$th ring relative to the vertical through the array center;

$$\psi_{\Phi m}^n = v_m^n \Delta_{\Phi s} \qquad v_m^n = \left[ \frac{\psi_{0m}^n - \psi_{\text{in } m}^n}{\Delta_{\Phi s}} + \frac{1}{2} \right]$$

$[\cdots]$ = integral part

$$\psi_{0m}^n = \kappa r_k^n((\cos \phi_m^n)\sin\epsilon_0' + (\sin \phi_m^n)\sin \beta_0')$$

$\Delta_{\Phi s}$ is the discrete phase shift step in the second level (fine) phase shifters which apply a phase to the entire module; $A^n$ is the amplitude of the signals at the output of the modules in the $n$th ring; $\delta_m^n$ is the random deviation of the amplitude of the $m$th module in the $n$th ring from the quantity $A^n$:

$\delta_m^n = \exp(-\Delta_m^n)$:

$$\Delta_m^n = \eta_{am}^n \Delta_a$$

$\Delta_a$ is the spread of the module signal amplitudes; $n_{am}^n$ is a random number normally distributed in the interval $(-1, 1)$, with zero mean value; $\psi_{\text{in } m}^n = \eta_{\psi m}^n \Delta_{\Phi s}$ is the initial phase shift of the $m$th module; $\eta_{\psi m}^n$ is a random number uniformly distributed in the interval $(-1, 1)$; $\Delta \psi_{\Phi m}^n = \eta_{\psi m}^n \rho_{\Phi s}$ is the phase error caused by the random phase distribution and imprecise setting of the phase shift at the $mn$th module; $\rho_{\Phi s}$ is the phase spread caused by the initial phase distribution and the errors in setting the module phase shifters:

$$\rho_{\Phi s} = \sqrt{\Delta_{\Phi s}^2 + \Delta_{sm}^2}$$

$\Delta \psi_m^n$ is the phase error caused by instabilities in the lines feeding the modules and inexact placement of the modules in the array; $\Delta_{sm}$ is the spread of the error in setting the module phase shifters; $t_m^n$ is the operational state of the $m$th module in the $n$th ring:

$$t_m^n = \begin{cases} 1 & \text{for } \xi_m^n = 1 \\ 0 & \text{for } \xi_m^n = \mu_m^n \end{cases}$$

$\mu_m^n = \eta_m^n \cdot 2\pi$ is the random phase, uniformly distributed in the interval $(-1, 1) \cdot 2\pi$:

$$\xi_m^n = \begin{cases} 1 & \text{for } \leqslant P_M \\ \mu_{nm} & \text{for } > P_M \end{cases}$$

$\eta_m^n$ is a random number uniformly distributed in the interval $(0, 1)$;

$$x = \begin{cases} 1 & \text{for } \xi_m^n = 1 \\ 0 & \text{for } \xi_m^n = \mu_m^n \end{cases}$$

and $P_M$ is the probability of a given module being inoperable.

To reduce the sidelobe levels caused by the discrete phasing, both in actual systems and in simulations (in (9.3) and (9.4)), the initial phase distributions $\psi_{\text{in } g}^p$ and $\psi_{\text{in } m}^n$ of the elements within the modules, and the modules within the array, are taken to be random, uniformly distributed within the discrete phasing limits $\Delta_{\Phi M}$ and $\Delta_{\Phi s}$. With such uniform distributions of the initial element and module phases, the sidelobes caused by the discrete phasing are uniformly "blurred" [1, 2].

With this model, the pattern in any direction may be calculated with a single pass through the algorithms. With the necessary number of passes, it is possible to estimate the statistical characteristics of the pattern.

## 9.3 MODELING THE SIGNAL PROCESSING IN THE RECEIVER

The basic signal processing functions performed in the radar receiver are linear filtering of the signals in RF and IF amplifiers, mixing down the carrier frequency, and amplitude, phase, or amplitude-phase detection. In most cases, mathematical modeling may be applied to the processes of linear filtering and mixing additive internal and external noises referenced to the receiver input, and further to the extraction of the envelope (or its square, logarithm, etcetera) and the phase of the filtered mixture.

We will now examine linear filtering of the signals. We will describe a single radar signal $s_0(t)$ at the input to the radar receiver [25]:

$$
\begin{aligned}
E^{1/2}s_0\,(t) &= \mathrm{Re}[s(t)\exp(-i\omega_0 t)] \\
&= \mathrm{Re}(S(t)\exp\{-i[\omega_0 t + \phi(t) + \phi_0]\}), \quad 0 \leqslant t \leqslant T
\end{aligned}
\tag{9.5}
$$

where $T$ is the observation time; $S(t)$ is the signal at time $t$ and $S(t) = 0$ outside the observation interval; $\omega_0 t + \phi(t) + \phi_0$ is the instantaneous signal phase; $\omega_0$ is the carrier frequency; $\phi(t)$ is the phase modulation, $\phi_0$ is the initial phase, and $s(t)$ is the complex signal envelope.

The signal $s_0(t)$ has energy:

$$
\int_0^T s_0^2(t)\,dt = E
\tag{9.6}
$$

We now write the impulse response of the linear filter:

$$
\begin{aligned}
h_0(t) &= \mathrm{Re}[h(t)\exp(-i\omega_0 t)] \\
&= \mathrm{Re}\{H(t)\exp[-i(\omega_0 t + \psi(t) + \psi_0)]\}
\end{aligned}
\tag{9.7}
$$

where $H(t)$ is the impulse response envelope; $\omega_0 t + \psi(t) + S_0$ is the instantaneous phase; $\omega_0$ is the filter frequency, which is selected to be equal to the signal's carrier frequency; $\psi(t)$ is the phase modulation of the impulse response; and $h(t)$ is the complex envelope of the filter's impulse response.

The signal $s_r(t)$ at the output of the filter is the convolution:

$$
s_r(t) = \int_{-\infty}^{\infty} s_0(\tau)h_0(t - \tau)\,d\tau
\tag{9.8}
$$

It is known [4] that the convolution in (9.8) may be written with the complex envelopes of the input signal and filter impulse response:

$$s_r(t) = \frac{\sqrt{E}}{2} \operatorname{Re}\{[\exp(-i\omega_0 t)] \int\limits_{-\infty}^{\infty} s(t)h(t - \tau)\,d\tau\}$$

$$= s_c(t)\cos(\omega_0 t) + s_s(t)\sin(\omega_0 t)$$

(9.9)

where

$$s_c(t) = \frac{\sqrt{E}}{2} \int\limits_{-\infty}^{\infty} S(t)H(t - \tau)\cos[\phi(\tau) + \psi(t - \tau) + \phi_0 + \psi_0]\,d\tau$$

$$-s_s(t) = \frac{\sqrt{E}}{2} \int\limits_{-\infty}^{\infty} S(t)H(t - \tau)\sin[\phi(\tau) + \psi(t - \tau) + \phi_0 + \psi_0]\,d\tau,$$

or

$$s_c(t) = \frac{\sqrt{E}}{2} \operatorname{Re}\left[\int\limits_{-\infty}^{\infty} s(\tau)h(t - \tau)\,d\tau\right]$$

$$s_s(t) = \frac{\sqrt{E}}{2} \operatorname{Im}\left[\int\limits_{-\infty}^{\infty} s(t)h(t - \tau)\,d\tau\right]$$

Equation (9.9), which describes the connection between the output of the linear filter and the convolution of the complex envelopes of the input signal and impulse response, is of importance in modeling radar filters, because it allows the direct modeling of high frequency oscillations to be avoided.

Analogous relations may be obtained for linear filtering of noise. We will examine white Gaussian noise $n(t)$ with zero mean and correlation function:

$$\overline{n(t_1)n(t_2)} = n_0\delta(t_1 - t_2)$$

where $n_0$ is the spectral noise density.

If the white noise is passed through a linear filter with the impulse response given in (9.7), there will result a normal random process:

$$M(t) = \int\limits_{-\infty}^{\infty} n(\tau)h_0(t - \tau)\,d\tau$$

with zero mean and correlation function:

$$K_M(t_1, t_2) = \overline{M(t_1)M(t_2)} = n_0 \int\limits_{-\infty}^{\infty} h_0(\tau)h_0(\tau + t_2 - t_1)\, d\tau \qquad (9.11)$$

It is possible to write an expression analogous to (9.9):

$$K_M(\tau) = \frac{1}{2}n_0 \, \text{Re}\{[\exp(-i\omega_0 t)] \int\limits_{-\infty}^{\infty} h(t)h^*(t - \tau)\, dt\}, \qquad (9.12)$$

where $\tau = t_1 - t_2$, and * denotes complex conjugate.

We will place the output noise in the form:

$$M(t) = M_c(t)\cos(\omega_0 t) + M_s(t)\sin(\omega_0 t) \qquad (9.13)$$

The noise given by (9.13) has a correlation function given by (9.12) if the correlation functions of the low frequency components $M_c(t)$ and $M_s(t)$ are determined by the equations:

$$\overline{M_c(t)M_c(t - \tau)} = \overline{M_s(t)M_s(t - \tau)} = \frac{1}{2}n_0 \, \text{Re}\left[\int\limits_{-\infty}^{\infty} h(t)h^*(t - \tau)\, dt\right]$$

$$\overline{M_c(t)M_s(t - \tau)} = \frac{1}{2}n_0 \, \text{Im}\left[\int\limits_{-\infty}^{\infty} h(t)h^*(t - \tau)\, dt\right] \qquad (9.14)$$

Thus, the components $M_c(t)$ and $M_s(t)$ are the results of filtering independent white noise samples $n_1(t)$ and $n_2(t)$ with spectral densities of $n_0$:

$$M_c(t) = \frac{1}{2} \, \text{Re}\left[\int\limits_{-\infty}^{\infty} n_1(\tau)h(t - \tau)\, d\tau\right] + \frac{1}{2} \, \text{Im}\left[\int\limits_{-\infty}^{\infty} n_2(\tau)h(t - \tau)\, d\tau\right]$$

$$(9.15)$$

$$M_s(t) = \frac{1}{2} \, \text{Re}\left[\int\limits_{-\infty}^{\infty} n_2(\tau)h(t - \tau)\, d\tau\right] + \frac{1}{2} \, \text{Im}\left[\int\limits_{-\infty}^{\infty} n_1(\tau)h(t - \tau)\, d\tau\right]$$

Thus, we obtain, in place of (9.10), a more easily modeled mathematical description of the process of filtering noise:

$$M(t) = \frac{1}{2} \, \text{Re}\{[\exp(-i\omega_0 t)] \int\limits_{-\infty}^{\infty} n_1(\tau)h(t - \tau)\, d\tau\}$$

$$+ \frac{1}{2} \, \text{Im}\{[\exp(-i\omega_0 t)] \int\limits_{-\infty}^{\infty} n_2(\tau)h(t - \tau)\, d\tau\} \qquad (9.16)$$

We now turn to the question of detection. If the signal and noise are passed through the linear filter, there results the process:

$$y(t) = s(t) + M(t) = Y_c(t)\cos(\omega_0 t) + Y_s(t)\sin(\omega_0 t) \qquad (9.17)$$
$$= Y(t)\cos[\omega_0 t - \phi(t)],$$

where $Y_c(t) = s_c(t) + M_c(t)$; $Y_s(t) = s_s(t) + M_s(t)$; $Y(t)$ is the process envelope; and $\phi(t)$ is the process phase modulation.

The envelope and phase modulation desired for modeling are calculated with the equations:

$$Y(t) = \sqrt{Y_c^2(t) + Y_s^2(t)} \qquad \phi(t) = \arctan[Y_s(t)/Y_c(t)] \qquad (9.18)$$

Expanding (9.9) and (9.15), we find the expression for the low-frequency signal and noise components at the output of the filter:

$$s_c(t) = \int_{-\infty}^{\infty} S_c(\tau)H_c(t - \tau)\, d\tau - \int_{-\infty}^{\infty} S_s(\tau)H_s(t - \tau)\, d\tau$$

$$S_s(t) = -\int_{-\infty}^{\infty} S_s(\tau)H_c(t - \tau)\, d\tau - \int_{-\infty}^{\infty} S_c(\tau)H_s(t - \tau)\, d\tau \qquad (9.19)$$

$$M_c(t) = \int_{-\infty}^{\infty} n_1(\tau)H_c(t - \tau)\, d\tau - \int_{-\infty}^{\infty} n_2(\tau)H_s(t - \tau)\, d\tau$$

$$M_s(t) = -\int_{-\infty}^{\infty} n_1(\tau)H_s(t - \tau)\, d\tau - \int_{-\infty}^{\infty} n_2(\tau)H_c(t - \tau)\, d\tau$$

where

$$S_c(t) = \sqrt{E/2}\,S(t)\cos[\phi(t) + \phi_0]$$
$$S_s(t) = \sqrt{E/2}\,S(t)\sin[\phi(t) + \phi_0]$$
$$H_c(t) = (1/\sqrt{2})H(t)\cos[\psi(t) + \psi_0]$$
$$H_s(t) = (1/\sqrt{2})H(t)\sin[\psi(t) + \psi_0]$$

The calculations in (9.18) and (9.19) are outlined in the block diagram of Fig. 9.1. The mathematical description of the processing performed by a radar receiver on the signals and noises allows it to be simulated on a computer.

We will now look at the question of representing continuous functions with discrete values. We will place the function $\lambda(t)$ in the form:

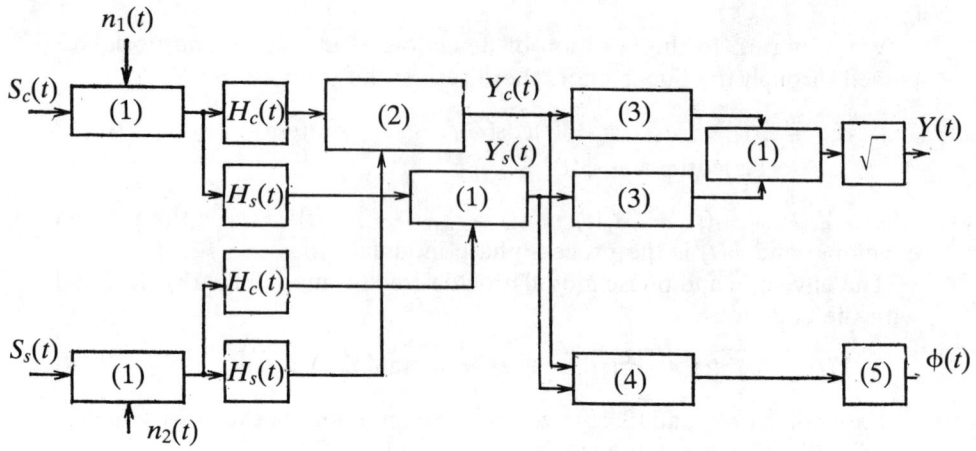

**FIGURE 9.1.** Block diagram of linear filtering and detection.
  (1) adder                    (4) division circuit
  (2) subtraction device   (5) $-\arctan$
  (3) square-law device

$$\lambda_\Phi(t) = \sum_{k=-\infty}^{\infty} \lambda_k \Phi_k(t) \tag{9.20}$$

with known functions $\Phi_k(t)$ and coefficients $\lambda_k$. If the condition:

$$\delta = \int_{-\infty}^{\infty} [\lambda(t) - \lambda_\Phi(t)]^2 \, dt = \min \tag{9.21}$$

is fulfilled, and the functions $\Phi_k(t)$ are orthogonal, then the coefficients $\lambda_k$ may be determined with the formula:

$$\lambda_k = \int_{-\infty}^{\infty} \lambda(t)\Phi_k(t) \, dt \left[ \int_{-\infty}^{\infty} |\Phi_k(t)|^2 \, dt \right]^{-1} \tag{9.22}$$

The set of functions $\Phi_k(t)$ is complete with respect to $\lambda(t)$ when $\delta = 0$. Such expansion by orthogonal functions is covered, for example, in [24, 39], where additional bibliographical data may be found.

Calculating the Fourier transform of both parts of (9.20), we obtain the expansion for the spectrum of the function $\lambda(t)$:

$$\tilde{\lambda}_\Phi(\omega) = \sum_{k=-\infty}^{\infty} \lambda_k \tilde{\Phi}_k(\omega) \tag{9.23}$$

with which it is not difficult to show that the following formula for determining the Fourier series coefficients is valid:

$$\lambda_k = \int_{-\infty}^{\infty} \tilde{\lambda}(\omega)\tilde{\Phi}_k^*(\omega) \, dw \left[ \int_{-\infty}^{\infty} |\tilde{\Phi}_k(\omega) \, d\omega| \right]^{-1} \tag{9.24}$$

Let the signal $x(t)$ with expansion $x_\Phi(t)$ be the input to a linear filter with impulse response $h(t)$, the expansion of which is

$$h_\Phi(t) = \sum_{k=-\infty}^{\infty} h_k \Phi_k(t).$$

We will assume that the result of the filtering $y(t)$ has the form:

$$y_\Phi(t) = \int_{-\infty}^{\infty} x_\Phi(\tau) h_\Phi(t - \tau) \, d\tau = \sum_{k=-\infty}^{\infty} y_k \Phi_k(t) \tag{9.25}$$

The coefficients $y_k$ must be the results of the discrete convolution:

$$y_n = \sum_{k=-\infty}^{\infty} x_k h_{n-k} \tag{9.26}$$

of the coefficients $x_k$ and $h_k$.

We will now find the constraints imposed on the functions $\Phi_k(t)$ [29]. From (9.25), we obtain

$$\sum_{k=-\infty}^{\infty} y_n \Phi_n(t) = \sum_{k=-\infty}^{\infty} \sum_{m=-\infty}^{\infty} x_k h_m \int_{-\infty}^{\infty} \Phi_k(\tau) \Phi_m(t - \tau) \, d\tau \tag{9.27}$$

We will calculate the Fourier transform of both parts of (9.27) in accordance with the requirement of (9.26):

$$\sum_{n=-\infty}^{\infty} y_n \tilde{\Phi}_n(\omega) = \sum_{k=-\infty}^{\infty} \sum_{m=-\infty}^{\infty} x_k h_m \Phi_k(\omega) \Phi_m(\omega)$$

$$= \sum_{k=-\infty}^{\infty} \sum_{n=-\infty}^{\infty} x_k h_{n-k} \tilde{\Phi}_k(\omega) \tilde{\Phi}_{n-k}(\omega)$$

$$= \sum_{k=-\infty}^{\infty} \sum_{n=-\infty}^{\infty} x_k h_{n-k} \tilde{\Phi}_n(\omega)$$

From this we find that $\tilde{\Phi}_n(\omega) = \tilde{\Phi}_k(\omega)\tilde{\Phi}_{n-k}(\omega)$, and, in particular,

$$\tilde{\Phi}_{k+1}(\omega) = \tilde{\Phi}_1(\omega)\tilde{\Phi}_k(\omega)$$

A non-trivial solution of this is

$$\Phi_k(\omega) = [\Phi_1(\omega)]^k \tag{9.28}$$

We will limit our example to the most commonly used set of functions satisfying (9.28):

$$\Phi_k(\omega) = \begin{cases} \exp(-i\omega k\Delta_M) & \text{for } |\Delta_M\omega| \leq \pi \\ 0 & \text{for } |\Delta_M\omega| > \pi \end{cases}$$

$$\Phi_k(t) = \frac{1}{\Delta_M} \frac{\sin[(\pi/\Delta_M)(t - k\Delta_M)]}{(\pi/\Delta_M)(t - k\Delta_M)} \tag{9.29}$$

Using (9.29) in (9.24), we find the expansion coefficients for the function $\lambda(t)$:

$$\lambda_k = \lambda(k\Delta_m)\Delta_M \tag{9.30}$$

The resulting expansion:

$$\lambda(t) = \sum_{k=-\infty}^{\infty} \lambda(k\Delta_M) \frac{\sin[\pi(t/\Delta_m - k)]}{\pi(t/\Delta_M - k)} \tag{9.31}$$

is called a Kotel'nikov series. It follows directly from (9.29) that it is sufficient to use the time interval $\Delta_M$ in a Kotel'nikov series to accurately represent a function whose bandwidth is limited by the expression:

$$|\omega| \leq \pi/\Delta_M$$

Simulations always involve processes which are of limited duration, and hence with infinite spectra. These processes are represented by series like (9.31) with errors, the acceptable value of which will be discussed later.

For the function $\lambda(t)$, $0 \leq t \leq \tau_\lambda$, the series (9.31) has a finite number of terms:

$$\lambda(t) = \sum_{k=0}^{N_\lambda} \lambda(k\Delta_M) \frac{\sin[\pi(t/\Delta_M - k)]}{\pi(t/\Delta_M - k)}$$

where $N_\lambda = \tau_\lambda/\Delta_M$.

When $x(t)$ is represented as a Kotel'nikov series and is limited to $0 \leq t \leq \tau_x$, and $h(t)$ is similarly represented in the interval $0 \leq t \leq \tau_h$, then the discrete convolution (9.26) may be expressed with the formula:

$$y(\Delta_M n) = \Delta_m \sum_{k=\max(0,n-N_h)}^{\min(N_x,n)} x(k\Delta_M)h(n\Delta_M - k\Delta_M) \qquad (9.32)$$

where $N_x = \tau_x/\Delta_M$, $N_h = \tau_h/\Delta_M$.

For convenience we will use the expression:

$$y(\Delta_M n) = \Delta_m \sum_{k=-\infty}^{\infty} x(k\Delta_M)h(n\Delta_M - k\Delta_M)$$

assuming that $x(t)$ and $h(t)$ are zero outside the given intervals.

Now, using the formulas just derived and (9.18) and (9.19), we may form an algorithm suitable for simulation to represent the linear filtering and detection processes:

$$\frac{1}{\Delta_M} Y_c(\Delta_M n) = \sum_{k=n-N_h}^{n} X_c(\Delta_M k)H_c(\Delta_M n - \Delta_M k)$$
$$- \sum_{k=n-N_h}^{n} X_s(\Delta_M k)H_s(\Delta_M n - \Delta_M k)$$

$$\frac{1}{\Delta_M} Y_s(\Delta_M n) = - \sum_{k=n-N_h}^{n} X_s(\Delta_M k)H_c(\Delta_M n - \Delta_M k)$$
$$- \sum_{k=n-N_h}^{n} X_c(\Delta_M k)H_s(\Delta_M n - \Delta_M k)$$

$$Y(\Delta_M n) = \sqrt{Y_c^2(\Delta_M n) + Y_s^2(\Delta_M n)}$$

$$\phi(\Delta_M n) = \arctan \frac{Y_s(\Delta_M n)}{Y_c(\Delta_M n)},$$

$$n = \pm N_{st}, N_{st} + 1, \ldots, N_{end} \qquad (9.33)$$

where, for $k = 0, 1, \ldots, N$, we obtain

$$X_c(\Delta_M k) = S_c(\Delta_M k) + N_1(\Delta_M k), \quad X_s(\Delta_M k) = S_s(\Delta_M k) + N_2(\Delta_M k)$$

and, for $N_{st} - N_h \leq k < 0$ and $N_c < k \leq N_{end}$, $N_c = \tau_c/\Delta_M$, $N_h = \tau_h/\Delta_M$, $N_{st} = t_{st}/\Delta_M$, $N_{end} = t_{end}/\Delta_M$, we have $X_c(\Delta_m k) = N_1(\Delta_M k)$ and $X_s(\Delta_M k) = N_2(\Delta_M k)$, where $t_{end} - t_{st}$ is the interval required to obtain the output data; $\tau_c$ and $\tau_h$ are the durations of the signal and filter's impulse response.

For $0 \leq n \leq N_x + N_h$ the algorithm output will consist of signal and noise, and, for $n < 0$ and $N_x + N_h < n$, only noise. The interval of input noise over $t_{st} - \tau_h \leq t < t_{st}$ is necessary to obtain the proper noise process at the filter output.

The sample interval $\Delta_M$ is chosen on the basis of the required accuracy. We will first examine a signal without noise, comparing the results of continuous (9.19) and discrete (9.33) filtering. In the continuous case, we have for a single component (dropping the subscripts):

$$Z(t) = S(\tau)H(t - \tau)d\tau \tag{9.34}$$

In the discrete case, using the index $\Delta_M$, we accordingly obtain the expressions:

$$Z_{\Delta M}(t) = \sum_{n=-\infty}^{\infty} Z_{\Delta M}(\Delta_M n)\frac{\sin[\pi(t/\Delta_M - n)]}{\pi(t/\Delta_M - n)} \tag{9.35}$$

where

$$Z_{\Delta M}(\Delta_M n) = \Delta_M \sum_{k=\min(0,n-N_h)}^{\max(N_c,n)} S(\Delta_M k)H(\Delta_M n - \Delta_M k) \tag{9.36}$$

Inserting $t = \Delta_M n$ in (9.34), we find

$$Z(\Delta_M n) = \int_{-\infty}^{\infty} S(\tau)H(\Delta_M n - \tau)\, d\tau$$

$$= \sum_{k=-\infty}^{\infty} \int_{k\Delta_M}^{(k+1)\Delta_M} S(\tau)H(n\Delta_M - \tau)\, d\tau \tag{9.37}$$

Thus, the error at the points of calculation are

$$\delta(\Delta_M n) = \sum_{k=-\infty}^{\infty} \delta_{nk}$$

where

$$\delta_{nk} = \Delta S(\Delta_M n - \Delta_M k) - \int_{k\Delta_M}^{(k+1)\Delta_M} S(\tau)H(n\Delta_M - \tau)\, d\tau \tag{9.38}$$

for which

$$Z_{\Delta M}(\Delta_M n) = Z(\Delta_M n) + \delta(\Delta_M n) \tag{9.39}$$

We will examine the mean square error which results when representing (9.34) with (9.35):

$$\rho_z^2 = \int\limits_{-\infty}^{\infty} |Z(t) - Z_{\Delta M}(t)|^2 \, dt = 2\pi \int\limits_{-\infty}^{\infty} |\tilde{Z}(\omega) - \tilde{Z}_{\Delta M}(\omega)|^2 \, d\omega$$

$$= 2\pi \int\limits_{-\infty}^{-\pi/\Delta M} |\tilde{Z}(\omega)|^2 \, d\omega + 2\pi \int\limits_{\pi/\Delta M}^{\infty} |\tilde{Z}(\omega)|^2 \, d\omega \qquad (9.40)$$

$$+ 2\pi \int\limits_{-\pi/\Delta M}^{\pi/\Delta M} |\tilde{Z}(\omega) - \tilde{Z}_{\Delta M}(\omega)|^2 \, d\omega$$

It follows from (9.40) that the error is limited from below by the quantity $\rho_{\Delta Mz}^2 \leq \rho_z^2$, where

$$\rho_{\Delta Mz}^2 = 2\pi \left\{ \int\limits_{-\infty}^{-\pi/\Delta M} |\tilde{Z}(\omega)|^2 \, d\omega + \int\limits_{\pi/\Delta M}^{\infty} |\tilde{Z}(\omega)|^2 \, d\omega \right\} \qquad (9.41)$$

It was shown in [64] that the upper bound on the estimate of the error is determined by the manner in which the spectrum $\tilde{Z}(\omega)$ decays, and for $\delta(\Delta_M n) = 0$, the following is true in many interesting cases:

$$\rho_{\Delta Mz}^2 \leq \rho_z^2 \leq 4\rho_{\Delta Mz}^2 \qquad (9.42)$$

It may be assumed (and actual modeling confirms this), that (9.42) may also be used to select $\Delta_M$, if the energy falling outside the interval $|\omega\Delta_M| \leq \pi$ is taken into account. The highest bandwidth found among the various processes is, of course, used to choose $\Delta_M$.

We will now examine the error in modeling the filtering of the noise. In the continuous case (9.19), we have for one of the components (dropping the subscripts):

$$P(t) = \int\limits_{-\infty}^{\infty} n(\tau)H(t - \tau) \, d\tau \qquad (9.43)$$

Accordingly, in the discrete case:

$$P_{\Delta M}(t) = \sum_{n=-\infty}^{\infty} P_{\Delta M}(n\Delta_M) \frac{\sin[\pi(t/\Delta_M - n)]}{\pi(t/\Delta_M - n)} \qquad (9.44)$$

where

$$P_{\Delta M}(n\Delta_M) = \Delta_M \sum_{k=-\infty}^{\infty} n(\Delta_M k)H(\Delta_M n - \Delta_M k)$$

and $n(\Delta_M k)$ are mutually independent normally distributed quantities with zero mean and variance equal to $\sigma^2$.

We will now calculate the correlation function of the noises $P(t)$ and $P_{\Delta M}(t)$:

$$R(t_2 - t_1) = \overline{P(t_1)P(t_2)} = n_0 \int_{-\infty}^{\infty} H(\tau)H(\tau + t_2 - t_1)\,dv$$

$$
\begin{aligned}
R_{\Delta M}(\Delta_M m - \Delta_M n) &= \overline{P_{\Delta M}(\Delta_M n)\,P_{\Delta M}(m\Delta_m)} \\
&= \Delta_M^2 \sum_{k=-\infty}^{\infty} \sum_{l=-\infty}^{\infty} \overline{n(\Delta_M k)n(\Delta_M l)}\,H(\Delta_M n - \Delta_M k)H(\Delta_M m - \Delta_M l) \\
&= D\Delta_M^2 \sum_{k=-\infty}^{\infty} H(\Delta_M n - \Delta_M k)H(\Delta_M m - \Delta_M k) \\
&= D\Delta_M^2 \sum_{k=-\infty}^{\infty} H(\Delta_M k)H(\Delta_M k + \Delta_M m + \Delta_M n)
\end{aligned}
$$

(9.45)

$$
\begin{aligned}
R_{\Delta M}(t_2, t_1) &= \sum_{n=-\infty}^{\infty} \sum_{m=-\infty}^{\infty} P_{\Delta M}(n\Delta_M)P_{\Delta M}(m\Delta_M) \\
&\quad \times \frac{\sin[\pi(t_1/\Delta_M - n)]}{\pi(t_1/\Delta_M - n)}\frac{\sin[\pi(t_2/\Delta_m - m)]}{\pi(t_2/\Delta_M - m)} \\
&= \sum_{n=-\infty}^{\infty} R_{\Delta M}(\Delta_M n)\frac{\sin[(\pi/\Delta_M)(t_1 - m\Delta_M)]}{(\pi/\Delta_M)(t_1 - m\Delta_M)} \cdot \\
&\quad \times \frac{\sin\{(\pi/\Delta_M)[t_2 - (m + n)\Delta_M]\}}{(\pi/\Delta_M)[t_2 - (m + n)\Delta_M]} \\
&= \sum_{n=-\infty}^{\infty} R_{\Delta M}(\Delta_M n)\frac{\sin[(\pi/\Delta_M)(t_2 - t_1 - n\Delta_M)]}{(\pi/\Delta_M)(t_2 - t_1 - n\Delta_M)}
\end{aligned}
$$

From this, in analogy with (9.37) and (9.38), we may write

$$D_n = \frac{n_0}{\Delta_M},$$

$$R_{\Delta M}(t) = \sum_{n=-\infty}^{\infty} R_{\Delta M}(\Delta_M n)\frac{\sin[(\pi/\Delta_M)(t - n\Delta_M)]}{(\pi/\Delta_M)(t - n\Delta_M)}$$

(9.46)

where

$$R_{\Delta M}(\Delta_M n) = R(\Delta_M n) + \delta(\Delta_M n), \qquad \delta(\Delta_M n) = \sum_{k=-\infty}^{\infty} \delta_{nk}$$

$$\delta_{nk} = \Delta_M H(\Delta_M k) H(\Delta_M n - \Delta_M k) - \int_{h\Delta_M}^{(k+1)\Delta_M} H(\tau) H(n\Delta_M - \tau)\, d\tau$$

The mean square error which results by expressing the correlation function in the form (9.46) may be estimated with (9.42).

In the particular case of matched filtering:

$$h_0(t) = s_0(-t), \qquad h(t) = s^*(-t)$$

We determine the signal-to-noise ratio by the formula:

$$q^2 = s^2(0)/2k_M(0) \tag{9.48}$$

Using (9.47) in (9.8) and (9.11), we obtain

$$S(0) = \frac{1}{\sqrt{E}} \int_{-\infty}^{\infty} S_0^2(t)\, dt = \frac{\sqrt{E}}{2} \int_{-\infty}^{\infty} S^2(t)\, dt = \sqrt{E}, \quad k_M(0) = n_0 \tag{9.49}$$

In this case:

$$s_c(0) = \sqrt{E}, \qquad s_s(0) = 0$$
$$\overline{M_c^2(t)} = \overline{M_s^2(t)} = n_0, \qquad \overline{M_c(0)M_s(0)} = 0$$

With discrete filtering according to (9.33), we have

$$s_{\Delta MC}(0) = \frac{\Delta_M \sqrt{E}}{2} \sum_{k=-\infty}^{\infty} S(\Delta_M k), \qquad s_{\Delta MS}(0) = 0 \tag{9.50}$$

$$R_{\Delta M}(0) = n_0 \frac{\Delta \sqrt{E}}{2} \sum_{k=-\infty}^{\infty} S^2(\Delta_M k) \tag{9.51}$$

The required signal-to-noise ratio is determined by (9.48). For continuous signals, through (9.49) we obtain

$$q^2 = E/(2n_0) \tag{9.52}$$

In the simulation with discrete signal and noise representations, the signal-to-noise ratio is found, with (9.50) and (9.51), to be

$$\frac{s_{\Delta MC}^2(0)}{2R_{\Delta M}(0)} = \frac{\Delta_M \sqrt{E}}{4n_0} \sum_{k=-\infty}^{\infty} S^2(\Delta_M k) \tag{9.53}$$

## 9.4 SIMULATING WIDEBAND SIGNAL PROCESSING IN RADAR RECEIVERS

We will now consider examples of processing signals transmitted with linear frequency modulation (LFM pulses). The input LFM pulse may be written [25] as

$$
\begin{aligned}
s_0(t) &= \sqrt{2E/\tau_c}\ \text{Re}[\exp(-i\{\omega_0 t + [\Omega/(2\tau_c)]t^2 + \phi_0\})] \\
&= \sqrt{E}\ \text{Re}[s(t)\exp(-i\omega_0 t)], \qquad |t| \le \tau_c/2
\end{aligned}
\tag{9.54}
$$

where $\tau_c$ is the length of the input pulse; $\Omega/(2\pi) = f_c$ is the frequency excursion, $\Omega \ll \omega_0$;

$$
s(t) = \sqrt{2/\tau_c}\ \exp(-i\{[(\Omega/(2\tau_c)]t^2 + \phi_0\})
$$

is the complex envelope of the input signal.

Let the processing consist of matched filtering followed by sidelobe suppression in the compressed pulse and detection. The impulse response of the matched filter is

$$
\begin{aligned}
h_0(t) &= \sqrt{2/\tau_c}\ \text{Re}(-i\{\omega_0 t - [\Omega/(2\tau_c)]\ t^2 - \phi_0) \\
&= \text{Re}[h(t)\exp(-i\omega_0 t)], \qquad |t| \le \tau_c/2
\end{aligned}
\tag{9.55}
$$

As given above, the filter is physically unrealizable, but this does not affect the issues being considered. The notation used in (9.55) is convenient in that the noise output is a maximum when $t = 0$.

The signal at the output of the matched filter, in accordance with (9.8), will be equal to

$$
\begin{aligned}
s_\Phi(t) &= \frac{\sqrt{E}}{\tau_c}\ \text{Re}\left( \int_{\max(-\tau_c/2,\, t-\tau_c/2)}^{\min(\tau_c/2,\, t+\tau_c/2)} \exp\left\{ -i\left[ \omega_0\tau - \frac{\Omega}{2\tau_c}\tau^2 - \phi_0 \right.\right.\right. \\
&\qquad \left.\left.\left. - \omega_0(t-\tau) + \frac{\Omega}{2\tau_c}(t-\tau^2) + \phi_0 \right]\right\} \right) d\tau \\
&= \sqrt{E}\ \frac{\sin[(\Omega t/2)(1 - |t|/\tau_c)]}{\Omega t/2}\ \cos(\omega_0 t), \qquad |t| \le \tau_c
\end{aligned}
\tag{9.56}
$$

The correlation function of the noise at the output of the matched filter will be, in accordance with (9.12):

$$
K_M(t) = \frac{n_0}{\sqrt{E}} s_\Phi(t) = n_0 \frac{\sin[(\Omega t/2)(1 - |t|/\tau_c)]}{\Omega t/2}\ \cos\omega_0 t
\tag{9.57}
$$

Calculating the Fourier transform of (9.57), we obtain the energy spectrum of the output noise:

$$\tilde{K}_M(\omega) = (n_0/\sqrt{E}) S_\Phi(\omega) \tag{9.58}$$

where $S_\Phi(\omega)$ is the spectrum of the signal at the output of the matched filter.

A Hamming filter is often used to suppress the range sidelobes, and has the impulse response:

$$\tilde{h}_1(t) = (1/\sqrt{1.36})\{\delta(t) + 0.425[\delta(t - 2\pi/\Omega) \tag{9.59}$$
$$+ \delta(t + 2\pi/\Phi)]\}$$

and frequency response:

$$h_1(\omega) = (1/\sqrt{1.36})[1 + 0.85 \cos (2 \pi\omega_0/\Omega)] \tag{9.60}$$

where $\omega_0/\Omega$ is an integer.

The signal at the output of the Hamming filter will be

$$s_1(t) = (1/\sqrt{1.36})\{s_\Phi(t) + 0.425[s(t - 2\pi/\Omega) \tag{9.61}$$
$$+ s(t + 2\pi/\Omega)]\}$$

and its spectrum will be $S_1(\omega) = S_\Phi(\omega)\tilde{h}_1(\omega)$.

We may obtain equations analogous to (9.57) and (9.58) for the energy spectrum of the noise at the output of the Hamming filter:

$$\tilde{K}_{M1}(\omega) = (n_0/\sqrt{E}) S_\Phi(\omega)|\tilde{h}_1(\omega)|^2 = (n_0/\sqrt{E}) S_1(\omega) \tilde{h}^*(\omega) \tag{9.62}$$

and the correlation function:

$$K_{M1}(t) = (n_0/\sqrt{E})\{s_\Phi(t) + 0.625[s_\Phi(t + 1/f_c)] \tag{9.63}$$
$$+ 0.133[s_\Phi(t - 2/f_c) + s_\Phi(t + 2/f_c)]\}$$

These expressions for the spectra allow us to use (9.42) to choose the sampling interval $\Delta_M$ necessary for modeling the LFM pulse processing in accordance with (9.33).

With large compression ratios $K_c = f_c$, $f_c \gg 1$, the signal at the output of the matched filter (9.56) will be well approximated by the equation:

$$s_\Phi(t) \approx \{\sqrt{E}[\sin(\Omega t/2)]/(\Phi t/2)\} \cos(\omega_0 t) \tag{9.64}$$

and from this it is not hard to find its spectrum:

$$S_\Phi(\omega) \approx \sqrt{E}(\pi/\Omega)\{\text{rect}[(\omega - \omega_0)/\Omega] + \text{rect}[(\omega + \omega_0)/\Omega]\} \tag{9.65}$$

Thus, the spectrum of the signal at the output of the matched filter is an even function which approximates a rectangle function of width $\Omega$ in the positive frequencies for large compression ratios.

The same is true for the energy spectrum of the noise at the output of the matched filter. The signal-to-noise ratio determined with (9.56) and (9.57) is

$$q^2 = s_\Phi^2(0)/[2K_M(0)] = E/(2n_0) \tag{9.66}$$

An analysis of (9.61)–(9.63) shows that the Hamming filter lowers the amplitude and increases the width of the peak signal response, while the overall spectrum of the signal is made narrower. The energy spectrum of the noise is narrowed even more, and the signal-to-noise ratio is reduced as follows:

$$q_1^2 = s_1(0)/[2K_{M1}(0)] \approx E/(2.72n_0) = q^2/1.36 \tag{9.67}$$

Similar effects result when other filters are used to reduce the range sidelobes.

The foregoing analysis allows us to conclude that when choosing the sampling interval $\Delta_M$ in a simulation of LFM pulse processing, the spectrum of the signal envelope at the output of the matched filter $S_{\mathrm{env}}(\omega)$ should be used. The mean square error (9.62), referred to the output signal energy, may be estimated with the inequality:

$$\frac{1}{\beta_\Phi} \int_{-\infty}^{\infty} |S_{\mathrm{env}}(\omega)|^2 \, d\omega + \frac{1}{\beta_\Phi} \int_{-\infty}^{\infty} |S_{\mathrm{env}}(\omega)|^2 \, d\omega$$

$$\leq \rho^2 \leq \frac{4}{\beta_\Phi} \left\{ \int_{\pi/\Delta_M}^{\infty} |S_{\mathrm{env}}(\omega)|^2 \, d\omega + \int_{-\infty}^{\pi/\Delta} |S_{\mathrm{env}}(\omega)|^2 \, d\omega \right\} \tag{9.68}$$

where $\beta_\Phi = \int_{-\infty}^{\infty} |S_{\mathrm{env}}(\omega)|^2 \, d\omega$.

With this sampling interval, it is possible to simulate the output signal envelope both at the output of the matched filter and at the output of various sidelobe suppression filters, with an error no worse than in (9.68).

If other detection schemes are simulated, then any resultant widening of the signal spectrum must be taken into account with a shorter sampling interval. Thus, for example, the normalized signal envelope (9.64) has a rectangular spectrum with bandwidth $\Omega$:

$$s_{\mathrm{env}}(t) = [\sin(\Omega t/2)]/(\Omega t/2), \qquad S_{\mathrm{env}}(\omega) = (2\pi/\Omega)\mathrm{rect}(\omega/2) \tag{9.69}$$

The mean square error will have the spectrum:

$$S_{env}^2(\omega) = \frac{4\pi^2}{\Omega^2} \int\limits_{-\infty}^{\infty} \left(\text{rect } \frac{\nu}{\Omega}\right)\text{rect } \frac{\omega - \nu}{\omega} \, d\nu$$

$$= \left(1 - \frac{|\omega|}{\Omega}\right)4\pi^2, \quad |\omega| \leq \Omega$$

(9.70)

occupying a bandwidth of $2\Omega$.

It should be noted that the actual signal processing performed in the radar receiver need not be simulated to model the output signals resulting from input signals of known form. Those signals or noises whose correlation functions are known beforehand may be simulated with much more economical methods. Therefore, the foregoing examples should only be considered as illustrative of the application of the simulation methods studied.

## 9.5 MODEL OF A MONOPULSE AUTOTRACKING RADAR

A simplified block diagram for a monopulse radar designed to perform automatic target tracking was presented in Fig. 4.10. With the help of the simulation to be described, it is possible to estimate the basic radar characteristics: the probability of detecting and tracking targets stably, and the tracking errors which are produced. In addition, it is possible to study the effects of various factors (RCS fluctuations, the parameters of various radar components, and so on) on the detection and tracking characteristics.

The most complex part of the simulation is developing a mathematical description of the angle discriminator. This description may be achieved by several methods. We will follow the method by which the signal and noise components are calculated for the various elements of the discriminator. This method provides the most complete description of the processes occurring in an actual discriminator, and, besides, allows us to calculate the effects of phase discrepancies in the channels.

In the modelled radar, the pattern is formed with a parabolic antenna of diameter $d_p$. With a uniform field distribution in the aperture, each of the four partial beams will be described by the expression:

$$F_i(\theta) = \frac{\sin(\frac{1}{2}k_\lambda d_p\theta_{rel} \pm \pi/2)}{\frac{1}{2}k_\lambda d_p\theta_{rel} \pm \pi/2}$$

(9.71)

where $\theta_{rel} = \theta_0/\theta_3$ is the angular coordinate in beamwidths; $\theta_3$ is the width of a partial beam in minutes; $\theta_0$ is the angular coordinate relative to the center of the beam in minutes; $k_\lambda = 2\pi/\lambda$ is the wave number; $d_p$ is the diameter of the antenna in meters; and $i = 1, 2, 3, 4$.

The sum pattern in one of the planes passing through the axis is described by the expression:

$$F_s(\theta) = \frac{(n/2)^2 \cos(\frac{1}{2}k_\lambda d_p \theta_{\text{rel}})}{(n/2)^2 + (\frac{1}{2}k_\lambda d_p \theta_{\text{rel}})^2} \tag{9.72}$$

The difference pattern in one of the planes has the form:

$$F_d(\theta) = \sqrt{F_1^2(\theta) + F_2^2(\theta) - 2F_1(\theta)F_2(\theta)\cos(\phi_1 - \phi_2)} \tag{9.73}$$

$$F_d(0) = \begin{cases} \arctan\left[-\dfrac{F_2(\theta)\sin\phi_1}{F_1(\theta) - F_2(\theta)\cos\phi_1}\right] \\ \qquad \text{for } F_1(\theta) > F_2(\theta)\cos\phi_1 \\[2ex] \arctan\left[-\dfrac{F_2(\theta)\sin\phi_1}{F_1(\theta) - F_2(\theta)\cos\phi_1}\right] + \pi \\ \qquad \text{for } \begin{cases} \phi_1 < 0, \\ F_1(\theta) < F_2(\theta)\cos\phi_1 \end{cases} \\[2ex] \arctan\left[-\dfrac{F_2(\theta)\sin\phi_1}{F_1(\theta) - F_2(\theta)\cos\phi_1}\right] - \pi \\ \qquad \text{for } \begin{cases} \phi_1 > 0, \\ F_1(\theta) < F_2(\theta)\cos\phi_1 \end{cases} \end{cases}$$

The reflected signal at the output of the antenna may be written in the form:

$$u_{\text{in}} = f(u_R, u_\phi)$$

where $u_R$ is the component of the reflected signal depending on the radar potential and target range, and $u_\phi$ is the component of the signal depending on the fluctuations in the target cross section.

It is convenient for the following analysis to convert the reflected signal to a relative quantity referred to the noise power in the receiver sum channel.

The signal-to-noise power ratio referred to the antenna input is

$$q_{\text{in}(0)}^2 = P_s \sigma_{\text{eff}}/R^4 \tag{9.74}$$

where $P_s$ is the radar potential, which is determined by the transmitter power, receiver sensitivity, pulse length, line losses, and so on; $\sigma_{\text{eff}}$ is the target's RCS; and $R$ is the target range.

In amplitude-sum-and-difference radars, the phase discrepancy between the high frequency channels may reach $10°$–$40°$ [21]. Such a discrepancy affects the angle accuracy but causes only a slight reduction in the amplitude of the sum signal. We will, therefore, neglect this factor when calculating the signal amplitude and signal-to-noise ratio in the sum channel. With these conditions, the sum signal and signal-to-noise power ratio at the output of the magic tee will have the form [25]:

$$u_s = u_{in} F_s^2(\theta) \tag{9.75}$$

$$q_s^2 = \frac{P_s \sigma_{eff}}{R^4} F_s^4(\theta) \tag{9.76}$$

The received sum signal passes through the linear portion of the receiver, where it is combined with noises, and on to an envelope detector.

The sum signal at the output of the linear portion of the receiver may be written as

$$u_{s\ out} = u_s + u_n \tag{9.77}$$

where $u_n$ is the noise voltage at the output of the linear portion of the receiver.

The amplitude and phase of the sum signal at the output of the linear portion of the receiver may be written in the form:

$$|u_{s\ out}| = \left|\sqrt{u_s^2 + u_n^2 + 2u_s u_n \cos(\phi_{s\ out} - \phi_n)}\right| \tag{9.78}$$

where $u_s$ and $u_n$ are the amplitude of the signal and noise in the linear portion of the sum channel; $\phi_n$ and $\phi_{s\ out}$ are the phases of the noise and sum signal at the output of the linear section of the receiver;

$$\phi_{s\ out} = \arctan \frac{|u_s| \sin \phi_s + |u_n| \sin \phi_n}{|u_s| \cos \phi_s + |u_n| \cos \phi_n} \tag{9.79}$$

where $\phi_s$ is the phase of the sum signal at the output of the tee, equal to

$$\phi_s = \arctan \frac{u(\theta_1)\sin \phi_1 + u(\theta_2)\sin \phi_{II}}{u(\theta_1)\cos \phi_s + u(\theta_2)\cos \phi_{II}} \tag{9.80}$$

$u(\theta_1)$ and $u(\theta_2)$ are the amplitudes of the sum signals in the first and second planes, respectively,

$$u(\theta_1) = u_{in} \left|\sqrt{F_1^2(\theta_1) + F_2^2(\theta_1) + 2F_1(\theta_1)F_2(\theta_1)\cos \phi_I}\right|$$

$$u(\theta_2) = u_{in} \left|\sqrt{F_3^2(\theta_2) + F_4^2(\theta_2) + 2F_3(\theta_2)F_4(\theta_2)\cos \phi_I}\right|$$

$\phi_I$ and $\phi_{II}$ are the phases of the sum signals in the first and second planes, respectively,

$$\phi_I = \arctan \frac{F_1(\theta_1)\sin \phi_1 + F_2(\theta_1)\sin \phi_2}{F_1(\theta_1)\cos\phi_1 + F_2(\theta_1)\cos \phi_2}$$

$$\theta_{II} = \arctan \frac{F_3(\theta_2)\sin \phi_3 + F_4(\phi_2)\sin \phi_4}{F_3(\theta_2)\cos \phi_3 + F_4(\theta_2)\cos \phi_4}$$

$\phi_1$, $\phi_2$ $\phi_3$, $\phi_4$ are the initial phases of the signals received by each of the partial beams, taking phase discrepancies into account.

The distribution of the phase of the noise will be considered uniform within the interval $(0, 2\pi)$.

For further analysis, the amplitude of the sum signal at the input to the detector will be taken as a relative value referred to the noise power of the receiver sum channel. Then

$$\left|\frac{u_{s \text{ out}}}{\sqrt{2}\sigma_n}\right| = \left| \sqrt{\frac{u_s^2}{2\sigma_n^2} + \frac{u_n^2}{2\sigma_n^2} + \frac{u_s u_n}{\sigma_n^2} \cos(\phi_{s \text{ out}} - \phi_n)} \right| \tag{9.81}$$

The signal-to-noise ratio at the output of the linear section of the receiver sum channel may be written in the form:

$$q_{s \text{ out}}^2 = [P_s\sigma_{\text{eff}}(\theta)F_s^4(\theta)]/(2\sigma_n^2 R^4) + u_{n \text{ rel}}^2$$
$$+ 2|\sqrt{[P_s\sigma_{\text{eff}}(\theta)F_s^4(\theta)]/(2\sigma_n^2 R^4)}| \, u_{n \text{ rel}} \cos(\phi_{s \text{ out}} - \phi_n) \tag{9.82}$$

where $u_{n \text{ rel}} = u_n/(\sqrt{2}\sigma_n)$ is the relative value of the noise amplitude in the receiver sum channel.

If the actual amplitude detector is approximated by an infinite bandwidth linear detector, then the voltage and signal-to-noise ratio at its output are determined by (9.78) and (9.79), respectively, multiplied by the detector gain. The difference signal at the receiver input is written as

$$u = u_d \sin(\omega t + \phi_2) \tag{9.83}$$

$$u_d = u_s\sqrt{F_1^2(\theta_1) + F_2^2(\theta)_1) - 2F_1(\theta_1)F_2(\theta_1) \cos \phi_d} \tag{9.84}$$
$$= u_s F_d(\theta)$$

where $\phi_d = \phi_1 - \phi_2$.

The phase of the difference signal at the output of the magic tee may be determined from the following relations:

$$
\phi_{dn} = \arctan \begin{cases} \dfrac{F_2(\theta_1)\,\sin\,\phi_d}{F_1(\theta_1)\,-\,F_2(\theta_1)\,\cos\,\phi_d} \\[4pt] \quad \text{for } F_1(\theta_1) > F_2(\theta_1)\,\cos\,\phi_d \\[8pt] \left[\dfrac{F_2(\theta_1)\,\sin\,\phi_d}{F_1(\theta_1)\,-\,F_2(\theta_1)\,\cos\,\phi_d}\right] + \pi \\[8pt] \quad \text{for } \begin{cases} \phi_d = 0, \\ F_1(\theta_1) < F_2(\theta_1)\,\cos\,\phi_d \end{cases} \\[8pt] \dfrac{F_2(\theta_1)\,\sin\,\phi_d}{F_1(\theta_1)\,-\,F_2(\theta_1)\,\cos\,\phi_d} - \pi \\[8pt] \quad \text{for } \begin{cases} \phi_d = 0, \\ F_1(\theta_1) < F_2(\theta_1)\,\cos\,\phi_d \end{cases} \end{cases}
\tag{9.85}
$$

The received difference signal is combined with the noise in the linear portion of the receiver. As a result, the difference signal at the output of the phase detector will be

$$
u_{d\ \text{out}} = u_d + u_{nd}
\tag{9.86}
$$

The amplitude of the difference signal is

$$
|u_{d\ \text{out}}| = |\sqrt{u_d^2 + u_{nd}^2 + 2u_d u_{nd}\,\cos(\phi_{nd} - \phi_{d\ \text{out}})}|
\tag{9.87}
$$

where $\phi_{nd}$ is the initial phase of the noise signal at the output of the linear portion of the difference channel.

The phase of the difference signal at the output of the linear section of the difference channel is determined from the following relations:

$$
\phi_{d\ \text{out}} = \phi_{dn} + \begin{cases} \text{arc sin}\,(|u_{nd}|/|u_d|)\,\sin(\phi_{dn} - \phi_{dn}) \quad \text{for } |u_d| > |u_{nd}| \\[4pt] \text{arc sin}\,(|u_{nd}|/|u_d|)\,\sin(\phi_{nd} - \phi_{dn}) \\[4pt] \quad \text{for } \begin{cases} \phi_{nd} - \phi_{dn} = 0/(\pi/2),\ |u_d| < |u_{nd}|, \\ \phi_{nd} - \phi_{dn} = (3\pi/2)/(2\pi) \end{cases} \\[8pt] \text{arc sin}\,(|u_{nd}|/|u_d|)\,\sin(\phi_{nd} - \phi_{dn}) + \pi/2 \\[4pt] \quad \text{for } \begin{cases} |u_d| < |u_{nd}| \\ \phi_{nd} - \phi_{dn} = (\pi/2)/\pi \end{cases} \\[8pt] \text{arc sin}\,(|u_{nd}|/|u_d|)\,\sin(\phi_{nd} - \phi_{dn}) - \pi/2 \\[4pt] \quad \text{for } \begin{cases} |u_d| < |u_{nd}| \\ \phi_{nd} - \phi_{dn} = \pi/(3\pi/2) \end{cases} \end{cases}
\tag{9.88}
$$

We will assume that the sum and difference signals are normalized with respect to the amplitude of the sum signal with an AGC system (see Fig. 3.26), using a control signal generator (CSG). The block diagram of the AGC system in the angle channel is shown in Fig. 9.2. The amplitude of the difference signal at the output of the detector, including the gain control, will have the following form:

$$|u_{dn}| = \left| \frac{u_{d\ out}}{u_{c\ out}} \right|$$

$$= \left| \sqrt{ \frac{u_d^2}{u_c^2\ _{out}} + \frac{u_{nd}^2}{u_c^2\ _{out}} + 2 \frac{|u_d|\ |u_{nd}|}{u_c^2\ _{out}} \cos(\phi_{nd} - \phi_{d\ out})} \right| \tag{9.89}$$

where $u_{c\ out}$ is the control signal voltage at the output of the linear section of the receiver, in accordance with which the gain of the difference channel is normalized.

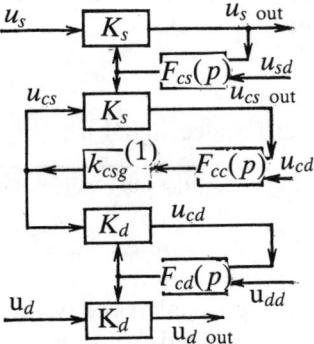

**FIGURE 9.2.** Block diagram of an AGC system.
(1) control signal generator

In relative quantities referred to the noise power in the difference channel, the normalized difference signal amplitude will be

$$|u_{dn}| = \left| \sqrt{ \frac{u_d^2 q_s^2}{q_d^2\ _{out}} + \frac{u_{nd\ rel}}{q_d^2\ _{out}} + \frac{2 u_d u_{nd\ rel} q_s^2}{q_d^2\ _{out}} \cos(\phi_{nd} - \phi_{d\ out})} \right| \tag{9.90}$$

where $q_d^2\ _{out}$ is the signal-to-noise power ratio of the control signal at the output of the linear section of the difference channel; $u_{nd\ rel}$ is the relative value of the difference channel noise. The sign of the error signal depends on the difference between the phases of the difference and sum signals. The phases of the sum-and-difference signal at the input to the phase detector are determined with (9.79) and (9.88).

The voltage at the output of the linear phase detector is

$$u_d(\theta) \approx 2K_{pd}u_{dn} \cos(\phi_{d \text{ out}} - \phi_{s \text{ out}}) \tag{9.91}$$

Then the error signal in the angle channel will be, for combined and separate formation of the amplitude and sign, respectively,

$$\Delta\theta_d = 2K_{pd} \left| \sqrt{\frac{u_d^2 q_s^2}{q_{d \text{ out}}^2} + \frac{u_{n \text{ rel}}}{q_{d \text{ out}}^2} + \frac{2u_d u_{n \text{ rel}} q_s^2}{q_{d \text{ out}}^2} \cos(\phi_{nd} - \phi_{d \text{ out}})} \right|$$
$$\times \cos(\phi_{d \text{ out}} - \phi_{s \text{ out}} \tag{9.92}$$

$$\Delta\theta_d' = K_{pd} \left| \sqrt{\frac{u_d^2 q_s^2}{q_{d \text{ out}}^2} + \frac{u_{n \text{ rel}}}{q_{d \text{ out}}^2} + \frac{2u_d u_{n \text{ rel}} q_s^2}{q_{d \text{ out}}^2} \cos(\phi_{nd} - \phi_{d \text{ out}})} \right|$$
$$\times \text{sign}[-\cos(\phi_{d \text{ out}} - \phi_{s \text{ out}})] \tag{9.93}$$

We will now find the signal-to-noise ratio of the control signal $q_d^2$ out in the receiver taking account of the operation of the AGC. When modeling the results of small changes in the output signal amplitude $\Delta u_{\text{out}}$ good results may be obtained with an exponential approximation to the AGC response to simulate its finite bandwidth. The requirement for small output signal variations corresponds to an actual condition when modeling monopulse radars, which perform automatic tracking. With this approximation to the AGC response, we have

$$K_c = K_{\max} \exp(-b_c u_c) \tag{9.94}$$

The amplitude of the control signal at the output of the receiver when the AGC is operating is

$$u_{c \text{ out}} = K_{\max}u_{d \text{ out}} \exp[-b_c(u_{d \text{ out}} - u_{dd})F_c(p)]$$
$$\text{for } K_{\max}u_{d \text{ in}} > u_{dd} \tag{9.95}$$

$$u_{c \text{ out}} = K_{\max}u_{d \text{ in}} \quad \text{for } K_{\max}u_{d \text{ in}} \leqslant u_{dd}$$

where $u_{d \text{ in}}$ is the voltage of the control signal at the input to the difference channel; $K_{\max}$ is the maximum gain of the difference channel; $u_{dd}$ is the AGC delay voltage in the difference channel; $b_c$ is the control response coefficient; and $F_c(p)$ is the open transfer function of the AGC filter.

Taking the logarithm of both sides of (9.95), and using the following assumptions:

$$u_{c \text{ out}} = u_{dd} + \Delta u_d, \quad \begin{cases} \ln(1 + \Delta u_d/u_{dd}) \approx \Delta u_d/u_{dd} \\ \ln[u_{d \text{ in}}/(\sqrt{2}\sigma_{nc})] = \ln q_{d \text{ in}} \end{cases} \tag{9.96}$$

we obtain

$$\ln q_{d \text{ out}(a)} = b_c F_c(p) u_{dd}/[1 + b_c F_c(p) u_{dd}]\ln q_{d \text{ in}} \qquad (9.97)$$

Expressions analogous to (9.97) may be obtained for the first (sum channel) and second (control signal generator) AGC loops:

$$\ln q_{d \text{ in}} = \{b_{cc} F_{cc}(p) u_{cd}/[1 + b_{cc} F_{cc}(p) u_{cd}]\}\ln q_c \qquad (9.98)$$

$$\ln q_c = \{b_{cs} F_{cs}(p) u_{sd}/[1 + b_{cs} F_{cs}(p) u_{sd}]\}\ln q_s \qquad (9.99)$$

where $q_{d \text{ in}}$ and $q_c$ are the signal-to-noise ratios for the control signal at the inputs of the difference and sum channels.

Solving (9.97) and (9.99) jointly, we can find the relation between the signal-to-noise ratio of the echo signal at the receiver input to the control signal signal-to-noise ratio at the output of the difference channel:

$$
\begin{aligned}
\ln q_{d \text{ in}(a)} = & \{b_{cs} u_{sd} F_{cs}(p)/[1 + b_{cd} u_{dd} F_{cd}(p)]\} \\
& \times \{b_{cc} u_{cd} F_{cc}(p)/[1 + b_{cc} u_{cd} F_{cc}(p)]\} \\
& \times \{b_{cd} u_{dd} F_{cd}(p)/[1 + b_{cd} u_{dd} F_{cd}(p)]\}\ln q_s
\end{aligned}
\qquad (9.100)
$$

The expression for the normalization error is also determined from (9.97) and (9.99) in the form:

$$
\begin{aligned}
\frac{\Delta u_{dd}}{u_{dd}} &= \ln q_s - \ln q_{d \text{ out}(a)} \\
&= \ln q_s \left[ 1 - \frac{b_{cs} u_{sd} F_{cs}(p)}{1 + b_{cs} u_{sd} F_{cs}(p)} \right. \\
&\quad \left. \times \frac{b_{cc} u_{cd} F_{cc}(p)}{1 + b_{cc} u_{cd} F_{cc}(p)} \frac{b_{cd} u_{dd} F_{cd}(p)}{1 + b_{cd}} \right]
\end{aligned}
\qquad (9.101)
$$

The error signal $\Delta\theta_p$, found with either (9.92) or (9.93), is fed to the subreflector drives from the output of the difference channel. The structure of the sub-reflector drive system and its model are shown in Figs. 9.3 and 9.4 [25]. The operation of the subreflector drives may be described with the following differential equations:

$$
\begin{aligned}
\Delta\theta_1 &= \Delta\theta_d \kappa_c, \quad \Delta\theta_2 = \Delta\theta_1 - \Delta\theta_3, \quad \Delta\theta_3 + \tau_c \frac{d\Delta\theta_3}{dt} \\
&= \tau_c v_c \frac{d^2\Delta\theta_c}{dt^2} + \tau_c r_c \frac{d^3\Delta\theta_c}{dt^3}
\end{aligned}
$$

$$
\begin{aligned}
\frac{d\Delta\theta_c}{dt} &+ (\tau_a + \tau_{ex} + \tau_{el}) + (\tau_a \tau_{ex} + \tau_{ac} \tau_{elc} + \tau_{ex\ c} \tau_{el\ c}) \\
&\times \frac{d^3\Delta\theta_c}{dt^3} + \tau_a \tau_{ex} \tau_{el} \frac{d^4\Delta\theta_c}{dt^4} = \Delta\theta_2
\end{aligned}
\qquad (9.102)
$$

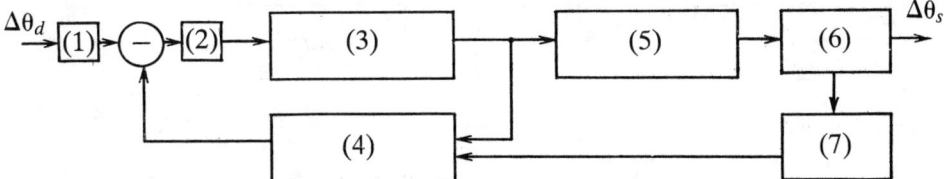

**FIGURE 9.3.** Block diagram of subreflector drive.

| | |
|---|---|
| (1) PTN | (5) auxiliary motor |
| (2) filter | (6) gears |
| (3) amplifier-oscillator | (7) TD |
| (4) correction circuit | |

The transfer function of the open generator-motor circuit is

$$k_2'(p) = 1/[p(1 + \tau_a p)(1 + \tau_{ex} p)(1 + \tau_{el} p)] \tag{9.103}$$

The transfer function of the parallel correction loop is

$$k_2''(p) = \tau_k p/(1 + \tau_k p) \tag{9.104}$$

The nonlinear function $\gamma_2(\theta)$ is approximated by the following expression:

$$\gamma_2(\theta) = \begin{cases} k_c \Delta\theta_0 - \Delta\theta_3 & \text{for } |k_c \Delta\theta_d - \Delta\theta_3| < \alpha_n \\ \alpha_n & \text{for } |k_c \Delta\theta_d - \Delta\theta_3| > \alpha_n \end{cases} \tag{9.105}$$

The notation in Fig. 9.4, and in (9.102) and (9.105) is as follows: $k_c$ is the gain of the control loop; $v_c$ is the gain of the output shaft speed feedback connection; $r_c$ is the gain of the output shaft acceleration feedback connection; $\tau_a$ is the electromechanical time constant for the motor acceleration; $\tau_{el}$ is the electrical time constant of the motor armature circuit; $\tau_c$ is the time constant of the parallel correction loop; $\tau_{ex}$ is the time constant of the generator excitation winding; $\Delta\theta_c$ is the measured value of the relative angle coordinate; and $\alpha_n$ is the linear portion of the amplifier.

In order to simulate the subreflector drive system, it is convenient to divide the nonlinear function $\gamma_2(\theta)$ into two parts: a filter for the uncorrected drive, and a filter for the correction circuit. We may then write

$$\Delta\theta_c = \Delta\theta_2[p(1 + \tau_a p)(1 + \tau_{ex} p)(1 + \tau_{el} p)] \tag{9.106}$$
$$\Delta\theta_{el} = \Delta\theta_c \tau_c p/(1 + \tau_c p) \tag{9.107}$$

The AGC-system filter is modeled as an aperiodic block with transfer function:

$$k(p) = [1/(pT_p)]/[1 + 1/(pT_p)] \tag{9.108}$$

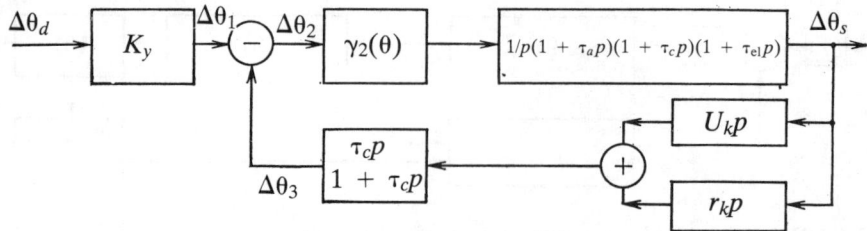

**FIGURE 9.4.** Block diagram of model for subreflector drive.

The search and detection processes are modeled with the beam following and Archimedean spiral in the plane perpendicular to the line of sight with respect to the indicated target data:

$$\rho_a = a_n \phi_n(t) = (h_a/2\pi)\phi_n(t) \tag{9.109}$$

where $h_a$ is the spiral step, determined by the beamwidth.

The block diagram of an automatic tracking radar simulation is presented in Fig. 9.5. The effective scattering surface $\sigma_{eff}$ is a function of the angle $\theta_t$ between the line of sight and the target direction, and is stored in look-up tables at intervals $\Delta\theta_t$. For the $i$th transmission event, the angle $\theta_t$ is determined, and the corresponding surface is retrieved from the look-up table to be converted to a reflected signal.

The algorithms simulating the antenna patterns and signals at the output of the tee comprise (9.71)–(9.73), (9.75), (9.77), (9.83), and (9.85). The sum signal at the output of the linear portion of the receiver is described by (9.78), (9.79), and (9.82). Equations (9.87) and (9.90) are used to describe the difference channel.

Equations (9.92) and (9.93) are used to describe the formation of the error signal with combined and separate formation of the amplitude and sign, respectively.

The normalization process performed by the AGC system is simulated in accordance with (9.95) and (9.97), the AGC filters with (9.108), and the subreflector drive filter with (9.106) and (9.107). The digital AGC and antenna drive filters are modeled digitally with discrete convolution, using the weighting function of the open system [25].

The receiver noises are modeled with the help of a random number generator. The random number generator produces a random sequence of numbers which are normally distributed.

The input data, for determining the dynamic and fluctuating angular errors, are realizations of the error signal produced by the antenna drive processing the input data, in the presence of receiver noises and reflected signal fluctuations.

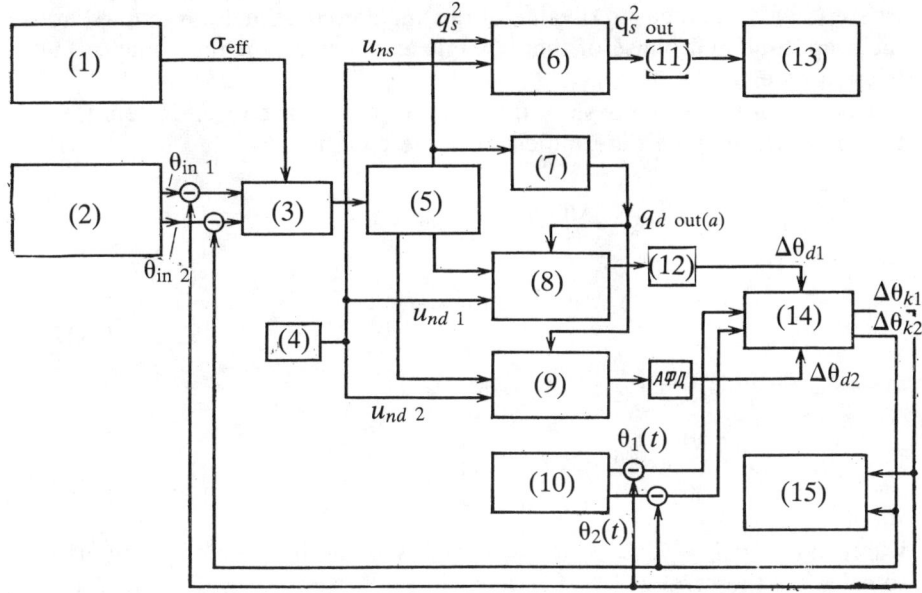

**FIGURE 9.5.** Block diagram of an autotracking monopulse radar model.

(1) reflected signal block
(2) target trajectory block
(3) antenna block
(4) random number generator
(5) sum/difference bridge
(6) sum signal receiver
(7) AGC block
(8) difference signal receiver
(9) difference signal receiver
(10) search block
(11) amplitude detector
(12) amplitude/phase detector
(13) detection
(14) subreflector drives
(15) statistical processor

The input is given in the form of the equivalent sinusoid:

$$\theta_{in}(t) = A_e \sin(\omega_e t) \tag{9.110}$$

where $A_e = \dot{x}/\omega_e$ is the amplitude; $\omega_e = \ddot{x}/\dot{x}$ is the frequency; and $\dot{x}$ and $\ddot{x}$ are the first and second derivatives of the polynomial used to approximate the target trajectory. The statistical processing is affected as an ensemble of realizations for the selected average values of the signal-to-noise ratio. The statistical processing problem consists of extracting the dynamic and fluctuating components from the general tracking error:

$$\Delta\theta(t) = \theta_{in}(t) - \theta_c(t) = \delta_d \sin(\omega_e t + \phi_d) + \delta_\Phi \tag{9.111}$$

where $\theta_c(t)$ is the measured value of the angular coordinate; $\delta_d$ and $\phi_d$ are the amplitude and phase of the dynamic error; and $\delta_\Phi$ is the value of the fluctuating error.

The amplitude and phase of the dynamic error have discrete representations which may be determined from the equations:

$$
\begin{aligned}
\delta_{di} &= \frac{2}{N}\left[\sin\phi_{di}\sum_{i=1}^{N}\Delta\theta_e(\Delta_t i)\cos(\omega_e\Delta_t i)\right. \\
&\quad \left. + \cos\phi_{di}\sum_{i=1}^{N}\Delta\theta_e(\Delta_t i)\sin(\omega_e\Delta_t i)\right] \\
\phi_{di} &= \arctan = \left[\sum_{i=1}^{N}\Delta\theta_e(\Delta_t i)\cos(\omega_e\Delta_t i)\right] \\
&\quad \times \left[\sum_{i=1}^{N}\Delta\theta_e(\Delta_t i)\sin(\omega_e\Delta_t i)\right]^{-1}
\end{aligned}
\tag{9.112}
$$

where $\Delta_t i = t$; $i = 1, 2, 3 \ldots , N$; and $N$ is the number of points used to calculate the errors:

$$
\Delta\theta_e(\Delta_t i) = \Delta\theta(\Delta_t i) - \overline{\Delta\theta(\Delta_t i)}, \qquad \Delta\theta(\Delta_t i) = \left[\sum_{i=1}^{N}\Delta\theta(\Delta_t i)\right]\bigg/N
$$

The average value of the dynamic error is given by

$$
\delta = \sum_{i=1}^{k_e}(\delta_{di}/k_e)
\tag{9.113}
$$

where $k_e$ is the number of realizations of the error signal; and $\delta_{di}$ is the dynamic error for fixed values of the average signal-to-noise ratio in each of the realizations.

After calculating the deterministic components:

$$
\delta_{di}\sin[\omega_e(\Delta_t i) + \phi_{di}]
$$

in each realization from $\Delta\theta_e(\Delta_t i)$, we obtain the realization of the fluctuating error:

$$
\delta_\Phi(\Delta_t i) = \Delta\theta_e(\Delta_t i) - \delta_{di}\sin[\omega_e(\Delta_t i) + \phi_{di}]
$$

The average value of the correlation function of the fluctuating error has the form:

$$
\overline{K(\Delta_t i)} = \sum_{i=1}^{k_e}[K_i(\Delta_t i)/k_e]
\tag{9.114}
$$

where

$$K_i(\Delta_t i) = \frac{1}{N - m} \sum_{i=1}^{N-m} \delta_\Phi(\Delta_t i)\delta_\Phi[\Delta_t(i + 1)]$$

$N$ is the total number of points used to calculate the correlation function; $m = 1, 2, \ldots, N/5$ are the points at which the correlation function is evaluated; $\Delta_t = T_p/N$ is the discrete time interval; and $T_p$ is the total duration of the process (integration time).

The average value of the spectral density of the fluctuating error is determined with the formula:

$$G(\Delta_f i) = 4\Delta_t \sum_{i=0}^{N-m} \overline{K(\Delta_t i)} \cos[\omega_e(\Delta_t i)] \qquad (9.115)$$

where $\Delta_f$ is the discrete frequency interval.

The average value of the variance of the fluctuating error is equal to

$$\overline{\sigma_{\Phi 0}^2} = \left\{ \sum_{i=1}^{k_e} \sum_{k_e=1}^{N} [\delta_\Phi(\Delta_t i)]^2 \right\} \bigg/ (Nk_e) \qquad (9.116)$$

The parameters $N$, $m$, $\Delta_t$, $\Delta_f$, and $k_e$ are selected on the basis of the accuracy required in the estimate of the dynamic and fluctuating errors.

To conclude this section we will examine some results of a simulation of the angle errors. We will look at the errors in determining the angular coordinates for two methods of forming the amplitude and sign of the error signal: separate, and combined. We will analyze the discriminator and fluctuation characteristics and calculate the dynamic and fluctuating errors directly.

We will first consider the effect of phase discrepancies in the high frequency channels on the discriminator and fluctuation responses. The fluctuation response is the variance of the error in a single measurement as a function of the offset angle. The results of a study on the effect of a phase discrepancy, treating both methods of forming the amplitude and sign of the error signal, are presented in Figs. 9.6–9.9, where the discriminator response is plotted in units of the beamwidth; the abscissa of the fluctuation response graphs is the offset angle in units of the beamwidth, and the unit along the ordinate is the maximum value of the variance which is obtained with separate formation of the amplitude and sign of the error signal for $q_s^2 = 10$ dB. It is apparent from the curves that the phase discrepancy has a significant effect on the discriminator and fluctuation responses when the amplitude and sign of the error signal are formed separately. The phase imbalance increases the slope of the discriminator response, and increases

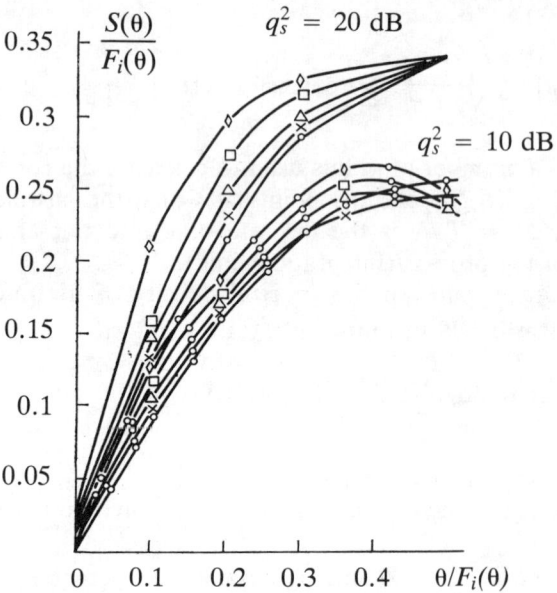

**FIGURE 9.6.** Discriminator response when forming amplitude and sign of error signal separately.

| Φ | |
|---|---|
| 0° | 30° |
| 10° | 40° |
| | 20° |

the variance of the fluctuation error. Thus, for example, with $q_s^2 = 20$ dB, the variance of the error increases along the axis by a factor of more than five, when the phase discrepancy goes from 0° to 40°. With large signal-to-noise ratios ($q_s^2 = 20$ $dB$) there are discontinuities in the null of the discriminator response (see Fig. 9.6).

If the amplitude and sign of the error signal are formed jointly, then the phase imbalance has a lesser effect on the discriminator and fluctuation responses and, consequently, causes less of an increase in the fluctuation angular errors. Thus, with $q_s^2 = 20$ dB, the variance of the fluctuation error increases by a factor of only 1.3 as the phase discrepancy goes from 0° to 30°. Furthermore, comparison of the fluctuation responses plotted in Figs. 9.7 and 9.9 show that the variance of the fluctuation errors at the output of the angle discriminator are about two times greater with separate formation of the amplitude and sign of the error signal than with the joint method. It should be noted that drops in the fluctuation responses for $\varphi = 0$ can arise only when the amplitude and sign of the error signal are formed jointly.

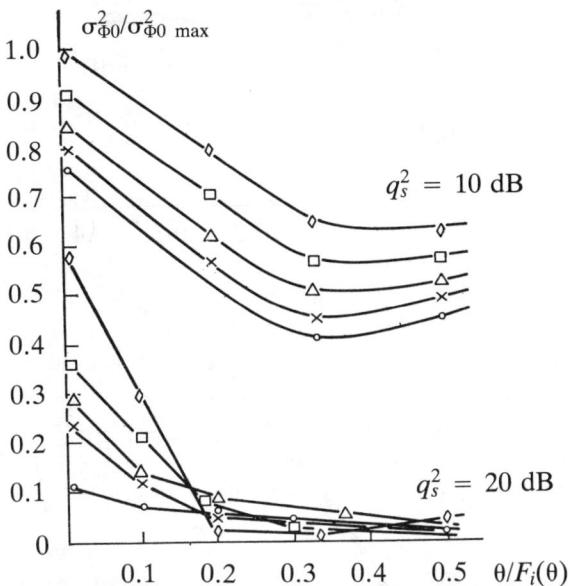

**FIGURE 9.7.** Fluctuation response when forming amplitude and sign of error signal separately.

| $\Phi$ | 20° |
|---|---|
| 0° | 30° |
| 10° | 40° |

The angle error fluctuations have been found to follow a log-normal distribution, with a correlation time which varies from realization to realization. The amplitude of the reflected signal fluctuations reaches 30 dB, and the correlation time varies from 0.1 to 5 seconds.

The fluctuations of the reflected signal cause an increase in the dynamic and fluctuation errors $\delta_d$ and $\sigma_{\Phi 0}^2$ (Table 9.1) (the plotted errors are normalized with respect to the maximum values of $\delta_d$ and $\sigma_{\Phi 0}^2$ which are obtained).

The dynamic error doubles in value for a signal-to-noise ratio $q_s^2 = 10$ dB, and increases by a factor of 1.5 for $q_s^2 = 20$ dB. The fluctuation error undergoes an even larger increase: by a factor of six for $q_s^2 = 10$ dB, and of 1.8 for $q_s^2 = 20$ dB. If the signal-to-noise ratio is on the order of 30 dB, the fluctuations in the reflected signal do not cause a noticeable increase in the angular error.

The angle errors in an autotracking monopulse radar depend on the properties of the AGC system. We will consider the effect of the AGC bandwidth on the angle error. The amplitude of the dynamic error $\delta_d$ and

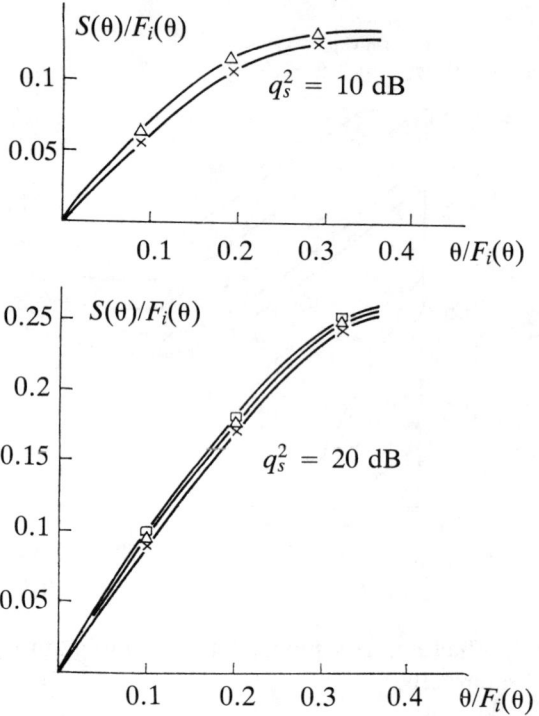

**FIGURE 9.8.** Discriminator response when forming amplitude and sign of error signal jointly.

$$\begin{array}{cc} \Phi & 20° \\ 10° & 30° \end{array}$$

**TABLE 9.1**

| modeling condition | $q_s^2 = 10$ dB | | $q_s^2 = 20$ dB | | $q_s^2 = 30$ dB | |
|---|---|---|---|---|---|---|
| | $\delta_d$ | $\sigma_{\phi 0}^2$ | $\delta_d$ | $\sigma_{\phi 0}^2$ | $\delta_d$ | $\sigma_{\phi 0}^2$ |
| with fluctuations | 1 | 1 | 0.5 | 0.18 | 0.34 | 0.03 |
| without fluctuations | 0.42 | 0.18 | 0.33 | 0.1 | 0.33 | 0.02 |

variance of the fluctuation error $\sigma_{\phi 0}^2$ are plotted as functions of the AGC time constant in Figs. 9.10 and 9.11. From these curves, it is evident that, for both fluctuating and nonfluctuating signals, the minimum dynamic and fluctuation errors result when the AGC time constant $t_a = 0.5 t_{cr}$. As the AGC time constant is increased, both of the errors increase.

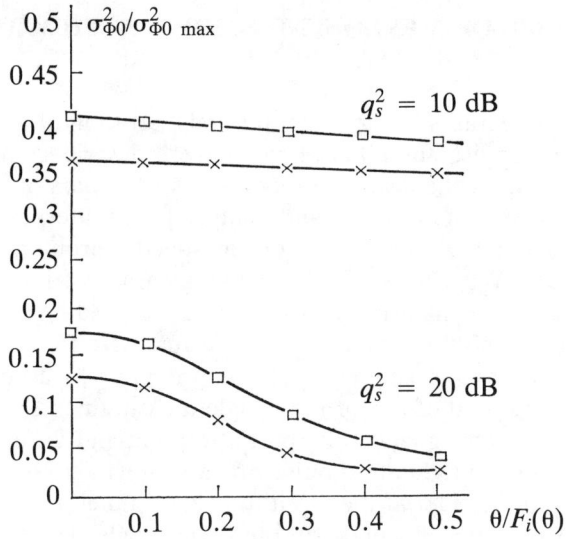

**FIGURE 9.9.** Fluctuation response when forming amplitude and sign of error signal jointly.

$$\begin{array}{cc} \Phi & 20° \\ 10° & 30° \end{array}$$

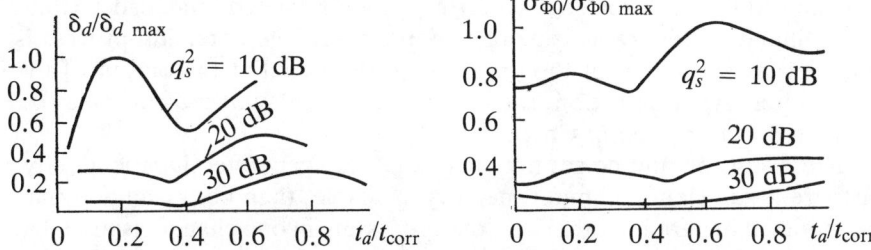

**FIGURE 9.10.** Error as a function of AGC time constant for a fluctuating signal.

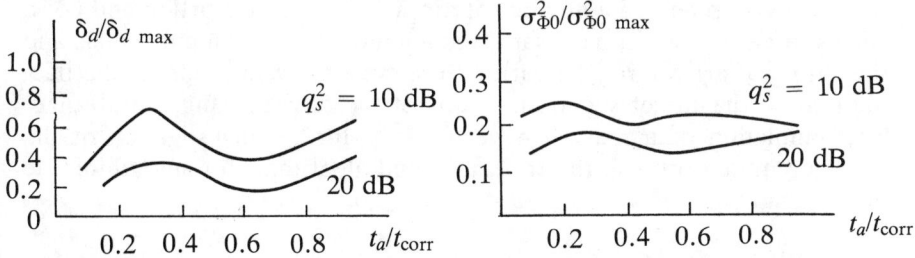

**FIGURE 9.11.** Error as a function of AGC time constant for a nonfluctuating signal.

## 9.6 A MODEL OF A FREQUENCY-MODULATED MONOPULSE SURVEILLANCE RADAR

The basic features of the structure and operation of an amplitude-comparison monopulse surveillance radar were described in Chapter 4. We will now consider the design of models of such radars, methods of making calculations with the models, and analyzing and interpreting the results.

We will put together some sort of specific problem for the sake of concreteness. We will assume that the accuracy of individual angle and time delay (range) measurements must be assessed. We will study models of amplitude-amplitude and amplitude-sum-and-difference radars for the case of measuring a single angular coordinate. In accordance with the monopulse method of performing angle measurements, the model should include two receiver channels (partial or sum-and-difference). Modeling linear frequency modulation pulse processing was covered in Section 9.4.

In the radar model at the input to each channel, a pulse is formed, the amplitude of which depends on the offset angle between the target and axis in accordance with the antenna patterns. Amplitude and phase distortions (random or regular) are added to the pulses, as required, for studying their influence. The distorted pulse is mixed additively with Gaussian white noise, the spectral density of which is modified to obtain the desired signal-to-noise ratio. A linear filter, whose impulse response is matched to the undistorted waveform, is then modeled, and also a Hamming filter to affect range sidelobe suppression. The detection process is modeled as extraction of the envelope of the signal at the output of the linear filter. After detection, the amplitude and delay of the pulse may be estimated as discussed in Chapter 4.

The distortion and noises in the channels are generated independently. If there is sufficient computer memory available, then both channels may be modeled in parallel; otherwise the processing in one channel is modeled and the results stored, and then the other channel is modeled and the target offset angle estimated. This is the basic cycle of the simulation.

To determine the statistics of the parameters measured in one cycle, the process is repeated a number of times, using new distortion and noise values in each cycle. If necessary, the amplitudes of the input signals and their time of arrival may be varied from cycle to cycle to model the fluctuations of the target signal and shifts of the target in angle and range. The simulation data may then be used for further investigation of the accuracy in determining the trajectory and algorithms for smoothing the output data.

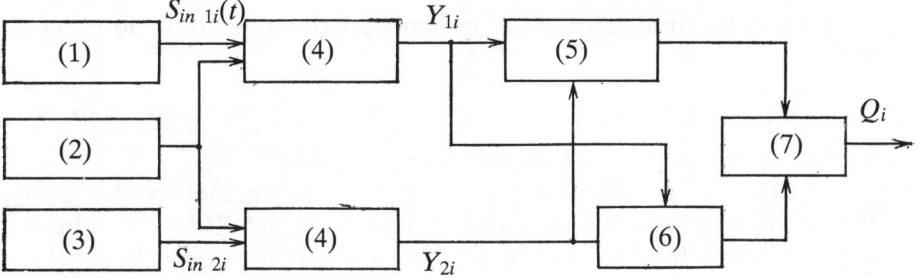

**FIGURE 9.12.** Block diagram for model of the amplitude-amplitude surveillance radar.

(1) signal 1 block        (5) subtraction device
(2) noise generator      (6) adder
(3) signal 2 block        (7) division circuit
(4) filtering and detection block

Let the normalized partial antenna patterns be described by the functions $F_1(\theta)$ and $F_2(\theta)$. Then in the amplitude-amplitude monopulse radar simulation (see Fig. 9.12), the amplitudes of the input signals 1 and 2 in the $i$th cycle are

$$A_{1i} = a_{qi}F_1(\theta), \qquad A_{2i} = a_{qi}F_2(\theta) \tag{9.117}$$

where $a_{qi}$ is a coefficient representing the signal-to-noise ratio, and $\theta$ is the target offset angle. The output signals may then be put in the form:

$$s_{in1i}(t) = A_{1i}[1 + \Delta s_{1i}(t)]\cos[\Phi_{1i}(t)]$$
$$s_{in2i}(t) = A_{2i}[1 + \Delta s_{2i}(t)]\cos[\Phi_{2i}(t)] \tag{9.118}$$

The amplitude and phase distortion may be introduced independently in both channels, or just in one channel. In the latter case, they will be the relative distortions of the response of one channel relative to the other. The simulation will then model the envelopes of the processes $Y_{1i}$ and $Y_{2i}$, which are the result of passing the signal and noise mixture through a matched filter and Hamming filter. They are evaluated at the moment when the envelope would be the greatest in the absence of noise. The methods and algorithms for such calculations were covered in Section 9.4.

To obtain an estimate of the target offset angle $\tilde{Q}_i$, the following ratio is used:

$$\tilde{Q}_i = \frac{Y_{1i} - Y_{2i}}{Y_{1i} + Y_{2i}} = \frac{F_1(\tilde{\theta}_i) - F_2(\tilde{\theta}_i)}{F_1(\tilde{\theta}_i) + F_2(\tilde{\theta}_i)} \tag{9.119}$$

In many practical cases, a linear approximation is justified close to the axis:

$$\tilde{Q}_i = \mu\theta_i \tag{9.120}$$

where $\mu$ is the slope of the angle-sensing response.

We will now examine a model of a sum-and-difference monopulse radar. It was shown in [43] that the patterns of the sum and difference channels are determined from the partial patterns with the matrix equation:

$$\begin{pmatrix} f_s(\theta) \\ f_d(\theta) \end{pmatrix} = \begin{pmatrix} \underline{n}_{11} & \underline{n}_{12} \\ \underline{n}_{21} & \underline{n}_{22} \end{pmatrix} \begin{pmatrix} F_1(\theta) \\ F_2(\theta) \end{pmatrix} \tag{9.121}$$

where the complex quantities $\underline{n}_{ij}$ ($i, j = 1, 2$) give the amplitude and phase characteristics of the sum-and-difference bridge. If we use the notation:

$$\underline{n}_{ij} = n_{ij} \exp(i\phi_{ij}) \tag{9.122}$$

then, from (9.121), we obtain

$$\begin{aligned} f_s(\theta) &= F_s(\theta) \exp[i\phi_s(\theta)] \\ f_d(\theta) &= F_d(\theta) \exp[i\phi_d(\theta)] \end{aligned} \tag{9.123}$$

where

$$\begin{aligned} F_s^2(\theta) &= \{n_{11} \cos[\phi_{11}F_1(\theta)] + n_{12} \cos[\phi_{12}F_2(\theta)]\}^2 \\ &\quad + \{n_{11} \sin[\phi_{11}F_1(\theta)] + n_{12} \sin[\phi_{12}F_2(\theta)]\}^2 \\ \phi_s(\theta) &= \arctan \frac{n_{11} \sin[\phi_{11}F_1(\theta)] + n_{12} \sin[\phi_{12}F_2(\theta)]}{n_{11} \cos[\phi_{11}F_1(\theta)] + n_{12} \cos[\phi_{12}F_2(\theta)]} \\ F_d^2(\theta) &= \{n_{21} \cos[\phi_{21}F_1(\theta)] - n_{22} \cos[\phi_{22}F_2(\theta)]\}^2 \\ &\quad + \{n_{21} \sin[\phi_{21}F_1(\theta)] - n_{22} \sin[\phi_{22}F_2(\theta)]\}^2 \\ \phi_s(\theta) &= \arctan \frac{n_{21} \sin[\phi_{21}F_1(\theta)] - n_{22} \sin[\phi_{12}F_2(\theta)]}{n_{21} \cos[\phi_{21}F_1(\theta)] + n_{22} \cos[\phi_{12}F_2(\theta)]} \end{aligned} \tag{9.124}$$

The bridge is strictly balanced if $n_{ij} = 1/2$ and $\phi_{ij} = 0$. The output of the sum and difference channels in the $i$th cycle, in accordance with the amplitude-sum-and-difference monopulse radar model shown in Fig. 9.13, are given by

$$A_{si} = a_{qi}F_s(\theta), \qquad A_{pi} \approx a_{qi}F_d(\theta) \tag{9.125}$$

and the input signals may be placed in a form analogous to those in (9.118), with the addition of the phase term from (9.124) to the argument of the cosine.

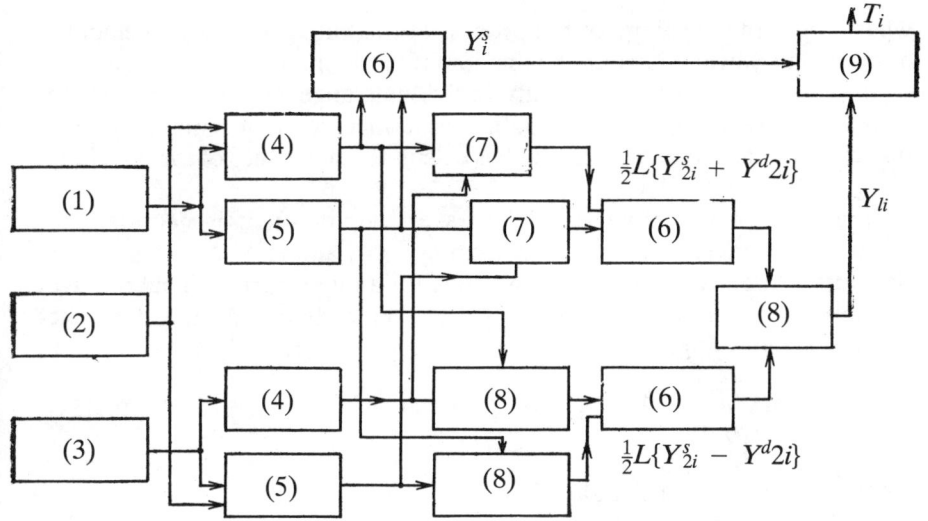

**FIGURE 9.13.** Block diagram for model of an amplitude-sum-and-difference surveillance radar.

(1) sum signal block
(2) noise generator
(3) difference signal block
(4) cosine filter block
(5) sine filter block

(6) detection block
(7) adder
(8) subtraction device
(9) division circuit

The input signals are mixed with noise and give, at the output of the linear filters, $y_{2i}^s$ and $y_{2i}^d$ (comprising two orthogonal components), each in accordance with the algorithms laid out in Section 9.3.

To estimate the target offset angle, the signals from the output of the phase detector and sum channel are used. To model the signals at the output of a square-law phase detector, it is necessary to calculate

$$Y_{si} = \tfrac{1}{4}L(y_{2i}^s + y_{2i}^d) - \tfrac{1}{4}L(y_{2i}^s - y_{2i}^d) \tag{9.126}$$

and, at the output of a linear detector,

$$Y_{1i} = \tfrac{1}{2}L(y_{2i}^s + y_{2i}^d) - \tfrac{1}{2}L(y_{2i}^s - y_{2i}^d) \tag{9.127}$$

where $L$ denotes the envelope.

The target offset angle may then be estimated with the ratio:

$$\tilde{T}_i = \frac{F_1(\tilde{\theta}) - F_2(\tilde{\theta})}{F_1(\tilde{\theta}) + F_2(\tilde{\theta})} = \frac{Y_{si}}{Y_i^s} = \frac{Y_{1i}}{Y_i^s} \tag{9.128}$$

where $Y_i^s$ is the envelope of the process at the output of the sum channel at the moment when it would be maximum for an undistorted input signal.

If it is necessary to study only individual range measurements (time delay measurements), then it is sufficient to include in the simulation only one channel, at the output of which is calculated the values of the envelope $Y_i$.

In order to determine the delay (see Chapter 4), it is necessary to evaluate the envelope at times in each cycle: $t_s$ and $t_s \pm \Delta\tau_s$, where $t_s$ is the shift of the maximum of the envelope for an undistorted signal relative to the center of the gate, and $\Delta\tau_s$ is the distance between the gates. The delay may be estimated with the calculation:

$$\tilde{t}_d = k_d[Y_i(t_s - \Delta\tau_s) - Y_i(t_s + \Delta\tau_s)]/[Y_i(t_s)], \qquad (9.129)$$
$$|t_s| \leq \Delta\tau_s/2$$

where $k_d$ is the dimensionality factor.

With the help of these models, it is possible to study the effect of the signal-to-noise ratio on the target angle accuracy of amplitude-amplitude and amplitude-sum-and-difference radars. To this end, the simulation is used to calculate $\tilde{Q}_i$ (9.119) and $\tilde{T}_i$ (9.128) for $i = 1, 2, \ldots , M_k$, for various signal-to-noise ratios and various target offset angles, as determined with $F_1(\theta)$ and $F_2(\theta)$.

We will present some simulation results for illustration. The angle-sensing response (9.120) was taken to be linear with unit slope for the calculations. The slopes of the discriminator responses were determined for an amplitude-amplitude radar:

$$\mu_Q = \frac{1}{M_k} \sum_{i=1}^{M_k} \frac{Q_i}{v_i}, \qquad v_i \neq 0$$

and for a sum-and-difference radar

$$\mu_T = \frac{1}{M_k} \sum_{i=1}^{\mu_k} \frac{T_i}{\theta_i}, \qquad \theta \neq 0$$

along with the variance of the angle error $\sigma_Q^2$ (9.119) for $\theta_i = 0$ (target on the axis). Linear and square-law detectors were treated in the amplitude-amplitude case. The number of cycles $M_k$ was chosen in order to obtain an error of less than 10% in the accuracy of the statistics. The results of the calculations are shown in Figs. 9.14 and 9.15. An inversely proportional relationship is shown for comparison with the dashed curve in Fig. 9.15. The curves are drawn as functions of the sum signal-to-noise ratio. The manner in which the discriminator response slope falls at lower

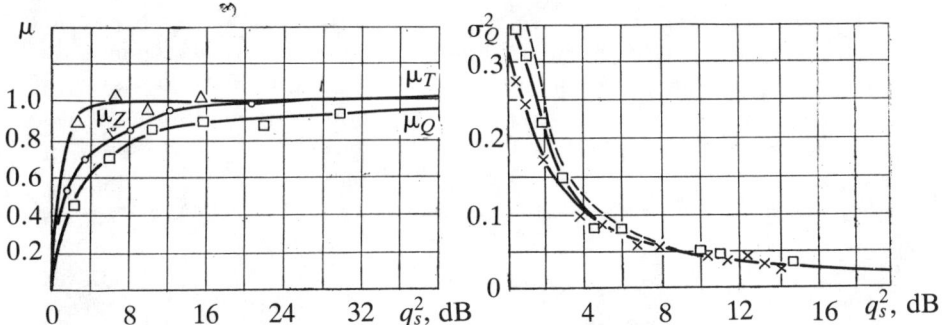

**FIGURE 9.14.** Change in the slope of the discriminator response.

**FIGURE 9.15.** Variance of angle estimate error:
  X   square-law detector
  ☐   linear detector

signal-to-noise ratios is clearly evident in Fig. 9.14, especially for the amplitude-amplitude radar. It may be seen in Fig. 9.15 that the widely-used formula by which the variance of the angular error is inversely proportional to the signal-to-noise ratio is valid for $q_s^2 \gg 8$ dB.

Analogous simulations were carried out for measuring the time delay (9.129). The slope of the discriminator response and the variance of the error $\delta^{2d}$ are plotted as functions of the signal-to-noise ratio in Figs. 9.14 and 9.16, respectively; the solid line in Fig. 9.16 corresponds to an inversely proportional relationship.

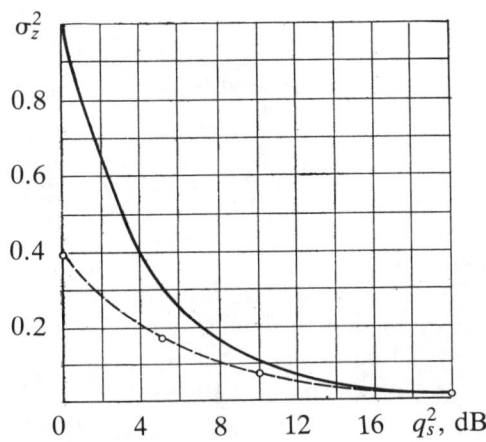

**FIGURE 9.16.** Variance of delay (range) error.

# Chapter 10

# The Application and Characteristics of Foreign Monopulse Radars

## 10.1 BALLISTIC MISSILE AND SPACE OBJECT TRACKING RADARS

In the West, monopulse radars are widely used in missile guidance and control systems, ballistic missile and space object intercept systems, satellite tracking facilities, and a number of other systems [21, 31, 51, 69, 99, 107].

The first monopulse radar deployed in the United States for tracking ballistic missiles and artificial satellites was the AN/FPS-16 [21]. The antenna of this system comprises a parabolic reflector and a four-horn feed (see Fig. 10.1). This automatic tracking amplitude-sum-and-difference monopulse radar performs tracking in two planes.

**FIGURE 10.1.** AN/FPS–16 radar antenna.

The performance of many AN/FPS-16 radars currently deployed has been significantly improved as a result of changes which have been made in their design. A polarization device has been placed between the feed and reflector to convert linear polarization to circular, and an improved crystal mixer and parametric amplifier are employed in the receiver, reducing the noise figure to 5 dB. The basic tactical-technical characteristics of the modernized system are presented in Table 10.1.

**TABLE 10.1**

| Basic parameters | Tactical-technical parameters | | | | |
|---|---|---|---|---|---|
| | AN/FPS-16 | AN/FPS-92 | AN/FPS-85 | AN/FPS-115 | AN/FPS -108 |
| operating range with $\sigma_{eff}$ = 1 m$^2$, km | 500 | 3700 | 4150 | 4800 | 3700 |
| operating frequency, MHz | 5400–5900 | 404–446 | 416–500 | 420–450 | 1000–1200 |
| peak power, MW | 3 | 5 | 50 | 0.6 | 16 |
| pulse repetition frequency, Hz | 142–1707 | 27 | 27.40 | 2–60 | 2–60 |
| pulse width, microseconds | 0.25–1.0 | 2000 | 5–250 | 250–1600 | 150–2000 |
| antenna dimension, m | 5 | 25.6 | receive 58.5 transmit 34 × 34 | 22 | 29 |
| beamwidth degrees | 0.7 | 2 | 0.5–1.5 | 2 | 0.55 |
| search sector, degrees: azimuth elevation | 360 0–90 | 360 0–90 | 120 0–90 | 240 3–85 | 120 0–80 |
| range measurement error, m | 4.6 | 300 | 700 | – | 4.6 |
| angle error, arc minutes | 0.7 | 1.8 | 3 | – | 3 |

The AN/FPS-49 monopulse radar, introduced in the BMEWS early warning system, was designed to detect and track missiles and artificial earth satellites. It is a pulsed Doppler system operating at two frequencies in the long and short wave portions of the decimeter waveband. A spiral search pattern is used for search. The radar can detect targets with an effective cross section $\sigma_{eff} = 1$ m$^2$ at a range of 4800 km. It then switches into a tracking mode in which it calculates the velocity, direction of flight, and, if the target is a missile, the point of impact [20].

The radar antenna has a 25-m diameter parabolic reflector, mounted under a radome with a diameter of 45 m. The antenna is rotated with a hydroelectric drive, and can achieve complete hemispherical coverage with 360° azimuth coverage and 90° elevation coverage.

In the detection mode, the AN/FPS-49 signals are passed from the receiver to a system which extracts and preprocesses the data. On the basis of range, angle, and Doppler information, the preprocessor computer uses signal correlation to sort out signals from nonballistic targets, and performs an initial estimate of the potential threat from targets with ballistic trajectories. Intercontinental ballistic missiles are distinguished from earth satellites on the basis of their range and azimuth rates, and also by comparing their motion parameters with known orbital element sets stored in the computer.

In 1966, the AN/FPS-92 was developed for the BMEWS system; it was a more reliable and jamming-immune radar based on the AN/FPS-49. Solid state circuitry was used widely in its design, along with a device to allow normal operation during periods when the Aurora Borealis is active [21]. The tactical-technical characteristics of the radar are presented in Table 10.1.

Serving the unified US air defense command NORAD is the AN/FPS-85, a more modern system for detecting and tracking space objects [99]. It uses a phased array antenna with programmed beam control. The radar frequency is 442 MHz. The operating range extends to several thousand kilometers. The system can detect, track, identify, and catalog objects in low-earth orbits and ballistic missiles, handling up to several hundred objects simultaneously. The outer appearance of the radar is shown in Fig. 10.2, and its basic operating parameters in Table 10.1. The length of the building is about 107 m, and ther height 47.3 m at the receiving antenna and 29 m at the transmitting antenna. The face is inclined at an angle of 45°.

**FIGURE 10.2.** Exterior appearance of AN/FPS–85.

The radar comprises nine subsystems: transmit antenna, receive antenna, beam control system, computer, tracking and control equipment, signal processor, testing and control, space object identification, and computer software.

The transmitting antenna consists of 5184 modules in a square array of $72 \times 72$ elements; the beamwidth is 1.4°; the peak power is 50 MW; and the pulse width ranges from 5 to 250 microseconds. The receiving antenna is made up of 19500 dipoles arranged in a circular network of squares. There are 4660 identical receiving modules, arranged in a spatial taper, less dense at the periphery. The inactive elements (dipoles) are terminated in matched loads.

The receiving beam-forming system creates nine beams, the axes of which are 0.4° apart. The positions of the transmitting and receiving beams are controlled by changing the phase at one frequency with low power levels. The computer system makes use of two IBM 360/651 processors.

The signal processor includes digital and analog components necessary for digitizing the intermediate frequency signals from each of the nine beams. There are three separate signal processors for search, track, and coherent processing.

A linear frequency-modulated pulse 250 microseconds long is used for both searching and tracking of targets at long ranges, and at shorter ranges an amplitude-modulated pulse 10 microseconds long is used.

Groups of pulses 5, 25, or 125 microseconds long are used for coherent processing. Such processing provides good range resolution (with pulse compression), and Doppler resolution (using pulses of long duration).

Other features of the AN/FPS-85 include sequential detection, dual-polarization reception, special methods for rejecting interference, and optimal energy management depending on the range and cross section $\sigma_{eff}$ of the target.

Toward the end of 1979 the AN/FPS-115 ("Pave Paws") radar was deployed to perform long range detection and tracking of ballistic missiles, including submarine launched ballistic missiles. In addition, this radar may detect and track artificial earth satellites. The exterior appearance of the radar is shown in Fig. 10.3, and its main parameters are presented in Table 10.1. The radar has two receive-transmit antennas, which provide azimuth coverage of 240°, and elevation coverage from 3° to 85° (which serves the interests of satellite tracking in the "Spacetrack" program) [107].

The AN/FPS-115 is the first radar in the United States to employ a solid state receive-transmit module, the dimensions of which are $30 \times 20$ c. Each phased array consists of 56 subarrays, and one subarray comprises 32 modules. The radar building (excluding the electrical power station) is trapezoidal, almost 36 m high and the same length on all three sides. The diameter of the phased arrays is nearly 22 m, within which are placed 1792

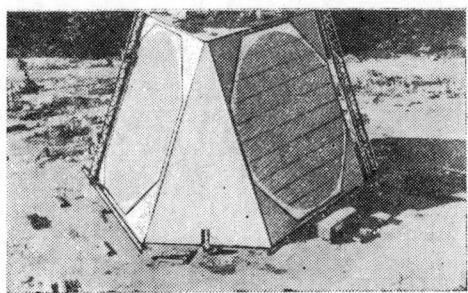

**FIGURE 10.3.** Exterior appearance of AN/FPS–115 radar.

active and 885 passive elements (2677 elements in all, in each array). To obtain the desired elevation coverage, the arrays are inclined at 20° to the vertical.

The computer power is provided by two Cyber 174-12s and two minicomputers. The main systems are the Cybers, one of which is operational while the other is held as a "cold" backup; if the radar goes into an early warning mode, the second computer is automatically switched to "hot" reserve.

The AN/FPS-108 *Cobra Dane* is a monopulse radar located on eastern Shemya (in the Aleutian Islands) and is tasked to gather reconnaissance information on Soviet ballistic missile tests conducted in the Kamchatka peninsula region. In addition, the radar may provide early warning of missile strikes and performs satellite detection and tracking as part of the "Spacetrack" system.

The Cobra Dane employs a phased array with 15,000 active elements 15.2 cm long. The azimuth coverage is 120°, and the coverage in elevation is 0° to 80°. The external appearance of the radar is shown in Fig. 10.4,

**FIGURE 10.4.** Exterior appearance of AN/FPS–108 radar.

and its basic parameters are presented in Table 10.1. The radar and auxiliary equipment are housed in a six-story building 30 m high. In an abutting
single-story building, there are control services and a precision measurement laboratory. In consideration of the severe conditions in the Aleutian
Islands, the radar is designed to withstand winds of up to 300 km per hour,
and earthquakes registering up to 4.0 *ball* (on a 12-point scale). The radar
uses a Cyber 74-18 computer [69].

## 10.2 RADAR SYSTEMS FOR LONG-RANGE TRACKING AND SPACE COMMUNICATIONS

The space tracking and communications systems TRAC(E) [21], which
tracks and communicates with earth satellites, the moon and other space
objects, has a monopulse amplitude-sum-and-difference tracking system
with a 25.5-m diameter parabolic reflector (see Fig. 10.5). The half-power
beamwidth of the sum pattern is about 1°. A spiral search pattern within
a 3° sector is used to detect moving targets.

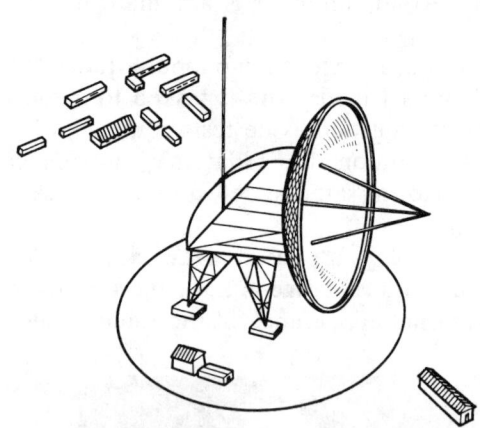

**FIGURE 10.5.** General form of TRAC(E) radar system.

The TRAC(E) system is one-way, allowing the measurement of angular
coordinates and Doppler frequency, and the reception of telemetry information. The earth station uses signals from the onboard transmitter. The
Pioneer IV spacecraft could be tracked out to 690,000 km using its onboard
0.2-W transmitter at 960 MHz.

The Telstar satellite detection and automatic tracking system uses the
monopulse method to measure angular coordinates [21]. This system transmits various coded information repeatedly over 118 channels. Transmission

of a complete cycle of information takes one minute. The Telstar satellite has the following characteristics: apogee—5600 km; perigee—1100 km; orbital inclination—45°. There are two beacons on the satellite operating at 126 MHz and 4080 MHz, along with telemetry and communications equipment. The uplink communications are transmitted at 6390 MHz, and the downlink at 4170 KHz. The telemetry information modulates a 136-MHz carrier transmitted to earth.

The earth station consists of a coarse guidance system, a fine guidance system, and an automatic tracking system (see Fig. 10.6), which operate on the beacon signals [21].

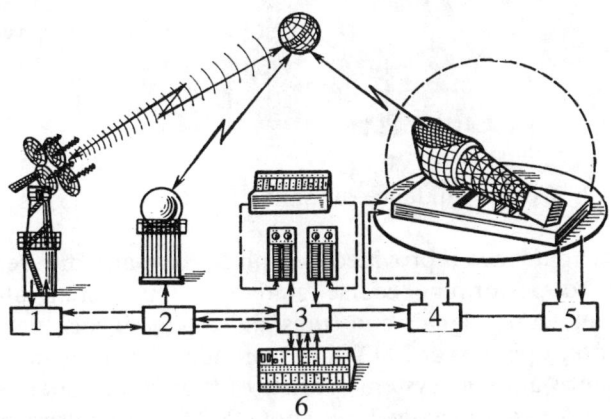

**FIGURE 10.6.** Functional diagram of earth station:
1: coarse guidance
2: fine guidance
3: computer
4: autotracking antenna
5: autotracking system
6: control console

Detection and tracking of earth satellites is performed as follows. Using data obtained from the launch, the operators of the ground control station direct the command-link antenna (see Fig. 10.7) used by the coarse guidance system to the point where the satellite is expected. After the beacon is received at 136 MHz, the coarse guidance system switches to an automatic tracking mode. The operator may then transmit a series of commands to enable the basic satellite systems. The onboard 4080-MHz beacon then begins transmitting, and is used by the earth station fine guidance system to acquire the satellite. After acquisition, the fine guidance system begins its automatic track, during which the fine guidance and tracking antennas work together.

**FIGURE 10.7.** Coarse guidance antenna.

The coarse guidance is provided by a phase-sum-and-difference system. In addition to performing coarse guidance, this system also receives microwave telemetry signals and transmits coded signals to the satellite. The transmitter operates at 200 W in a continuous transmission mode. The coarse guidance antenna system consists of four spiral antennas (see Fig. 10.7). The half-power beamwidth is about 20°. The signals received by the spiral antennas go through four waveguide bridges, where they are converted to sum and difference signals in elevation and azimuth. The sum and difference signals pass through high-pass filters to a three-channel phase-sensitive receiver.

Fine guidance is performed by an amplitude-sum-and-difference system. The antenna is composed of a 2.4-m diameter parabolic reflector, a 0.9-m hyperbolic subreflector, and a four-horn feed. The beamwidth is 2.1°. The signal from the output of the antenna goes to four magic-tees. The output sum and difference signals go on to three parametric amplifiers, and further to the tracking receiver, in which the azimuth and elevation error signals are extracted. The error signals are then passed to the azimuth and elevation antenna drive systems.

The automatic tracking system is also an amplitude-sum-and-difference system, operating at the onboard beacon frequency of 4080 MHz. A parabolic horn antenna is used (see 4 in Fig. 10.6), which consists of a parabolic reflector illuminated with a conical horn. The top of the horn coincides with the focus of the paraboloid, and the horn axis is perpendicular to the paraboloid axis. The antenna has the following basic characteristics: conical horn aperture angle—31.5°; focal length—18.2 m; aperture diameter—20.6 m; aperture area—334 m$^2$; beamwidth (at 4080 MHz)—0.24°. The antenna is enclosed in a 65-m diameter housing.

The US Haystack radar system is an extremely powerful monopulse radar designed to track space equipment and to support radioastronomy studies, including studies of such distant planets as Mars, Mercury, and Jupiter [21]. The antenna system comprises a 37-m diameter parabolic reflector and a 2.85-m subreflector, with horn feeds. The beamwidth is about 3 minutes of arc. The antenna is housed in a spherical enclosure with a 45-m diameter.

The transmitter transmits either continuously or long pulses, and has an average output power of 100 kW, operating at 7750 MHz. The signals are amplified in the receiver by several parametric amplifiers, cooled to the temperature of liquid helium.

## 10.3 ANTIMISSILE DEFENSE RADARS

The American *Safeguard* antiballistic missile defense system (see Fig. 10.8) consists of PARs (perimeter acquisition radars), MSRs (missile site radars), Sprint short-range and Spartan long-range interceptors, computing facilities, and communications facilities [3, 22].

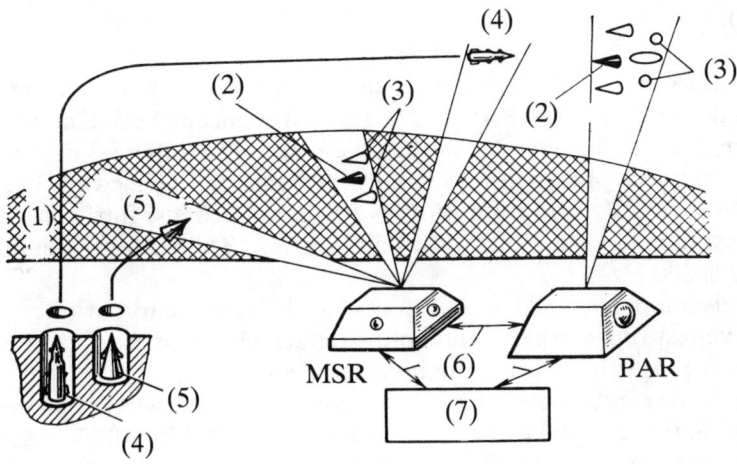

**FIGURE 10.8.** Components of the Safeguard ABM system.

    (1) atmosphere    (5) Sprint
    (2) warhead       (6) communications
    (3) decoys        (7) computer
    (4) Spartan

The PAR (see Fig. 10.9) is an amplitude-sum-and-difference radar tasked to perform long-range target detection and tracking, to hand off targets to the appropriate MSR, and to perform guidance of the Spartans during the initial portion of their flight. The radar operates in the UHF band at 442 MHz. The radar is housed in a building which looks like a four-sided pyramid with its top removed. The radar antenna is a highly modularized phased array, 35.4 m in diameter, consisting of 6600 elements, each of which has an average power of 1.1 kW. The arrays are situated on two sides so as to provide 180° azimuth coverage. The antenna faces are inclined at 30° to the vertical. The beamwidth is 1.2°. The pulses are linearly frequency-modulated with a width of 5 to 250 microseconds, and a compressed pulse width of one microsecond. The pulse repetition frequency is 10 Hz.

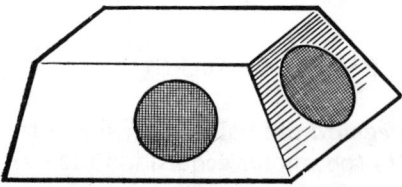

**FIGURE 10.9.** General appearance of the PAR (Perimeter Acquisition Radar).

The MSR (Fig. 10.10) is a monopulse radar designed to detect and track targets designated by the PAR and to guide Spartan and Sprint interceptors to them. The radar operates in the 10-c band. The antennas are circular receive-transmit phased arrays 4.1 m in diameter, inclined at 30°. The beamwidth is 1.7°. The pulse width is 100 microseconds. Intrapulse phase shift keying is used to affect pulse compression. The pulse repetition frequency is 200 Hz.

The development and improvement of antimissile radars always requires extensive test-range experimentation. All radars built for use on antimissile radar test ranges have been monopulse radars.

The Tradex radar was designed to study complex target and processing algorithms for identifying payloads at ranges to 3200 k [20]. The general form of the antenna system is shown in Fig. 10.11. The radar employs a dual frequency scheme in which two transmitters operate simultaneously. The first transmitter has a peak power of 4 MW and operates at 425 MHz, while the second operates at 1320 MHz with a peak power of 1.25 MW. The pulse repetition frequency is 1500 Hz.

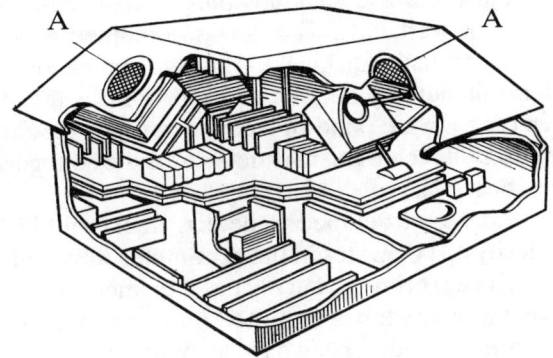

**FIGURE 10.10.** General appearance and location of equipment in MSR (Missile Site Radar) building.

A   antenna array

**FIGURE 10.11.** Tradex radar antenna.

Search and target detection are performed at 1320 MHz. The use of frequency-modulated pulses provides a high average power level, and after pulse compression, good range resolution. Targets are tracked at 425 MHz. The data obtained during the tracks are accumulated and presented to the operator in analog form, and to the computer after being digitized.

The Hapdar radar has a phased array receive-transmit antenna consisting of 4300 elements, of which 2165 are active. This radar can detect and track targets with an effective cross section of 1 $m^2$ to a range of 370 km. The range measurement error is in the vicinity of 20 m [31].

The Altair radar is designed to study the physical phenomena associated with the re-entry of warheads. The data so obtained is necessary for developing methods of distinguishing warheads from decoys. The radar antenna is a 45-m diameter parabolic reflector with a five-horn feed. The radar operates over a wide band of frequencies (60, 425, and 1320 MHz), and, according to its developers, may determine the configuration of targets and their velocities [31].

The Adar radar was developed to detect, track, and identify warheads among many decoys. It operates in the centimeter waveband with a phased array antenna. The experimental array has a diameter of 1.5 m and several thousand travelling wave tubes mounted on its surface. Ten independent receive beams are formed. The Adar radar has a large operating range, wide instantaneous bandwidth, the capability to alter its waveform, and good throughput capability. The range resolution is 1.5 m [31].

## 10.4 SURFACE-TO-AIR MISSILE GUIDANCE RADARS

Monopulse radars are widely used in American systems for guiding surface-to-air missiles, including Nike-Ajax, Nike-Hercules, and Sam-D (Patriot), and in shipboard surface-to-air fire control systems (Tartar, Talos, Terrier, and others).

The Nike-Ajax system includes systems for surveillance, target tracking and missile guidance; the latter two are monopulse systems [21]. The radar operates at 9100 MHz ($\lambda = 3.3$ c), and has a 1.9-m diameter lens antenna with a beamwidth of 1.2°. The transmitted peak power is 200 kW. The pulse length is 0.25 microseconds, and the pulse repetition frequency is 2000 Hz.

The Nike-Hercules system [21] includes a target detection radar and a monopulse guidance radar. The guidance system employs a spherical antenna system with a double lens (see Fig. 10.12) contained within a silicone resin radome.

The guidance radar was modified with increased transmitter power and an antenna twice as large as the first (i.e., 8 m). The modernized Nike-Hercules system may engage not only aircraft, but also missiles, which may be detected out to 300–1000 km.

The SAM-D (Patriot) air defense system uses a multifunction monopulse phased array radar, which detects and tracks targets, and also tracks the missiles during the final stage of their guidance to the target. The Patriot system may simultaneously process information on 100 targets and guide eight missiles, and may perform terminal guidance on three missiles simultaneously. The use of a multifunction radar in Patriot allows it to replace nine separate radars found in the Nike-Ajax and Nike-Hercules surface-to-air systems.

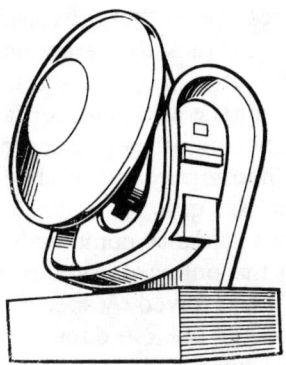

**FIGURE 10.12.** Nike-Hercules guidance antenna.

The multifunction radar and its computer are located in two separate structures, which in a military operation are joined to one another and a third unit housing four 60-kW turbine generators. The radar operates in the 7-cm waveband. The antenna comprises several separate phased arrays (see Fig. 10.13). The largest array, 1 (2.4-m diameter), contains 5161 elements, and is tasked to perform target search and tracking for both targets and missiles. Array 2, composed of 251 elements, is a receive-only array which obtains information from the missiles during tracking. The other four or five 51-element arrays, 3, serve to suppress the antenna sidelobes and to increase the jamming immunity when subject to electronic countermeasures. In the simplest version of the Patriot system these antennas are not used.

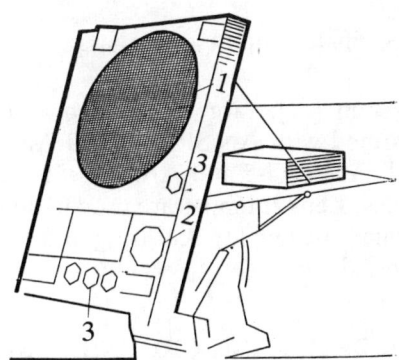

**FIGURE 10.13.** Patriot radar antenna.

The Patriot radar makes use of "target via missile" (TVM) guidance. The radar uses antenna 1 to track the target and missile during the final acquisition stage. The 30-cm monopulse tracking antenna and homing system, mounted on the missile with an inertial unit and autopilot, are controlled from the ground to increase the probability of detecting the signals reflected from the illuminated target. The signals from the onboard tracker are transmitted to the ground, where they are received through antenna 2. The radar forms guidance commands while tracking the target, which are transmitted to the onboard systems with phase-coded signals. The phase-coded signals are received by special antennas on the missile— four in the nose and two in the tail. No data are processed on the missile.

Monopulse radars have been widely used on shipboard surface-to-air guidance systems.

The monopulse fire control AN/SPG-51 system (Tartar) (Fig. 10.14) performs automatic target acquisition and tracking [21]. The illuminating beam is turned on several seconds before launch, and the missile is guided by the signals reflected from the target. A single parabolic reflector antenna with a diameter of 2.4 m is used for track and illumination.

**FIGURE 10.14.** AN/SPG-51 fire control radar.

In the Talos system [21], target detection and surface-to-air missile guidance are performed with AN/SPG-49 and AN/SPG-56 radars. In the Terrier system [21], AN/SPG-5 and AN/SPG-55 radars are used to guide surface-to-air missiles. Lens antennas are used in these monopulse systems, simultaneously forming beams for acquiring and tracking targets, and for tracking the missiles.

## 10.5 RADARS FOR DETECTING GROUND TARGETS IN CLUTTER

An experimental phase-sum-and-difference monopulse radar was developed in 1958 in the United States to detect and determine the range to targets in vegetative clutter [21].

## Basic Radar Parameters

Operating frequency..................................................35GHz
Pulse power .......................................................35kW
Pulse length ......................................................0.06µs
Pulse repetiton frequency ........................................ 4000Hz
Intermediate frequency.............................................60MHz
Bandwidth..........................................................20MHz
Noise figure, each receiver channel............................... 18dB
Diameter of parabolic reflectors ..................................61cm
Antenna gain.......................................................44dB
Bandwidth: azimuth ................................................ 3°
            elevation................................................ 1°

The antenna system (see Fig. 10.15) consists of twin parabolic reflectors, the centers of which are separated by 61.4 cm for phase comparison, each with a feed horn. The target signals leave the antenna and enter a magic tee in which sum and difference signals are formed, after which the converted signals enter a two-channel receiver, at the output of which appear the sum and difference video signals. An electronic switch at the output of the receiver maintains simultaneous indication of the sum and difference video signals on the display.

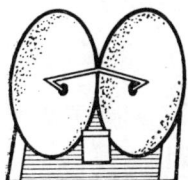

**FIGURE 10.15.** Ground target detection radar antenna.

The oscillograms of the video sum (b) and difference (c) signals resulting from reflections from a vehicle in dense trees and shrubbery are shown in Fig. 10.16. The output resulting from a conical scanning radar located next to the monopulse radar shown for comparison (a). The range from the radar to the vehicle was 750 m. As may be seen in Fig. 10.16(a), it is almost impossible to detect the target with the conical scanning radar due to the interfering reflections. The monopulse radar detects this target without error by the peak in the sum signal (d) for which there is no difference signal. The signal at (e), following the target signal, was caused by reflections from the road shoulder behind the vehicle.

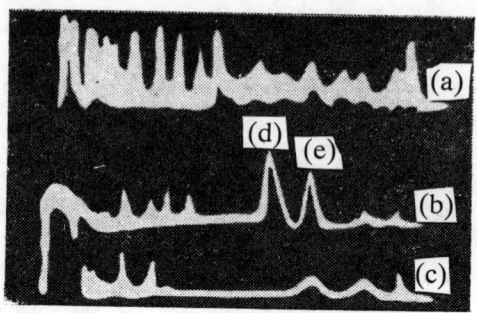

**FIGURE 10.16.** Oscillograms of video signal at radar output:
- (a) conical scan
- (b) monopulse sum channel
- (c) different channel
- (d) vehicle return
- (e) road shoulder return

# Bibliography

1. Ayzin, F.L., Dolzhenkov, A.A., Zimin, D.B., and Sedenkov, E.G. *O podableniy parastinykh lepestkov v diagramme naprablennosti kommutatsionnykh antennykh reshetok (Suppression of unwanted sidelobes in the pattern of switched antenna arrays).* Radiotekhnika i elektronika, vol. 11 (1971), no. 7, pp. 1268–1271.

2. Bakhrakh, L.D. and Voscresenkiy, D.I. (Editors). *Antenny: Sovremennoe sostoyanie i problemy (Antennas: Modern technology and problems),* vol. 16. Moscow: Sovietskoe Radio, 1979. 207 pp.

3. Anureev, I.I. *Oruzhie protivoraketnoy i protivokosmicheskoy oborony (Missile and space defense weapons).* Moscow: Voenizdat, 1971. 304 pp.

4. Bykov, V.V. *Tsifrovoe modelirovanie v statisticheskoy radiotekhnike (Digital processing in statistical radio engineering).* Moscow: Sovietskoe Radio, 1971. 326 pp.

5. Bakin, S.A. and Shustov, L.N. *Osnovy radioprotivodeystviya i radiotekhnicheskoy razvedki (Fundamentals of electronic countermeasures and radio reconnaissance).* Moscow: Sovietskoe Radio, 1968. 444 pp.

6. Volkov, V.M. *Logarifmicheskie usiliteli (Logarithmic amplifiers).* Kiev: Gostekhizdat, 1962. 244 pp.

7. ———. *Logarifmicheskie usiliteli na transistorakh (Transistor logarithmic amplifiers).* Kiev: Tekhnika, 1965. 266 pp.

8. Tartakovskiy, G.P. (Editor). *Voprosy statisticheskoy teorii radiolokatsii (Problems in statistical radar theory),* vols. 1, 2. Moscow: Sovietskoe Radio, 1964. 424 pp., 1079 pp.

9. Glagolevskiy, V.G. and Shishov, Yu. A. *Antenny radiolokatsionnykh stantsiy (Radar antennas).* Moscow: Voenizdat, 1977. 111 pp.

10. Rabiner, L.R. and Rader, I. *Digital signal processing.* New York: IEEE Press, 1972.

11. Druzhinin, V.V. and Kontorov, D.S. *Konfliktnaya radiolokatsiya: Opyt sistemnovo iccledovaniya (Conflict radar: Experience from systems research).* Moscow: Radio i Svyaz, 1982. 124 pp.

12. Efremov, Yu.G. and Nevgasimiy, A.F. *Poteri preobrazobaniya smecitelya prifazovom metode podavleniya zerkal'novo kanala (Mixer conversion losses with phase suppression of image channel).* Radiotekhnika i elektronika (1976), no. 2, pp. 404–405.

13. Maksimov, M.V. (Editor). *Zashchita ot radiopomekh (Radar anti-jamming techniques).* Moscow: Sovietskoe Radio, 1976. 496 pp. (English Trans., Artech House, 1979).

14. Zelkin, E.G. *Postroenie izluchayushey sistemi po zadannoy diagramme naprablennosti (Design of radiating systems for a given radiation pattern).* Moscow: Gocenergoizdat, 1963, 272 pp.

15. Zufrin, A.M. *Metody postroeniya sudovykh avtomaticheskikh uglemernykh sistem (Design methods for shipboard automatic angle measurement systems).* Leningrad: Sudostroenie, 1970. 404 pp.

16. Kanareykin, D.B., Pavlov, A.F. and Potekhin, V.A. *Polyarizatsiya radiolokatsionnykh signalov (Radar signal polarization).* Moscow: Sovietskoe Radio, 1966. 470 pp.

17. Kinber, V.E. and Tishchenko, V.A. *Polyarizatsiya izlucheniya osesimmetrichnykh zerkal'nykh antenn (The polarization of the radiation of axially symmetric antennas).* Radiotekhnika i elektronika, vol. 17 (1972), no. 4, pp. 680–687.

18. Krasenko, N.P. and Rybakov, B.S. *Analiz tochnosti pelengovaniya sluchaynykh poley dvukhkanal'nymi monoimpul'snymi sistemami (Analysis of the accuracy of angle measurement on random fields with two-channel monopulse systems).* Izvestiya vuzov SSSR. Radioelektronika, vol. 19 (1976), no., 4, pp. 10–12.

19. Kuzin, E.V. *Monoimpul'snaya radiolokatsiya (Monopulse radar).* Moscow: Voenizdat, 1969. 100 pp.

20. Leonov, A.I. *Radiolokatsiya v protivoraketnoy oborone (Radar in anti-missile defense).* Moscow: Voenizdat, 1967. 136 pp.

21. Leonov, A.I. and Fomichev, K.I. *Monoimpul'snaya radiolokatsiya (Monopulse radar).* Moscow: Sovietskoe Radio, 1970. 392 pp.

22. Lovenar, N.N. *RLS protivoraketnoy oborony: mif ili real'nost' (Anti-missile defense radar: myth or reality).* Zarybezhnaya radioelektronika (1970), no. 10, pp. 3–16.

23. Lukoshkin, A.P. *Radiolokatsionnie usiliteli s bol'shim diapazonom vkhodnykh signalov (Radar amplifiers with wideband input signals)*. Moscow: Sovietskoe Radio, 1964. 255 pp.

24. Mishchenko, Yu.A. *Radiolokatsionnie tseli (Radar targets)*. Moscow: Voenizdat, 1966. 140 pp.

25. Leonov, A.I. (Editor). *Modelirovanie v radiolokatsii (Modeling radars)*. Moscow: Sovietskoe Radio, 1979. 264 pp.

26. Khetagurov, Ya.A. (Editor). *Mul'tiprotsessornye vychislitel'nye sistemy (Multiprocessor computer systems)*. Moscow: Energiya, 1971. 320 pp.

27. Narbut, V.P. and Khmel'nitskaya, N.S. *Raschet kross-polyarizatsionnykh diagramm ocesimmetrichnykh parabolicheskikh antenn s sinfaznym raspredeleniem polya (Calculating the cross-polarization pattern of an axially symmetric parabolic antenna with a coherent field distribution)*. Voprosy radioelektroniki. Ser. OT, vol. 21 (1967), pp. 43–55.

28. Nemilikher, Yu.A. *et al. Postroenie skhem diodnykh SHF preobrasobateley chastoty s fazovym podavleniem zerkal'novo kanala (Design of diode microwave frequency converters with phase suppression of the image channel)*. Poluprovodnikovye Pribory i Tekhnika Elektrosvyazi (Semiconductor Devices and Communications Technology; I.F. Nikolaevskiy, Ed.). Moscow: Svyas', 1974. 146 pp.

29. Oppenheim, Johnson. *Diskretnoe predstablenie signalov (Discrete signal representation)*. TIIER (1972), no. 6, pp. 102–107.

30. Paliy, A.I. *Radioelektronnaya bor'ba (Radioelectronic conflict)*. Moscow: Voenizdat, 1981. 320 pp.

31. Galkin, V.I., Zakharchenko, I.I., and Mikhaylov, L.V. (Editors). *Radiotekhnicheskie sistemy v raketnoy tekhnike (Radio systems in missile technology)*. Moscow: Voenizdat, 1974, 340 pp.

32. Rhodes, R.S. *Vvedenie v monoimpul'snuyu radiolokatsiyu (Introduction to monopulse)*. Moscow: Sovietskoe Radio, 1960. 159 pp. (Trans. from English, Artech House, 1980).

33. Sergievskiy, B.D. *Opredelenie uglovykh koordinat sovokupnosti izluchateley (Determining the coordinates of an ensemble of radiators)*. Radiotekhnika (1967), no. 4., pp. 83–89.

34. ———. *Metody i sredstva protivodeystviya protivoraketnoy oborony (Methods and means of anti-missile defense)*. Zarùbezhnaya radioelektronika (1966), no. 1, pp. 3–31.

35. Sviridov, E.F. *Sravnitel'naya effectivnost' monoimpul'snykh radiolokatsionnykh sistem pelengatsii (Comparative effectiveness of monopulse angle measurement systems)*. Leningrad: Sudostroenie, 1964. 116 pp.

36. Skolnik, M. (Editor). *Spravochnik po radiolokatsii (Radar Handbook)*. Vols. 1, 2, 3. Moscow: Sovietskoe Radio, 1979. 456 pp., 406 pp., 528 pp. (Trans. from English, McGraw-Hill, 1970).
37. Tartakovskiy, G.P. *Dinamika sistem avtomaticheskoy regulirovki (Dynamics of automatic control systems)*. Moscow: Gosenergoizdat, 1957. 191 pp.
38. Dulevich, V.E. (Editor). *Teoreticheskie osnovy radiolokatsii (Theoretical radar fundamentals)*. Moscow: Sovietskoe Radio, 1964. 732 pp.
39. Ufimtsev, P.Ya. *Metod kraevykh zadach v fizicheskoy teorii difraktsii (Methods of solving boundary value problems in physical diffraction theory)*. Moscow: Sovietskoe Radio, 1962. 243 pp.
40. Fal'kovich, S.E. *Priem radiolokatsionnykh signalov na fone fluktuatsionnykh pomekh (Reception of radar signals against a background of random fluctuations)*. Moscow: Sovietskoe Radio, 1961. 311 pp.
41. Fradin, A.A. *Antenno-fidernie ustroystva (Antenna-feed devices)*. Moscow: Svyas', 1977. 400 pp.
42. Hannan, P. *Microwave antennas derived from the Cassegrain telescope,* IEEE Trans., vol. AP-9 (1961), no. 2, pp. 140–53.
43. Hellgren, G. *On the theory of monopulse radar,* Goteborg, 1960.
44. Khurgin, Ya.N. and Yakovlev, V. *Metody teorii tselykh funktsiy v radiofizike, teorii svyazi i optike (Methods of the theory of integer functions in radiophysics, communications theory and optics)*. Moscow: Fizmatgiz, 1962. 220 pp.
45. Tsar'kov, N.M. *Mnogokanal'nie radiolokatsionnie izmeriteli (Multi-channel radar measuring devices)*. Moscow: Sovietskoe Radio, 1980. 190 pp.
46. Adatia, N.A. and Keen, K.M. *Radiation characteristics of a double offset antenna, theory and experiment* (Antennas and Propag. Internat. Sympos., Quebec, 1980). Symposium Digest, vol. 2 (1980), pp. 545–548.
47. Aerospace Daily, vol. 73 (1975), no. 39, pp. 309–311.
48. Andrews, R.S. *Antenna and other systematic effects on amplitude comparison monopulse systems.* Electronic Circuits and Systems, vol, 3 (1979), no. 3, pp. 103–108.
49. Asseo, S.J. *Detection of target multiplicity using monopulse quadratic angle.* IEEE Trans., vol. AES-17 (1981), no. 2, pp. 271–280.
50. Aviation Week, vol. 89 (1968), no. 1, pp. 24–28.

51. Aviation Week, vol. 102 (1975), no. 20, pp. 34–39.
52. Barton, D.K. *Radiolokatsionnoe soprovozhdenie tseli pri malykh uglakh mesta (Low-angle radar tracking)*. TIIER, vol. 62 (1974), no. 6, pp. 37–61 (see Proc. IEEE, vol. 62 (1974), no. 6, pp. 687–704).
53. Bielli, P. *Axially symmetrical reflectors* (IEEE Internat. Sympos., Session 14, May 19, 1978). Symposium Digest (1978), pp. 337–380.
54. Birkemeier, W.P. and Wallace, N.D. *Radar tracking accuracy improvement by means of pulse-to-pulse frequency modulation.* IEEE Trans. Commun. Electron., vol. 81 (1962), pp. 571–575.
55. Bodnar, D.G. *Cross-polarisation characteristics of monopulse difference patterns* (Antennas and Propag. Internat. Sympos., Quebec, 1980). Symposium Digest, vol. 1 (1980), pp. 477–480.
56. Pat. 3927406 (SSHA).
57. Budinasky, J.H. *Solid state phased arrays for ECM applications* (Conf. Nat. Aerospace Electron., Dayton, Ohio, May 15–17, 1972). Proc. (1972), pp. 162–166.
58. Cooper, D.C. *Zero-steering antenna system for receiving a signal close to the direction of strong interference.* Electronics Letters, March 22, 1973, pp. 140–141.
59. Pat. 3806925 (SSHA).
60. Davis, R. *Design-to-price takes on the cruise missile.* Electronic Warfare, vol. 9 (1977), no. 4, pp. 52–54, 56, 60.
61. Davis, W.A. *Principles of electronic warfare. Radar and EW.* Microwave J., vol. 23 (1980), no. 2, pp. 52–59.
62. Dax, P.R. *Keep track of that low flying attack.* Microwaves, vol. 15 (1976), no. 4, pp. 32–36, 40–42, 47–53.
63. ———. *Accurate tracking of low elevation targets over the sea with a monopulse radar. Radar present and future.* (IEEE Conf., London, October 23–25, 1973). Proc. (1973), pp. 160–165.
64. Defense Electronics, vol. 13 (1981), no. 1, p. 75.
65. Pat. 2943318 (SSHA).
66. Dicken, L.W. *Report on the use of null steering in suppressing main beam interference* (Radar-77, Internat. Conf., London, Oct. 25–28, 1977). Proc., pp. 226–231.
67. Dilpare, A.L. *Chaff primer.* Microwave J., vol. 9 (1970), no. 12, pp. 46–47.
68. Dijk, J. *et al.* *The polarization losses of offset paraboloid antennas.* IEEE Trans., vol. AP-22 (1974), no. 4, pp. 513–520.
69. Elektronika, vol. 48 (1975), no. 16.
70. Guili, D. and Tiberio, R. *A modified monopulse technique for*

*radar tracking with low-angle multipath.* IEEE Trans., vol.
AES-11 (1975), no. 5, pp. 741–747.

71. Ghobrial, S.J. and Fututh, M.M. *Cross-polarisation measurements
    using a composite feed with parasitic elements.* Electronics
    Letters, vol. 11 (1975), no. 20, pp. 481–482.

72. Ghobrial, S.J. *The effect on the polarisation characteristics of the
    receiving feed on cross-polarisation of receiving reflector
    antennas.* Radio and Electronic Engineering, vol. 45 (1975),
    no. 7, pp. 346–350.

73. ———. *Gain and cross-polarisation of reflector antennas with
    surface errors* (Antennas and Propag. Internat. Sympos.,
    Washington, D.C., 1978). Symposium Digest, pp. 253–256.

74. Ghobrial, S.J. and Jarvas, J.A. *Effects of off-set antenna profile
    errors on their cross-polarisation performance* (Antennas and
    Propag. Internat. Sympos., Quebec, 1980). Symposium Digest,
    vol. 1 (1980), pp. 248–251.

75. Gonzalez, D.G. and Pollon, G.E. *Report on cross-polarisation
    limitation on accuracy of reflector monopulse antennas* (Internat.
    Colloq. on Radar, Paris, 1977), pp. 333–336.

76. Hoisington, D.B. *Monopulse Doppler and fuse jamming primers.*
    *International Countermeasures Handbook,* 5th ed., 1979–1980,
    pp. 349–354.

77. Hockhman, G.A. and Olver, A.D. *Cross-polarized performance of
    small corrugated feeds* (IEEE Internat. Sympos. on Antennas
    and Propag., Symposium Digest (1978), pp. 431–434.

78. Howard, D.D., Sherman, S.M., Thomson, D.N., and Campbell,
    J.J. *Experimental results of the complex indicated angle
    technique for multipath correction.* IEEE Trans., vol. AES-10
    (1974), no. 6, pp. 779–787.

79. Howard, J.E. *A low angle tracking system for fire control radars*
    (Rec. IEEE Internat. Radar Conf., Arlington, N.J., 1975),
    pp. 412–417.

80. Van Brundt, L.B. *International Countermeasures Handbook,* 7th
    ed., EW Communications, Dunn Loring, VA., 1981–1982,
    pp. 344–347.

81. Interavia Air Letter, vol. 20/II (1960), pp. 14–16.

82. Jones, E.M.T. *Paraboloid reflector and hyperboloid lens antenna.*
    IRE Trans., vol. AP-2 (1954), pp. 119–127.

83. Johnson, G.R. *Jamming passive lobing radars.* Electronic Warfare,
    vol. 9 (1977), no. 2, pp. 75–85.

84. Johnson, M.A. and Stoner, D. *ECCM from the radar designer's
    view-point.* Microwave J., vol. 21 (1978), no. 3, pp. 59–63.

85. Johnston, S.L. *Guided missile ECM/ECCM.* Microwave J., vol. 21 (1978), no. 9, pp. 20–26.
86. Johnson, G.R. *Jamming CW radar.* Electronic Warfare, vol. 9 (1977), no. 3, pp. 95–105.
87. Johnston, S.L. *Radar electronic ECCM.* Trans. on Aerospace and Electronic Systems, (1978), no. 1, pp. 109–117.
88. Klaus, D.E. and Hollins, R.P. *Monochannel direction finding improves monopulse technique.* Defense Electronics, vol. 14 (1982), pp. 35–43.
89. Kitsuregava, T. *Report on a tri-reflector antenna with no cross-polarized components* (IEEE Internat. Sympos. on Antennas and Propag., 1979). Symposium Digest, pp. 92–95.
90. Pat. 4143375 (SSHA).
91. Kumar, A. *Waveguide feed reduces cross-polarization levels.* Microwave J., vol. 21 (1978), no. 3, pp. 86–88.
92. Loomis, R. *Threat and ECM techniques.* Internat. Defense Rev. (1976), no. 1, pp. 55–58.
93. Martin, A.G. *Symmetrical low cross-polarization from a dielectric sphere loaded horn.* IEEE Trans., vol. AP-27 (1979), no. 6, pp. 862–863.
94. Mendelovicz, E. and Oestreich, E.T. *Report on the phase-only adaptive nulling with discrete values* (IEEE Internat. Sympos. on Antennas and Propag., 1979). Symposium Digest, pp. 193–198.
95. Microwave J., vol. 24 (1981), no. 9, p. 45.
96. Miller, B. *US penetration capability erodes.* Aviation Week, vol. 88 (1967), no. 17, pp. 90–109.
97. Microwave System News, vol. 9 (1979), no. 2, p. 28.
98. Olver, A.D. *et al. Design of corrugated feed for low cross-polarization* (Internat. Conf. on Antennas and Propag., London, 1978). Proc. (1978), part 1, pp. 355–359.
99. Rid. Radiolokatsionnaya sistem AN/FPS-85. TIIER, vol. 57 (1969), no. 3, pp. 78–92.
100. Sakamoto, H. and Peebles, P. *Conopulse radar.* IEEE Trans., vol. AES-14 (1978), no. 1, pp. 199–208.
101. Skolnik, M. *Comment on the angular resolution of radar.* Proc. IEEE, vol. 63 (1975), no. 9, pp. 1354–1355.
102. Sherman, S.M. *Complex indicated angle applied to unresolved radar targets and multipath.* IEEE Trans., vol. AES-7 (1971), pp. 160–170.
103. Simens, R.F. *History and principles of electronic warfare,* III. Frequency Technology, January 1970, pp. 29–33.

104. Sims, R.J. and Craf, E.R. *The reduction of radar glint by diversity techniques.* IEEE Trans., vol. AP-19 (1971), no. 4, pp. 462–468.
105. Smith, P.G. and Mrstic, A.V. *Multipath tracking errors in elevation scanning and monopulse radars.* IEEE Trans., vol. AES-15 (1979), no. 6, pp. 765–776.
106. Steinber, B.D. *A proposed approach for increasing the azimuthal resolution in H.F. radar.* IEEE Trans., vol. AP-20 (1972), no. 5, pp. 613–618.
107. Stein, K.J. *Pave Paws phased array radar nears acceptance.* Aviation Week, vol. 110 (1979), no. 15, pp. 60–65.
108. Spellman, M. *Spread spectrum radios thwart hostile jammers.* Microwaves (1981), no. 9, pp. 85–90.
109. Van Brunt, L.B. *Applied ECM,* vol. 1, *E.W. Engineering.* EW Communication, Dunn Loring, VA 1978, 973 pp.
110. Watson, P.A. and Ghobrial, S.J. *Off-axis polarization characteristics of Cassegrainian and frond-fed paraboloid antennas.* IEEE Trans., vol. AP-20 (1972), pp. 691–698.
111. White, E.G. ITT vendor's partner, Microwave J., vol. 18 (1975), no. 7, pp. 16–19, 61.
112. Pat. 3221328 (SSHA).
113. Dver, F.B. *Millimeter waves—the EW challenge,* Microwave J., vol. 23 (1980), no. 9.

# Index